SMITSWINKEL PEAK

SWARTKOP MOUNTAINS

CAPE P

242

An aerial view of Cape Point, looking north. [Photograph by Charles Field, Wynberg, Capetown.]

LIFE BETWEEN TIDEMARKS ON ROCKY SHORES

T. A. Stephenson

Anne Stephenson

With a Foreword by C. M. Yonge

University of Edinburgh

W. H. FREEMAN AND COMPANY
San Francisco

Printed in the United States of America

International Standard Book Number: 0-7167-0687-3 (cloth)
                                    0-7167-0698-9 (paper)

Library of Congress Catalog Card Number: 79-152055

9 8 7 6 5 4 3 2 1

Dedicated to C. F. A. Pantin

T. A. Stephenson

FOREWORD

Following the death in 1961 of our friend Alan Stephenson, Professor Carl Pantin, who had a common interest with him in the study of animal distribution, suggested that he and I make joint application to the Department of Scientific and Industrial Research for a grant (later to be taken over by the Natural Environment Research Council) to enable Anne Stephenson to complete—indeed effectively to write—the book on the life of the rocky shores of the world that her husband had planned but of which he had written only the two introductory and the final chapters. She was the only person who could have done this. She had worked alongside Alan in Australia, when they were my colleagues on the Great Barrier Reef Expedition in 1928 and 1929, and then for ten years in South Africa, when he occupied the chair of Zoology in the University of Cape Town. There, he organized a far-ranging survey of the shores of South Africa, from temperate to near-tropical waters, from the Atlantic round into the Indian Ocean. They also viewed together the shores and fringing reefs of Mauritius. On Alan's subsequent return to Great Britain as Professor of Zoology at Aberystwyth, with the aid of grants from the Department of Scientific and Industrial Research, they jointly undertook intensive surveys of selected shore areas along the Atlantic and Pacific coasts of North America and also worked in Bermuda.

It was on the basis of knowledge so widely gleaned, a unique wealth of intertidal experience, that Alan had planned this book. Many of his observations had been published in a variety of scientific journals over a long period of years, although some of the Pacific surveys, notably those of the shores around Pacific Grove and La Jolla, make their first appearance in the pages of this book. But in any case, what had been so widely published obviously needed to be brought together and coordinated with the relevant work of others, many of them inspired by Alan's example.

It seemed tragic to Carl and me that this book might never appear. Anne agreed to do all she could, and our subsequent application for a grant, which was to include money for her support as well as for expenses, was successful. A final total of some £1600 was received over a period of eight years. I wish to thank the former Department, and the Natural Environment Research Council, for their indispensable help and also the Finance Officer of the University of Glasgow for his disinterested labours in administering this grant. Amongst other things, this money permitted the preparation of a significant number of new text figures on a standard as near as possible to the perfection of those left by Alan, superb artist as well as magnificent zoologist.

In the initial years, Carl Pantin and I visited and corresponded with Anne and assisted her with the planning of the book and with the resolution of the many problems that Alan's death had left. A further blow, of equal import to Anne and to me, came with Carl's death in 1967. Mistakenly, as I now feel, I carried on alone doing what I could, in the intervals of extensive world travel, to act in a dual capacity. Over the years I have read all that has been written but, of course, without that prime interest that Carl had in the problems of animal distribution; I have also discussed with Anne the numerous figures that were specially prepared for this book to supplement the incomparable ones that Alan had done. There was the final pleasure of seeing the manuscript eventually completed—a major labour nobly maintained by Anne over what I now realize, from start to final publication, involved a period of ten years. We can only hope that Alan would have approved.

I would wish, in conclusion, to emphasize that this is not just another in the long series of books about intertidal life. It is the unique record of personal studies over a surprisingly wide extent of the rocky shores of the world supplemented by some account of other such shores, notably in the Arctic, West Africa, northern Australia and New Zealand. The nomenclature was certainly valid when the original studies were made, but in certain cases it may now be outdated; in no case, however, should there be any doubt about what species is being described. No attempt has been made to refer to any but very strictly relevent literature. The theme is the distribution of animals and plants on rocky shores and it is on this plane that its merits must be assessed.

July 1971   C. M. YONGE

CONTENTS

PREFACE

This book must, to some extent, be regarded as a salvage operation. There were wider issues that were formed in my husband's mind of which, naturally, I cannot write. His un-rivalled knowledge of the animals and seaweeds of the intertidal coasts in so many parts of the world was not mine, nor his unparalleled intuitive perception of what was relevant; but the field work we did together, his approach to problems, certain of his conceptions, his whole attitude to ecological studies—with these I was entirely familiar. His extraordinarily high standard of care and accuracy in all he did, for the achievement of which nothing was too much trouble, has been before me all the time I have summarised his work. Whether what I have done would merit his approval I can never know, but that has been my aim during all these years of laborious writing.

ANNE STEPHENSON

July 1971

1

PERSONAL INTRODUCTION

The nature of a book is determined chiefly by its author's personality and experience, but also by the epoch during which it is written. The publisher makes another contribution, by producing it well or badly. The readers of the book can see the contents in a better perspective if they have some knowledge of these productive agencies, than if they read it *in vacuo* against no known background. For this reason I think that a short personal introduction will not be irrelevant.

The personal circumstances that led to the production of this book are as follows. My father was a botanist and I absorbed a knowledge of the British flora from him as I grew up; my earliest publications were short papers on British orchids, written in collaboration with him. I began to paint birds and flowers at a very early age, and my dual interest in plants and in painting has persisted and developed throughout life, but since about 1928 the botanical side of my work has been concerned almost entirely with seaweeds. I first took an interest in animals when, just before becoming a university student, I was attracted by sea anemones. During my training as a zoologist and my early professional years, I studied these animals in considerable detail, and most of my publications at that time referred to them. While collecting anemones on the British coasts I began to notice the distribution of plants and animals between tidemarks, but it was not until 1928 that an event occurred that caused this growing interest in shore populations to fuse with an increasing consciousness of geographical distribution. This event was the opportunity, provided by Dr (later Sir) Maurice Yonge, to spend a year on the Australian Barrier Reef as a member of the Great Barrier Reef Expedition of 1928-29. During this expedition my work first became definitely ecological. It so happened, moreover, that almost as soon as the expedition was over I was appointed Professor of Zoology at the University of Cape Town, and a ten-year sojourn in South Africa (1931-40) provided an opportunity for the ecological survey of an almost unknown coastline of a particularly fascinating type. My fate as an ecologist was now sealed. After leaving Cape Town I began to work consciously and deliberately towards collecting information that would make it possible to prepare a preliminary sketch of the distribution of plants and animals between tidemarks in the world as a whole. In pursuit of this aim I have visited, apart from the coasts already mentioned, the Atlantic and Pacific coasts of North America, various parts of Britain, the oceanic islands of Bermuda and Mauritius, and points on the coast of East Africa. While no one is more conscious than myself that the number of coasts I have never seen far exceeds those that I have seen, I believe it is also true that I have worked on more seashores than most other ecologists, and it is certainly true that the ones visited were very carefully chosen. By this I mean that I have tried to see *examples* of all the main types of rocky shore, subject to all sorts of variations in the temperature of the adjacent sea, from the tropics at one extreme to icy regions at the other.

Throughout this work certain conditions had to be fulfilled. No one can satisfactorily examine such long stretches of coastline by himself, at least on the scale that this particular problem demanded. Most fortunately, I have had the unfailing support of my wife Anne, whose name has stood by my own in many publications. For 36 years she has acted as my research assistant, has accompanied me on all my expeditions and has taken an active part in every phase of the investigation. I have also had many other assistants, for whose help I am profoundly grateful.

One problem that faces the authors of scientific books becomes increasingly pressing and difficult with every passing year of the twentieth century. What is to be done about the existing literature covering the subject of the book? The traditional solution is for the author to read every item that has been published, and to compile a colossal bibliography including them all. This may run to hundreds if not thousands of references. In his text he tries to summarise the content of this vast array, and to mention as many authors as possible. This method has certain disadvantages. While the bibliography is useful up to a point, nobody looks up all the contained items, and it would probably be more useful still if most of them were eliminated. This would save money, for the publisher, and would limit the list to those sources most likely to be useful to readers other than the author. Further, the method often tends to turn the book into a "compilation," consisting of endless pages of short sentences such as "Caterpillars have been seen to stand on their heads (Pettigrew, 1915). Phillimore (1821), however, attaches no importance to this." Such a text is usually wearisome, and while it does contain a great many facts—too many for anyone to assimilate—it often fails to present the essence of the subject in a digestible form. One of the most valuable contributions an author can make to his presentation is a living comprehension of the whole subject, which illuminates it throughout, and which perhaps he alone can provide. This impalpable witchcraft can itself be submerged or destroyed by too much preoccupation with matter that, in ceasing to be subordinate, has become insubordinate. We may also remember that the mere volume of scientific literature has now become so enormous that if an author reads everything that has ever been written in his field, he may not survive to do anything more.

Reviewers of scientific books have not always been very helpful in this connection. Repeatedly one sees such comments as: "This book is justified if only by its bibliography, which makes the literature of the subject available to students for the first time." This is a stupendous bit of snobbery that is not only untrue (students can always find references if they want them) but that greatly exaggerates the value of bibliographies. It is often so phrased as to imply that the *only* value of the book lies in its bibliography. This may be quite unjust to the author; and in any case no book can have been worth publishing if its bibliography were more valuable than its text. If so, why not publish the list of references alone?

What is the solution of this problem? I can answer this question no more satisfactorily than anyone else. I am convinced that we must make a change, but there will be a transitional period during which we experiment with other methods, and only after this period can the best solution be found. As far as this book is concerned, my wife and I have tried to work towards such a solution in the following ways: (1) by limiting the bibliography to those items most likely to be serviceable to others; (2) by reducing the number of references to authors in the text to a reasonable level; and (3) by presenting the material, so far as we are able, in a readable form, and dwelling on essentials and broad outlines rather than on detail. We have never subscribed to the view that a simple statement need be a superficial one, but have believed, rather, that only where there has been a mastery of detail can a statement emerge that is both true and simple.

While we may fail in all these objectives, we shall at least have attempted to advance. By mentioning our aims, we do not imply for a moment that many scientific books of the past have not achieved them—of course they have—simply that such objectives have not, on the whole, been a recognised aim of scientific authors, and that the time has come to adopt them more widely.

There has also been a widespread assumption that scientific information should be presented in a very impartial and impersonal manner. This has been likely to result in presentations that are merely colourless and dull, and that do not necessarily *eliminate* the personal predilections of the author, but merely conceal them, so that the reader cannot readily detect them and allow for them. There is a growing feeling that no presentation can be fully impersonal, and that possibly a personal presentation is more valuable, after all, than too detached a survey, so long as it respects the facts and selects as fairly as it can. I believe firmly in looking at both sides of a question and in evaluating pros and cons as judiciously as may be possible; but I also feel that a scientific book will not be the success that it might be unless it is so informed by its author's personality that it represents a coherent individual comment on the whole situation. Whether this be true or not, any contribution to the subject of shore ecology that I have been able to make has been essentially a personal one; and what is personal can hardly be formulated in completely impersonal terms. I am an admirer of the works of some of the leading naturalists of the late nineteenth and early twentieth centuries. These men would visit a country new to them, and would produce an account of its geology, climate, fauna and flora, often admirably illustrated, that showed a mastery in all these fields and was lively and readable without being unscientific. Perhaps it should be one of our aims to emulate such authors, making due allowance for the improvement in scientific methods and the increase in scientific knowledge that has taken place since their day.

PUBLICATIONS IN MARINE BIOLOGY BY T. A. STEPHENSON

1931 [with Anne Stephenson, G. Tandy, and M. A. Spender]. The structure and ecology of Low Isles and other reefs. In *Scientific Reports of the Great Barrier Reef Expedition, 1928-1929*, vol. 3, pp. 17-112. London: British Museum (Natural History).

1934 [with A. Zoond and E. J. Eyre]. The liberation and utilisation of oxygen by the population of rock-pools. *J. Exp. Biol.* 11: 162-172.

1936. The marine ecology of the South African coasts, with special reference to the habits of limpets. *Proc. Linnean Soc. London* 148: 74-79.

1937 [with Anne Stephenson and G. A. du Toit]. The South African intertidal zone and its relation to ocean currents, I. A temperate Indian Ocean shore. *Trans. Roy. Soc. S. Afr.* 24: 341-382.

1938a [with Anne Stephenson and K. M. F. Bright]. The South African intertidal zone and its relation to ocean currents, IV. The Port Elizabeth district. *Ann. Natal Mus.* 9: 1-20.

1938b [with E. J. Eyre]. The South African intertidal zone and its relation to ocean currents, V. A sub-tropical Indian Ocean shore. *Ann. Natal Mus.* 9: 21-46.

1938c [with N. A. H. Bokenham]. The colonisation of denuded rock surfaces in the intertidal region of Cape Peninsula. *Ann. Natal Mus.* 9: 47-81.

1938d [with N. A. H. Bokenham and F. L. M. Neugebauer]. The vertical distribution of certain intertidal marine gastropods in False Bay, with notes on the development of two of them. *Ann. Natal Mus.* 9: 113-138.

1939. The constitution of the intertidal fauna and flora of South Africa, I. *J. Linnean Soc. London Zool.* 40: 487-536.

1940 [with Anne Stephenson and J. H. Day]. The South African intertidal zone and its relation to ocean currents, VIII. Lamberts Bay and the west coast. *Ann. Natal Mus.* 9: 345-380.

1943. The causes of the vertical and horizonal distribution of organisms between tide-marks in South Africa. *Proc. Linnean Soc. London* 154: 219-232.

1944. The constitution of the intertidal fauna and flora of South Africa, II. *Ann. Natal Mus.* 10: 261-358.

1947. The constitution of the intertidal fauna and flora of South Africa, III. *Ann. Natal Mus.* 11: 207-324.

1949. The universal features of zonation between tide-marks on rocky coasts. *J. Ecol.* 37: 289-305.

1950 [with Anne Stephenson]. Life between tide-marks in North America, I. The Florida Keys. *J. Ecol.* 38: 354-402.

1952 [with Anne Stephenson]. Life between tide-marks in North America, II. Northern Florida and the Carolinas. *J. Ecol.* 40: 1-49.

1954a [with Anne Stephenson]. Life between tide-marks in North America, IIIA. Nova Scotia and Prince Edward Island: description of the region. *J. Ecol.* 42: 14–45.

1954b [with Anne Stephenson]. Life between tide-marks in North America, IIIB. Nova Scotia and Prince Edward Island: the geographical features of the region. *J. Ecol.* 42: 46–70.

1961a [with Anne Stephenson]. Life between tide-marks in North America, IVA. Vancouver Island, I. *J. Ecol.* 49: 1–29.

1961b [with Anne Stephenson]. Life between tide-marks in North America, IVB. Vancouver Island, II. *J. Ecol.* 49: 227–243.

2

SCIENTIFIC INTRODUCTION

One of the most vital of natural rhythms is the ebb and flow of the tides. Eternally they surge up and down the coasts of the world, many of which are as remote and secret as they were before the appearance of man. One can sail past these lonely coastlines for days at a time, without seeing anyone; and nothing about them is more impressive than their vast emptiness. It is no wonder that marine biologists, newcomers in the world's history and limited in numbers, have as yet made their little sketches of shore communities on a mere fraction of the endless miles of coastline. This thought would be too daunting if it were not true of the shore, as of so many other things, that "plus ça change, plus c'est la même chose." We need not interpret this as meaning that all seashores are one seashore and therefore boring; but in the sense that although the tapestry formed by the variations is kaleidoscopic in its permutations, there is a theme underlying the variations.

PRESENT TRENDS IN SHORE ECOLOGY

If fewer seashores are known than unknown, one might well suppose that it is premature to attempt any general description of their inhabitants on a worldwide scale. This is not so, for several reasons. The first (the existence of a theme) has already been mentioned. The second is that progress towards a fuller understanding of seashore life must be limited in scope until a preliminary picture for the whole world exists. We cannot tell what is true or false until this sketch has been made for future correction and extension. A third source of reassurance may be found in the comparative histories of oceanography and of shore ecology. The oceanographers began to study the world as a whole much earlier than did the shore ecologists. Their broader outlook dates especially from the year 1872, when the *Challenger* expedition set sail. But if the earth's coastlines cover so many thousands of miles as to seem almost infinite, how much vaster are the oceans themselves. Where the tidal fringe of the shores is narrow and relatively accessible, the ocean bed is wide, deep, and extremely inaccessible, hidden by darkness and sub-merged under unimaginable tons of water; and water, as much as the sea floor, is the subject of oceanographical studies. This being so, the oceanographers have been obliged to base their conclusions on *samples* of all kinds, on observations made at numerous points or along narrow lines of the ocean's vastness; for them also, more has always been unseen than seen. Furthermore, as their work must be done chiefly from ships, their technical difficulties greatly exceed those of students of the shore. Despite all these drawbacks, the oceanographers have described their results in worldwide terms for nearly a hundred years, and no one doubts the soundness of this approach. Their results, progressively revised, became increasingly precise, but could never have reached their present comprehensiveness by means of a more piecemeal approach. It is one of the curiosities of biological history that the shore ecologists, with their less unmanageable task, concentrated at first so much on detail that it was not until 1949 that any attempt at a large scale picture began to be made. But the history of oceanography surely indicates that if exponents of this science were laying widespread foundations in 1872, it can hardly be premature for the shore ecologists to do the same thing nearly eighty years later.

The history of shore ecology has not only differed from that of the rest of oceanography, but has passed through several phases. After it had emerged from the stage of casual natural history and had become more systematic and scientific, there was a tendency for its students to undertake very precise and detailed examinations of small areas—a single rock-pool, a strip of shore a yard wide, or at most an estuary or a hundred yards of beach. These investigations were attractive because of the degree of accuracy attainable, and because many of them dealt with numbers or weights of animals and algae, and were thus quantitative, conveying an air of mathematical precision. Such studies produced excellent results, but on the whole these were interesting in themselves rather than productive of a wider understanding of the shore. It may be added that botanists tended to develop an ecological outlook on the problems involved earlier than did the zoologists, but that the latter have now completely caught up with them. It was rather a long time before anyone began to describe the ecology of long stretches of coastline, much less that of whole countries. But such broad surveys began to be developed in the 1930s in the hands of a limited group of observers, and that phase in the maturation of the subject is now in full swing. Attempts to explain the *reasons* for the distributions that we see on the shores have continually been made, along with the descriptive work; but this aspect of the subject still requires much further development.

Shore ecology is still a young science, although an enormous bulk of information has already been amassed. In thinking of the future we may profitably compare the present state of knowledge about seashores with our present knowledge of evolution. It is now over a hundred years since Darwin published his views on evolution by means of natural selection. These years have been most fruitful. Our knowledge has advanced immensely, genetics and mathematics have come to the rescue, and many biologists are now very happy about natural selection. They feel that it explains the mechanism of evolution almost completely. However, it seems that before long entirely new discoveries in the evolutionary field may well be made, and that these may wholly undermine the present orthodoxy. Natural selection, no doubt an effective agency within its own limits, may then appear in a truer perspective and no longer occupy the whole stage. Shore ecology, perhaps luckily for its own prospects, has not yet reached any such stage. It is more in the phase during which the more we see of the shore, the less we feel we know about it, and this seems a healthier condition than one prevailing in the evolutionary field. Nonetheless, shore ecology is probably on the verge of having gone as far as its present impetus will carry it, and here too we must hope for new discoveries that will impart a new direction. There is still too much about the shore that appears magical, or at least beyond present explanation.

THE IMPORTANCE OF STUDYING PLANTS AND ANIMALS TOGETHER

As the shore between tidemarks is inhabited both by animals and by seaweeds, our knowledge of its inhabitants has come from botanists as well as from zoologists. Sometimes the botanists have ignored the animals, or the zoologists have ignored the plants; but the best results have emerged when, as has happened fairly often, the algae and animals have been considered together. For general purposes this is almost essential. It is so important that, at the risk of stressing the obvious, we shall dwell upon it briefly; and a few examples will perhaps emphasize our meaning more clearly than a broader statement.

Let us suppose that we wish to study the occurrence of the fucoid algae on the British coasts. This can hardly be done effectively unless we realise that the distribution of these seaweeds depends not only on the plants themselves, and on the climatic conditions in which they live, but also upon animals. An area of rock can be free from fucoids simply because limpets, in the course of their browsing, have removed the spores or sporelings—that is, a plant can be *absent* from an area it might otherwise have colonised, because of an animal.

A second example may be taken from South Africa. There are species of *Gelidium* in that country that, if left to themselves, form short bushes of erect branches a few inches high. But very often an area of *Gelidium* forms the feeding-ground for a limpet, which browses over it so continually that, although the alga does not disappear, it is able to produce nothing more than small flat fronds growing horizontally over the rock, and these make a thin layer only a few millimetres thick. The shape of these prostrate fronds is so unlike that of the normal ecrect ones that it was not possible to identify them until they had been allowed to grow up into a more normal form by removing the limpets. Here we have an example of the production, by an animal, of a completely different growth form in a plant.

My last example will be taken from the Bermuda Islands. Here, the whole appearance of the tidal belt on the shore can be changed by the activities of animals directed against plants. Some of these animals, such as parrotfish and crabs, peck at the algae growing on the rock, or nip at them with their claws. Other animals, such as snails, limpets, or chitons, rasp the algae from the rock by means of their filelike radulae. These depredations have more than one effect. The rock of Bermuda consists of an aeolian limestone, which weathers into the most intricate surface forms. It is spiky, with innumerable points and sharp edges, hollows, crevices, and cavities, sometimes almost fretted like lace. The scraping activities of the animals mentioned tend to smooth off many of these surface irregularities, so that between tidemarks the rock becomes increasingly smooth as one goes from higher to lower levels. Here we have a case where the rock itself is affected, as well as the covering of algae, which becomes a short turf. But here and there another similar effect is visible in Bermuda. We shall see from the later parts of this book that one of the most widespread features of seashores is the existence near high-water level of a horizontal strip in which the surface of the rock is discoloured to a grey or blackish tint by a filmlike incrustation of myxophycean algae or of marine lichens. There are places in Bermuda where one is surprised to find, at the right level, a strip of rock *paler* in colour than the rest, rather than darker. This appears mysterious until it is discovered that the blackish film has indeed developed in such places, as usual, but that the bulk of it has been browsed away, on an unusually comprehensive scale, by littorinid snails. Such examples as these, which could be multiplied, seem to demonstrate the futility of attempting to understand any intertidal population of plants and animals unless both are considered together.

We have always thought the conclusion just stated to be so obvious as hardly to need mentioning. That this impression is wrong was sharply illustrated at an international congress that we attended in 1958. One of us was invited, on this occasion, to introduce the subject of shore ecology at the beginning of a symposium on that topic. With some hesitation it was decided to stress the need for considering plants and animals together, when studying seashores. There need have been no hesitation. After the session was over, we lunched with a young botanist who had had the opportunity of examining remote seashores never before described (ecologically) by biologists. Not only had he ignored the animals during his work, but he showed no appreciation whatever of the seriousness of the omission. He had even, it would seem, failed to interpret the distribution of the algae, because the key to it lay in their relation to the animals, which had been ignored. Not only had the remarks of the preceding morning been anything but superfluous, but they had fallen on deaf ears. So much for the futility of human effort!

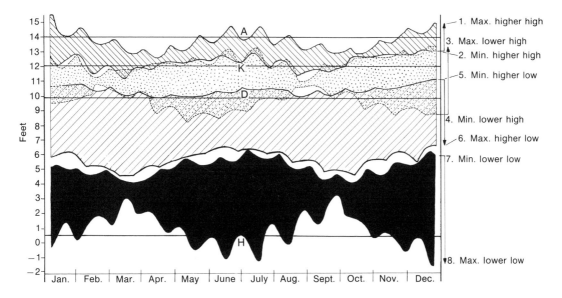

FIGURE **2.1**

The behaviour of the tide at Sand Heads, on the Strait of Georgia, British Columbia, during the year 1947, based on tide tables for that station: *left*, heights in feet above and below datum; *right*, the arrows and numbers 1–8 refer to the various high and low waters. The amplitudes of the high and low waters are shown in bands: *striped left-to-right*, higher high waters; *stippled*, lower high waters; *striped right-to-left*, higher low waters; *black*, lower low waters. The four horizontal lines labelled *A, K, H*, and *D* represent some of the more important zone boundaries on Brandon Island, which is some twenty miles from Sand Heads (see Figure 2.4).

TIDAL RANGES   *(Revised by Sir George Deacon)*

It is a general feature of tidal behaviour that the two high waters of the day do not rise to the same height and that the two low waters do not fall to the same low level. This variation is known as diurnal inequality.*

In Atlantic waters, especially in the North Sea and the English Channel, the diurnal inequality is insignificant; but in other oceans it may be large, as at Sand Heads (Figure 2.1), and sometimes so great that the half-daily tides are almost suppressed and there is only one obvious tide a day, as at some places in the Pacific Ocean, particularly in the South China Sea. The diurnal inequality also varies throughout each month according to the declination of the moon about the plane of the equator. When there is a marked diurnal inequality, a so-called mixed tide, it is useful—if admittedly somewhat confusing—to refer to the different high and low waters as higher high water, lower high water, higher low water and lower low water. Such fortnightly variations of tidal height are shown at Sand Heads in the Strait of Georgia where there is a large inequality as shown in Figure 2.1, and at Halifax, Nova Scotia, where the inequality is relatively small (see Figure 2.2).

---

*For a general account of tides, see Marmar (1930) and King (1962).

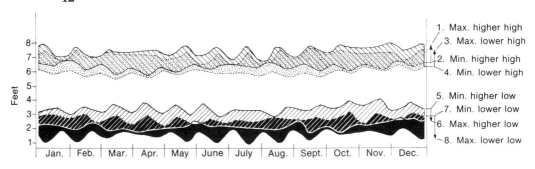

FIGURE **2.2**

A diagram similar to Figure 2.1 for Halifax, Nova Scotia, where the diurnal inequality is insignificant.

It is of importance to the shore ecologist to realise that, with a mixed tide as at Sand Heads, where the varying lower high waters and higher low waters (see arrows between *3* and *4* and between *5* and *6* in Figure 2.1) fall so often between the shoreward limit of the minimum higher high waters (*2*) and the seaward retreat of minimum lower low waters (*7*), only over the middle part of the shore are there *two* exposures every day. The higher part of the shore is covered, and the lower part uncovered, only once a day. At Sand Heads, the range of variation of the higher high waters (upper band of oblique lines) is much less than that of the lower low waters (shown black), but in other regions—in Bermuda, for example—the diurnal inequality is featured principally in the high waters (see Figure 2.3). Examples may also be found in the Hawaiian Islands, the Philippines, and South Australia.

One approach to the investigation of the possible biological significance of the varying tidal ranges might be to map the maxima and minima of the higher and lower high waters and of the higher and lower low waters. In the figures these eight varying limits are indicated by lines and arrows on the right-hand side. In Figure 2.1, the four horizontal lines marked *A, K, D,* and *H* represent some of the more important zone

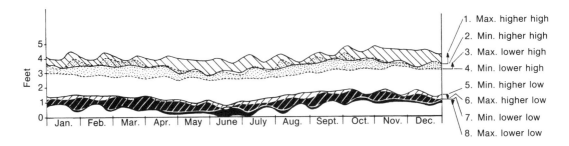

FIGURE **2.3**

A diagram similar to Figure 2.1 for Bermuda, where the diurnal inequality is mainly among the high waters.

boundaries at Brandon Island in Departure Bay, about twenty miles from Sand Heads. Figure 2.4 demonstrates the relation of zone boundaries to some of the eight tide levels. The following is a comparison of Figures 2.1 and 2.4.

Line *A* marks the lower limit of terrestrial lichens on both coasts and lies very close to level *1* (the maximum higher high water). Lines *D* and *K* show the upper boundaries of barnacles on the south and north coasts, respectively. Line *D* lies between levels *4* and *5*, the minimum of higher low waters and the minimum of lower high waters, while line *K* corresponds with levels *2* and *3*, which coincide as the maximum lower high waters and the minimum higher high waters. This demonstrates the different heights of the barnacle limit on the two shores. Line *H* indicates the upper boundary of the beard zone on the south coast, which is almost the same as that of the *Metridium* zone on the north coast, both at the tide level *8*, the maximum of lower low waters.

These variations in zone levels and species cannot be attributed to tidal action, since the tide levels affect both coasts alike, nor to marked differences in wave action, which

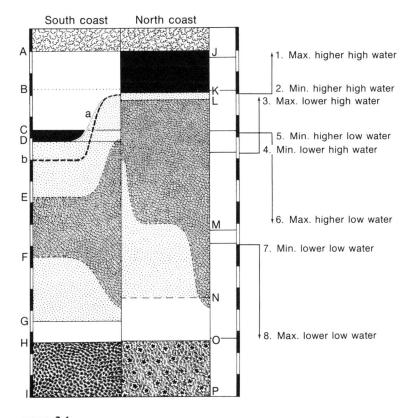

FIGURE **2.4**

The relation of some zone boundaries to the tide levels at Brandon Island, in Departure Bay, British Columbia. Compare with Figure 2.1.

do not appear to exist. They can only be due to differences in the coasts themselves: the south coast is sloping and sunny, and the north coast is cliffed and shady (see Figures 10.4 and 10.5). The apparent relation between zone boundaries and tide levels at two other places, Pacific Grove and La Jolla, are shown in Figures 10.4 and 10.22.

Thus annual means of the different tidal ranges can be worked out graphically by drawing "envelopes" to include the different maxima, minima and means; more diagrams covering a year are shown for St John and Key West (Figures 2.5 and 2.6). This graphical method, though simple, is a very time-consuming task, and it could be done more quickly by digital computation using harmonic constituents of the nearest standard port, though this would require the help of a mathematician with some knowledge of tidal theory.

Important factors are at what time in the 24 hours low waters are reached in winter and summer, and for how long the exposure lasts. The effect of long hours of bright light, heat, and drought on sea shores is very great, particularly on the biota of lower zones, which require constant wetting. At San Francisco, for example, the lowest low waters occur mainly after 4 PM in winter and before 8 AM in the summer; thus the lowest zones are not uncovered during the greatest heat of the day. Similarly, at San Diego, the maximum low waters occur during the afternoon in winter, but before 6 AM in the summer. Other examples are given throughout the book, as at places in South Africa, at La Jolla and Pacific Grove, and in Chile. The biological consequences of the conditions given above require closer study, because much remains to be done before we can elucidate the nature of all the relationships between tidal influences and zonation.

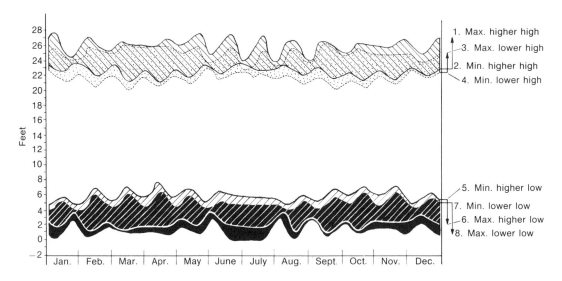

FIGURE **2.5**

A diagram similar to Figure 2.1 for St John's, New Brunswick, where the tidal ranges are very great.

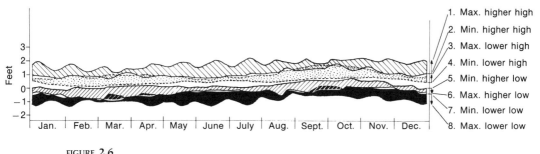

FIGURE **2.6**
A diagram similar to Figure 2.1 for Key West, Florida, where the tidal range is very small.

ZONATION BETWEEN TIDEMARKS

It has been well known for a long time that the seaweeds and animals inhabiting rocky shores are not arranged haphazardly, but that as we proceed from the higher and drier parts of the shore to the lower and wetter parts, the assortment of species present not only changes, but on the whole increases in number and variety. This results in a curious effect: the different species come to inhabit horizontal strips or belts of the shore that lie one above the other. In many places, these strips, commonly called *zones,* are brightly coloured by the organisms that populate them, and are sharply marked off from each other, often producing a spectacular appearance along the shore. In other places, they are less conspicuous and less readily distinguishable, but are probably rarely, if ever, completely absent. They have caused a great deal of trouble to biologists, because they present an intriguing phenomenon that has attracted many investigators; yet, while they reveal some of their secrets to any persistent student, they conceal others, so that the whole problem remains baffling beyond a certain point, and very susceptible to different explanations that are not necessarily compatible with one another.

This is partly because the zonation of the various coasts of the world presents a bewildering series of variations. Furthermore, there is often a real difficulty in giving an accurate description of the zonation of some shores. If an author is dealing with smooth rocky surfaces with an even slope, he may find that the zonation is diagrammatically clear and that it can be described with precision. But if the shore presents a tumbled chaos of broken rocks and boulders lying at all angles, the task of interpreting the zonation (much less describing it) becomes a very special one. (It is as essential to learn to "read" such a shore as it is to learn to read printed matter. Unfortunately, there have been authors who have attempted to *write* about such shores without ever having been able to read them.) In these difficult cases it is essential to realise that all the different facets of the rock have their own appropriate zonation. Just as one expects a selection of plants and animals under a stone to be different from that occurring on top of it, or one selection under a shaded overhanging rock and another on a sunlit slope, so one must expect one zonation on rocky faces directed landward, another on faces directed seaward, a third on surfaces at right angles to the shoreline, and yet others on slopes inclined at different angles. The zonation on a thoroughly broken shore has to be described in terms of the prevailing arrangement of species on all the rocky facets of the same type at the same level. Attempts have been made to achieve this by means of very careful instrumental surveys along narrow strips of shore, crossing the tidal region at different angles. However, it remains probable that the best instrument for appreciating these interrupted zonations is the suitably trained human eye.

As this problem of zonation is one of the central features of shore biology, this book will dwell on it in considerable detail. Although zonation was noted and studied before 1900, our substantial knowledge of it belongs to the twentieth century. Despite the fact that authors in various parts of the world had described the zonation on a considerable number of shores, for a long time it was very difficult to correlate the description of a New Zealand shore with that of another in Japan, a third in Britain, and (even worse) that of a coral reef in the Maldive Islands. This difficulty was due partly to the fact that many of the authors were acquainted with the shores of one country only, or even of a small part of that country's shores. They described what they saw in whatever terms seemed suited to it, and applied names to the zones that they described. Some of them worked out "systems" of zonation that applied to the shores they had seen and might perhaps apply to others. Since zonation is by no means precisely the same everywhere, it is not surprising that the descriptions given for one part of the world were very difficult to compare with those given for another part. As nobody knew whether all the shores of the world had anything in common, far less what the common features might be if they existed, no coherent picture could be built up of zonation in general and of its mode of variation from place to place.

In 1949, having studied seashores in many parts of the world for some thirty years, we decided to risk publishing a conclusion to which we had been led during our various wanderings. This was that, in spite of the almost endless variations in zonation that do undoubtedly occur, there is nevertheless a theme, or basic plan, underlying the variations. We described what we believed the basic plan to be, and indicated what seemed to be the major variations. We also introduced a terminology that we hoped might prove adequate for general use.

At this point, it will be relevant to explain the meaning of a remark that was made in the first chapter of this book, to the effect that our contribution to shore ecology has been a "personal" one. An essential feature of it has been that our own eyes—the same two pairs of eyes—have seen many shores of many varieties in widely separated parts of the world. Not until the *same* eyes had seen *enough* shores could there be any real possibility of recognising the most widespread features of zonation. The fact that the shores were seen by two pairs of eyes, rather than one, has been an added advantage, as we could check and correct each other's impressions. Naturally, if one uses human eyes and brains in this way, as if they were scientific instruments (which, in a sense, they undoubtedly are), one realises clearly that, like other implements, they have their defects and may introduce errors. But this is offset by the fact that they secure a degree of consistency which can hardly be attained otherwise, and make genuine comparisons possible.

WIDESPREAD FEATURES OF ZONATION

Our paper entitled "The universal features of zonation between tide-marks on rocky coasts" appeared in the *Journal of Ecology* in December 1949 (vol. 37, p. 289). Since then, other workers have had time to test and discuss our conclusions. With few exceptions, they have agreed with our conception wholly or in part; and we have had the pleasure of meeting some of them personally, and of reconsidering our own ideas in the light of theirs.

The account that follows represents the point of view we put forward in 1949, a little modified in the light of subsequent developments. If it differs only in minor respects from the original account, this is not because of any desire on our own part to cling to any particular thesis, but because on the whole the work that has been done since 1949 confirms and extends our view. We are not in the least averse to revising our scheme if it proves to need revision, but, so far, it seems to be serving its purpose; and, as a good deal of it is concerned with fact rather than theory, that part will require revision only if significant new facts come to hand.

If we consider how variable is the behaviour of the tides from one place to another, and how the plant and animal populations change between the tropics and the poles, we might reasonably expect that there would be a corresponding diversity of zonations showing little in common. This expectation is partly misleading, because (as we shall see later in this chapter) it appears that zonation, although undoubtedly related to tides, is not directly caused by them. That zonations vary remarkably is an undoubted fact; they change not merely from tropics to poles, but within a few yards on the same beach, clearly responding to local variations in the incidence of wave action and other controlling agencies. In spite of this, the more one sees of the world's shores, the more evident it becomes that, however various the zonations may be, they are all variations on a fairly simple theme. In other words, zonation has certain features that are very widespread.

Having stated this, we may point out immediately one of the ways in which it has been misunderstood. One can enumerate the widespread features of tendencies verbally, or summarise them by means of diagrams. Either will result in a mental picture of a sort of "average" or "typical" zonation. Such a picture is useful up to a point, as it serves as a kind of standard against which one can compare actual shores, and many of them, in fact, can be interpreted quite readily with its help. But if anyone, especially one with little experience of typical shores, visits a region such as the arctic, where conditions are very different from those in the warm and temperate parts of the earth, he may find little that seems, at first sight, to correspond with any "average" or "typical" shore, and may conclude that none of the widespread features of zonation are recognisable on the shores before him.

This conclusion is not necessarily correct. The truth may be, rather, that the exceptional conditions have *modified* the widespread features so much that they are not easily recognised in their changed form. Some of them may have been suppressed altogether. But this does not alter the fact that the widespread tendencies exist. Where they can (and they can in many places), they become fully manifest; elsewhere, they suffer interference; but even where they are least in evidence, some traces of their influence can usually be found. It may perhaps be useful at this point to mention a completely different example for comparison. The various parts of the vertebrate skeleton vary almost endlessly from one group to another, in connection with the evolutionary history of any particular vertebrate species and with its mode of life. But because we sometimes find vertebrates without limbs, or with wings or flippers instead of legs, we cannot conclude for a moment that the vertebrate skeleton has no general features or tendencies that recur whenever there is nothing to change them or to suppress them.

Perhaps part of the difficulty has arisen because of the word "universal." In our original paper we referred to the general features of zonation as being very widespread, frequently recurrent, and "perhaps even universal in occurrence." At no time did we suppose that any feature of zonation was literally universal in the sense that sheets of barnacles, for example, could be expected near the north and south poles, any more than they could be expected in the Himalayas. What we had in mind was that the tendency of some species of barnacles to form sheets in the middle part of the shore would lead them to do so anywhere in the world where something, such as ice, did not prevent them. There is undoubtedly something world-wide about such a tendency. But seeing that our original usage of the word "universal" can be misunderstood, and may therefore confuse the issue, we speak in this book of "widespread" features of zonation, rather than of "universal" ones.

The widespread features or tendencies are as follows:

1) Near high-water level there is an arid zone subject to conditions transitional between those of land and sea. It is affected by spray, but wetted by waves only in heavy weather or at the higher spring tides, when at least its lower parts become washed or submerged. The number of species inhabiting the zone is relatively small, and includes snails adapted to arid conditions and belonging to the genus *Littorina* and to related genera, or to genera of snails containing similarly adapted species.

2) The surface of the rock in the zone just described, or in its lower part, is commonly blackened by encrusting myxophyceans, or lichens of the *Verrucaria* type, or both. The blackening may exist either as separate patches or as a continuous belt, often overlapping into the zone below; but this "black zone" is one of the most persistent features of shores.

3) The middle part of the shore is the most fully intertidal region and part of it at least is covered and uncovered by the sea every day. The number of species inhabiting the zone is greater than in the drier one above, and typically includes numerous balanoid (acorn) barnacles belonging to such genera as *Balanus, Chthamalus* and *Tetraclita*. The upper limit of the zone is marked by the disappearance of barnacles in quantity. There is a strong tendency for the zone to become differentiated into subzones, commonly two or three in number. In the upper of these the barnacles tend to be maximal in quantity and are either the dominant forms or among these. In the lower, they are usually in competition with other sedentary forms, which may overshadow or obliterate them, though this by no means always happens.

4) The lowest part of the shore is uncovered at spring tides and not at neap tides,

and in some places, where wave action is persistently strong, only at the lowest spring tides in calm weather. It supports the fringe of the infralittoral population (that is, the population of the region below tidemarks that is never uncovered), is exposed to the air for a relatively small proportion of the time, and in many places rarely if ever dries off. The population is extremely variable, but is usually rich. In many cold-temperate regions, it consists of a forest of large brown algae (laminarians, for example) with animals and an undergrowth of smaller plants among their holdfasts. On coral reefs, it commonly contains the upper edge of the rich growth of coral that extends down the reef face below low-water level. In warm-temperate regions, it may support (*a*) a dense covering of simple ascidians (*Pyura*, for example); (*b*) a dense growth of rather small mixed algae, primarily rhodophyceans; or (*c*) other communities. Large brown algae may or may not occur here, but if they do, they tend to be less strongly developed than in cold-temperate regions.

5) In the wet zone just described, the encrusting, nonjointed calcareous algae of the family Corallinaceae (referred to broadly as "lithothamnia") tend to encrust all surfaces suitable to them, often forming continuous pinkish sheets. This encrustation not only continues below tidemarks, but may also extend above the zone described in paragraph 4. But it does not usually form a continuous growth above the lowest strip of the barnacle zone of paragraph 3, except in pools, caves or other specially damp places.

We may summarise the foregoing paragraphs by saying that seashores show a widespread tendency to develop three principal zones, a *Littorina* zone above, a barnacle zone in the middle, and a wetter zone of variable population below. In addition, there is typically a blackish myxophycean or lichen zone at a high level, and an incrustation of lithothamnia over all the lower parts of the rock. Perhaps it should be added that, above the *Littorina* zone, there are often other zones distinguished by coloured lichens (orange, grey, green, white, brown) belonging to species other than those characteristic of the black zone. But as these higher zones belong to the maritime land flora rather than to the tidal zone, we shall be little concerned with them here.

These widespread zones are often best displayed on fairly continuous rocky slopes (whether steep or gradual) subject to wave action that is strong without being maximal. But the zones do not necessarily "hang together." They vary independently, so that it does not follow that all of them are at their best under the same conditions. Consequently one does not necessarily see all the widespread features at the same time on the same shore. Places can be found where these features and no others are displayed; but more frequently one finds shores on which, in addition to the widespread features, there are additional communities that are local in distribution, which naturally complicates the whole picture.

TERMINOLOGY

Like everything else, a terminology can be good or bad. If it is good, it may be a valuable tool; if bad, it darkens counsel and is an unmitigated pest. The several branches of biology seem to call for terminologies in different degrees. Anatomy, for example, obviously needs a great many names for the structures of which it speaks: particular blood-vessels and their branches have to be named, and so on. Although anatomists have by no means always shown discretion in their choice of these names, the system works well on the whole and a multiplicity of technical labels is regarded as inevitable. Ecology, on the other hand, seems to us to be a subject that has little need for technical terms, and in which any great multiplication of them becomes a positive disadvantage. The subject itself, despite a spurious air of simplicity, is exceedingly complex. This being so, the only reason for burdening it with a complicated terminology would be that the terminology had a simplifying or clarifying effect, that it made the inherent complexities of the subject less difficult to understand.

Unfortunately, ecologists have not always remembered this. Some of them have described their results in simple language that readers can easily follow, but others have indulged in a terminological debauch that can have few parallels in scientific history. We have studied the works of these authors with care, but we have never been able to see either that the terms were necessary or that they advanced understanding of the subject. To the contrary, they seem to us to confuse it intolerably, and also to be calculated to produce an impression that unsolved problems have been solved, whereas, in fact, they have merely been shelved by attaching an unnecessary label to them. We believe that Dr Singer's comment, which we quoted elsewhere, is justified: "From the beginning, however, it [ecology] has been cursed, more than most sciences, by a horde of technical terms equally hideous, unnecessary and obfuscating!" (Singer, 1931, p. 279).

We realise, of course, that the authors who have put forward or developed ecological terminologies have been actuated by a genuine desire to advance the subject. But it has been necessary to mention the matter here, if only to make it clear that in this book we are adopting as few special terms as we can, not because we are unfamiliar with them, but because we believe that we can do much better without them. On the other hand, clearly *some* special terms are necessary, and we have even, in the interest of clarity and accuracy, been guilty of inventing a few ourselves.

To return to the zones of the seashore, a great many systems of nomenclature have been invented for these. It might be expected that in a book of this sort we should fill many pages by enumerating and comparing these systems, both verbally and by means of diagrams. We intend to do nothing of the sort, because we feel that the endless repetition of such comparisons resembles those "post-mortems" that so often follow games of bridge, and that in the interests of that development of scientific writing of which we have spoken in our first chapter (p. 2), it is better to resist the temptation to include them. It is doubtful whether readers would derive substantial assistance from them, and in any case they can be found in Hedgpeth (1962), Hodgkin (1963), Lewis (1964), and Southward (1958).

In this book, we shall use the terminology that we proposed in 1949, because the comments on it that have appeared since that date hardly seem to justify a change, unless perhaps in the case of one or two details (Figure 2.7).

We propose to call the *Littorina* zone of the shore the "supralittoral fringe," because it represents the upper fringe of the tidal area, the boundary strip between land and sea. The term "supralittoral zone" we apply (as it has often been applied by others) to the maritime region above high water, which may extend several miles inland; our supralittoral fringe follows immediately to seaward of this.

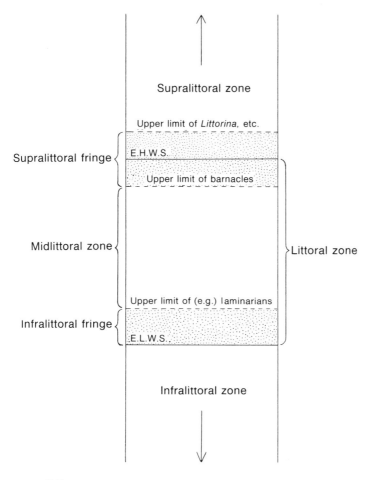

FIGURE **2.7**

A generalised scheme of intertidal zonation. The terminology introduced in
the text is given on the left-hand side. *E.H.W.S.*, extreme high water of spring
tides; *E.L.W.S.*, extreme low water of spring tides.

For the barnacle zone of the middle part of the shore we propose the name "midlittoral
zone." Some commentators have given reasons for preferring to call it simply the
midlittoral, without the word zone. This seems to us quite unobjectionable.

For the lowest of the three main intertidal zones we propose the name "infralittoral
fringe," because again it represents the intertidal fringe of the "infralittoral zone," which
lies below low-water mark altogether and is permanently submerged. Many authors prefer
the adjective sublittoral to infralittoral, and this again seems perfectly reasonable; our
own preference for infralittoral is merely that infra- and supra- make a contrasting pair,
and have the substantial advantage that they are not readily confused when seen repeatedly on a printed page. Supralittoral and sublittoral, on the other hand, are easily
confused.

The zone of myxophyceans, or dusky lichens, or both, can simply be called the "black zone"; and the covering of lithothamnia on the lower part of the shore hardly needs a more specific name.

It may be noticed that in naming the three major zones of the shore, we have used combinations of the word littoral without using this word by itself, devoid of any prefix. This is because the word has been used in such fundamentally different senses by different authors that it has no standard meaning, whereas if we use the terms "tidal zone" or "intertidal zone" for the strip of coast between extreme high water and extreme low water, there is no ambiguity.

This tendency to confusion, which so readily besets the human mind, has sometimes led to the impression that, because authors disagree among themselves about what the main zones of the shore should be called, there is some doubt about the existence of the zones themselves. We must therefore emphasize that the zones themselves are a part of nature; they exist, whatever biologists think about them. It is the authors, not the zones, that are uncertain.

There remains the problem that each of the three main zones may be subdivided into subzones, and that names are often required for these. But whereas the principal zones occur over a great part of the earth's shores, their populations changing from place to place, the subzones are often characterized by very locally distributed organisms. Consequently, while one seeks general names for the main zones, which refer to no particular organisms, it is quite practicable to use the names of characteristic organisms for subzones. This means, of course, that the subzones can have different names in different parts of the world; in practice, this has proved to be no inconvenience. We can therefore speak, for example, of a "*Pyura* subzone" or "*Pyura* community" occurring in the broader zone known as the infralittoral fringe.

SELECTED REFERENCES

Anand, P. L., 1937. An ecological study of the algae of the British chalk-cliffs, I and II. *J. Ecol.* 25: 153-188, 344-367.

Berner, L., 1931. Contribution à l'étude sociologique des algues marines dans le Golfe de Marseilles. *Ann. Mus. Hist. Nat. Marseille.* 24(1): 3-84.

Broekhuyser, G. J., 1940. A preliminary investigation of the importance of desiccation, temperature and salinity as factors controlling the vertical distribution of certain intertidal marine gastropods in False Bay, South Africa. *Trans. Roy. Soc. S. Afr.* 28: 255-292.

Chapman, V. J., 1946. Algal zonation in the West Indies. *Ecology* 27(1): 91-93.

Colman, J. S., and Anne Stephenson, 1966. Aspects of the ecology of a "tideless" shore. *In* H. Barnes, ed., *Some Contemporary Studies in Marine Science,* pp. 163-170. London: Allen and Unwin.

Cranwell, L. M., and L. B. Moore, 1938. Intertidal communities of the Poor Knights Islands, New Zealand. *Trans. Roy. Soc. N. Z.* 67: 375-407.

du Rietz, G. E., 1925. Die Hauptzüge der Vegetation der Insel Jungfrun. *Svensk Bot. Tidskr.* 19(3): 323-346.

Feldman, J., 1938. Recherches sur la végétation marine de la Méditerranée. *Rev. Algol.* 10: 1-339.

Fritsch, F. E., 1931. Some aspects of the ecology of fresh-water algae. *J. Ecol.* 19: 233-272.

———, 1945. *The Structure and Reproduction of the Algae,* vol. 2. Cambridge: Cambridge University Press.

Geitler, L., 1930-32. Cyanophyceae. *In* G. L. Rabenhorst, ed., *Kryptogamenflora von Deutschland, Österrich und der Schweiz,* vol. 14. Leipzig: Akademische Verlagsgesellschaft.

Ginzberger, A., 1925. Der Einfluss des Meerwassers auf die Gliederung der süddalmatischen Küstenvegetation. *Österr. Bot. Z.* 74(1): 1-14.

Grubb, V. M., 1936. Marine algal ecology and the exposure factor at Peveril Point, Dorset. *J. Ecol.* 24: 394-423.

Hedgpeth, J. W., 1962. Intertidal zonation and related matters. *In* E. F. Ricketts and J. Calvin, *Between Pacific Tides* (3d ed.), pp. 345-516 (appendix). Stanford, California: Stanford University Press.

Hodgkin, E. P., 1960. Patterns of life on rocky shores. *J. Roy. Soc. W. Aust.* 43(2): 35-43.

Johnson, D. S., and A. F. Skutch, 1928. Littoral vegetation on a headland of Mt. Desert Island, Maine. *Ecology* 9(3): 188-215, 307-338.

King, C. A. M., 1962. *Oceanography for Geographers.* London: Arnold.

Kylin, H., 1937. Über die Farbstoffe und die Farbe der Cyanophyceen. *Förhandl. Kungl. Fisiogr. Sällsk. Lund* 7(12): 1–28.

Marmar, H. A., 1930. *The Sea.* New York: Appleton.

Singer, C., 1931. *A Short History of Biology.* Oxford: The Clarendon Press.

Smith, A. L., 1921. *Lichens.* Cambridge: Cambridge University Press.

Southward, A. J., 1958. The zonation of plants and animals on rocky sea shores. *Biol. Rev. Cambridge Phil. Soc.* 33: 137–177.

3

DISTRIBUTION OF
SHORE POPULATIONS

SOUTH AFRICA

We begin this survey of world shores with the results of the ten years of study of the South African coasts from which it sprang.

COMPONENTS OF THE SOUTH AFRICAN BIOTA

During our work on the South African coasts we experimented with several systems of categories for classifying the species on them, and found that the one postulated in Figure 3.1 gave the most useful results. In this figure the coast of South Africa has been represented as a strip divided into three sections, the West Coast (Port Nolloth to Cape Point), South Coast (Cape Point to Cape Padrone), and East Coast (Cape Padrone to Durban). (See Figures 8.1 and 8.3.) Related to these three divisions are shown various hypothetical types of distribution, numbered 1–13, showing all the main categories, as well as some that are relatively localised.

In fact, species are found in South Africa that fit most if not all of these categories, but some categories contain more species than others. For the purposes of an analysis of the population, it is convenient to combine some of these categories into groups.

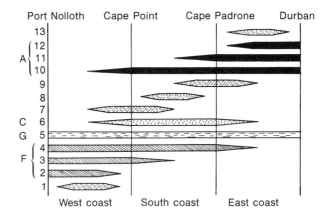

FIGURE **3.1**

Thirteen ways in which organisms may be distributed round the South African Coast. The coast is divided into three sections; distances are not drawn to scale. Category *12*, for example, includes those warm-water species that do not extend as far south as Cape Padrone; category *11*, those that reach beyond Cape Padrone to the south coast; and category *10*, those that stretch beyond Cape Point—the other categories can be interpreted similarly. Categories *1*, *7–9*, and *13* are various local components, each of which comprises species restricted to particular localities; the cold-water component (*F*) includes categories *2–3*; the ubiquitous component (*G*) is coextensive with category *5*; the south-coast (or warm-temperate) component (*C*) is coextensive with category *6*; and the warm-water component (*A*) includes categories *10–12*.

Thus categories 2–4 (all containing species characteristic of the whole of the west coast) may be grouped as a cold-water component, while categories 10–12 (all containing species characteristic of the whole of the warmest part of the coast) may be grouped as a warm-water component. Some of the constituent species of both components react very sharply to changes of temperature, and their ranges are limited by the temperature of the coastal waters; others, more eurythermic, persist for varying distances along the coast beyond the region where they are most typical. Using this classification, we find that there are four main components on the South African coast: warm-water, cold-water, south-coast (warm-temperature), and ubiquitous; and that, in addition to this, there are certain more local components, notable among which are those of the overlapping regions in the southwest and southeast. For the actual proportions and rates of fall-off of these components, see Figure 3.2, which demonstrates very clearly the reality of the control exercised by water masses of different temperatures on coastal populations. In Figure 3.3 are shown the main currents that affect the coasts of South Africa.

The Warm-Water Component (*Categories 10–12*)

The warm-water component is the largest in the South African biota. It is represented in our lists by 159 species (54 plants and 105 animals). It includes those species whose distribution extends from Durban westward, though not as far as Port Nolloth. Their

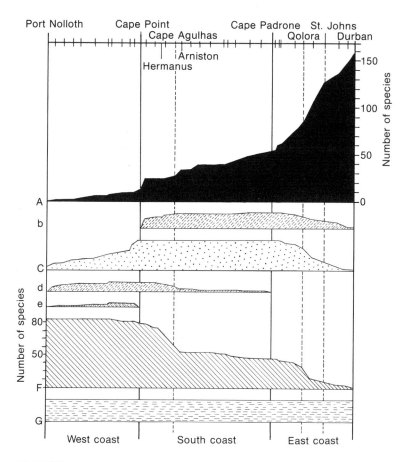

FIGURE **3.2**

The distribution of the geographical components of the intertidal biota of the coast of South Africa. For details and scale, see Figures 8.26 and 8.27 (pp. 129 and 130), in which the distributions of fauna and flora are illustrated separately.

range, however, does not begin at Durban: many of these species, if not all, extend into warmer water still, but Umpangazi, Natal, was the eastern limit of our survey. In a westerly direction these species gradually disappear; and in accord with the point at which they disappear, they can be placed roughly into three categories. Category 12 contains the majority (109 species), which apparently die out sooner or later on the east coast, none reaching Cape Padrone; category 11 contains the small number (58 species) that pass that point and extend varying distances along the south coast; and category 10 contains a few (12 species) that continue as far as the southern part of the west coast.

Category 12 contains the more stenothermic forms, and constitutes a very typical tropical and subtropical collection. There are algae of such typical coral reef genera as *Caulerpa, Sargassum, Microdictyon, Valonia, Padina, Turbinaria,* and *Galaxaura.*

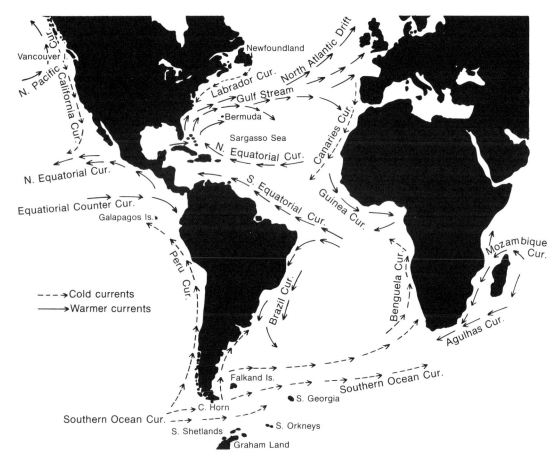

FIGURE **3.3**

The position and flow of the main currents along the coasts of Africa and along the east and west coasts of North and South America. A comparison with Figure 3.4 shows how closely the changes in the intertidal biota of South African shores are correlated with the temperatures of currents that bathe the African coasts. The distribution of the cold-water fauna that creeps down the east coast of North America is controlled by the Labrador Current, which passes south to Cape Cod, and the marked change of species at Cape Hatteras coincides with the course of the Gulf Stream, which at this section of the coast swings away eastwards into the Atlantic Ocean (see Figures 3.9-3.11). The cool California Current influences the constituents of the temperate biota on the west coast, (see Figures 3.6-3.8); on the east, the warm Gulf Stream carries semitropical species into the temperate regions.

Tropical coelenterates are strongly represented: *Xenia coerula;* an abundance of zoanthids of the genera *Palythoa, Zoanthus,* and *Isarus;* and a fair representation of reef corals themselves—species of *Anomastrea, Favia, Favites, Goniastrea, Pocillopora, Psammocora, Stylophyllia,* and *Symphyllia.*

Among the remaining groups of animals, the Mollusca, Crustacea, and Echinodermata are particularly well exemplified, and there are frequent examples of reef-colonising species. These include the oyster *Ostrea cucullata;* the gastropods *Monodonta australis,*

Littorina scabra, Turbo coronatus, and members of the genera *Drupa* and *Nerita;* the sea urchins *Diadema savignyi, Echinometra matthaei, Echinostrephus molare, Stomopneustes variolaris,* and *Tripneustes gratilla;* the bêches-de-mer *Holothuria cinerascens* and *H. vagabunda;* the prawn *Alphoeus crassimanus;* and the crabs *Grapsus maculatus, G. strigosus, Ocypoda ceratophthalmus, O. kuhlii, Trapezia cymodoce* (which is a commensal of the coral *Pocillopora*), and *Zozymus aeneus.*

There are, however, no coral reefs off the shores of Natal. The substratum is rock (sandstone, dolerite, and so forth) and the corals are not sufficiently vigorous to create a true coral reef. The coast is an example of the marginal belt of coral seas. It exhibits some very definite coral-reef associations; some species of these associations, perhaps freed from coral dominance, are able to grow exuberantly, notably the zoanthids. But the temperature is not high enough to support a really flourishing growth of corals: compared with a coral reef proper, the growth is sparse and the number of species very limited. Few seem common, the most plentiful being the curious *Anomastrea irregularis.* There are hardly any reef-builders, such as species of *Acropora* and *Porites,* and the coral growth occurs for the most part in rock pools—not, as on a coral reef, in the infralittoral fringe, which in Natal is often dominated by algae. There are surprisingly large numbers of simple corals in pools, caves, and underhangs.

The South-Coast Component *(Category 6)*

The south-coast component, an important group that includes 27 species of animals and 5 of algae, has its headquarters on the south coast between Cape Point and Cape Padrone. The constituent species range beyond these points for varying distances along the southern parts of the east and west coasts where some of them remain plentiful, but they rarely reach either as far as Port Nolloth in the west or Durban in the east. Many of the species are common, so that it is difficult to choose examples, but the following are among the most interesting: *Patella cochlear,* which is so abundant that it characterises one of the subzones of the intertidal belt; *P. oculus,* which is accompanied throughout most if not all of its range by the small commensal polyclad *Notoplana ovalis;* the giant *P. patriarcha;* the smaller *Helcion pruinosa;* the very large snail *Turbo sarmaticus* and the smaller ones *Oxystele sinensis* and *Cominella cincta;* the abundant prickly chiton *Acanthochiton garnoti;* the large ormer *Haliotis midae;* and the red seaweeds *Gelidium pristoides* and *G. cartilagineum,* each characteristic of certain levels on the shore.

The Cold-Water Component *(Categories 2–4)*

The occurrence of the cold-water component begins not at Port Nolloth, but north of that point in regions we were unable to visit. The constituent species of the component discussed here are only those that extend from Port Nolloth along the coast to the east; few of them range as far as Durban.

On the Natal coast, mean annual temperatures are in the neighbourhood of 20°C and above (21.8°C for Durban); but at Port Nolloth, which lies in the region of minimum temperature of the surveyed area, they are nearer 12°C. On the west coasts of South Africa there is a temperature gradient from north to south; Port Nolloth lies in the region of minimum temperature, whence there is a slight rise towards Cape Point.

There is no parallel here to the great abundance of species inhabiting the east coast, but we have listed five forms as living on the west coast only; with further knowledge they may be found to extend at least a little beyond it, but they are undoubtedly species that are virtually confined to the west coast. A good example is the giant kelp *Laminaria pallida*, which reaches its greatest development in regions of minimal temperature in the north, but becomes overshadowed by *Ecklonia buccinalis* southward; east of Cape Point, *L. pallida* does not occur in significant amounts. There is a rapid falling off of west-coast species in the Agulhas region between Hermanus and Arniston, which can be compared with the diminution of warm-water forms occurring between Port St John and Qolora.

The "personnel" of the cold-water component is very unlike that of the warm-water one. We find a group of 42 species (28 algae and 14 animals) that are more or less limited to the west coast, or continue for varying distances along the south coast, but never reach the east one. The largest number of these consists of what may be described as typical west-coast algae, including not only the giant species *Ecklonia buccinalis* and *Laminaria pallida*, but a number of other relatively large forms, such as *Aeodes orbitosa*, *Schizymenia obovata*, *Pachymenia carnosa*, *Grateloupia longifolia*, *Iridophycus capensis*, and *Gigartina stiriata*. The ecologically important *Champia lumbricalis* belongs to this group, as do the three principal epiphytes of the kelp, *Polysiphonia virgata*, *Carpoblepharis flaccida*, and *Suhria vittata*. Most of these seem to die out west of Cape Agulhas, or near it. The animals, although fewer than the plants, include such very important species as *Mytilus crenatus*, *M. meridionalis*, *Patella compressa*, *P. granatina*, *Argobuccinum argus,* and *Cominella delalandii*.

The Ubiquitous Component   *(Category 5)*

It is remarkable that no fewer than 23 of the species in our list (17 animals and 6 algae) are distributed round the entire coastline of South Africa from Port Nolloth to Durban, undoubtedly extending beyond these localities at both ends of their range. At the beginning of the survey, we were doubtful whether we should find any species able to adapt themselves to so wide a series of temperature ranges as those that prevail at Port Nolloth and in Natal, but undoubtedly such species do exist. Some of them seem to be distributed fairly evenly round the coast, showing nothing more than minor local variations from one place to another; but many show a definite change in colour, size, or abundance as they are traced round the coast. The following are examples.

Gunnarea capensis is a polychaet, building coarse sandy tubes. It is very abundant in many places on the west coast, where its tubes may form enormous aggregations covering huge areas of rock to a depth of 15 inches, and are able to change the whole configuration of a beach. The species continues to be important on the south coast, although the large colonies tend to become less frequent, but in Natal it tends to construct small groups of tubes, or only isolated ones, rather than large masses.

Patella barbara is a extremely variable limpet. It remains abundant all along the coast, attaining a larger average size in the colder water, and developing a curious variety in Natal that superficially resembles *P. tabularis.*

Patella granularis is another limpet numerous throughout the length of the coast, but it undergoes a distinct change in size from east to west, attaining a much larger size in the colder water.

Parechinus angulosus is a sea urchin plentiful on the west and south coasts, becoming much scarcer in Natal. On the south coast, and on the southern part of the east coast as well, it may be scarlet, violet, lilac, or various other colours. On the southern part of the west coast, however, it is uniformly pallid; in Natal, we saw only one rather dingy colour form.

Tetraclita serrata and *Octomeris angulosa* are large barnacles that form a most important element in the fauna of the east and south coasts. These become less plentiful on the west coast, and are fairly rare at Port Nolloth, being confined to crevices and other restricted habitats.

Asterina exigua is a small starfish abundant all along the coast. In the east and south, it is varied in colour: individuals are commonly marked by a variegated pattern, and the colours may be very bright. On the west coast, they are uniform in tint or little marked, and assumes such dull shades as drab, greyish, brownish, and dark green.

Siphonaria capensis, a pulmonate limpet, is plentiful everywhere, but it develops a variety with a blackish lining in the shell on westward coasts.

Gelidium cartilagineum (a tough, wiry, red seaweed) and *Littorina knysnaënsis* (a small periwinkle) belong, strictly speaking, in this category; but their centres of distribution are such that they have been considered as belonging more truly to the south-coast and the cold-water components, respectively.

The Local Components *(Categories 1, 7, 8, 9, and 13)*

There appear to be certain species that do not belong to any of the categories so far described. Their distributions differ from the type of occurrence already outlined, chiefly in that they seem to be more restricted—some of them to regions coinciding with none of those as yet defined.

For instance, there are a few species (category 7) that extend onto the west and south coasts, but not as far as Port Nolloth in the west or Cape Padrone in the east; the best examples are two coarse brown algae, *Bifurcaria brassicaeformis* and *Pycnophycus laevigatus,* ecologically important in their restricted region. Species belonging to category 9 are more numerous (11 animals and 6 algae) and include such important forms as the small prickly kelp *Ecklonia radiata, Sargassum heterophyllum* (with its headquarters on the south and east coasts), and the ormer or abalone *Haliotis sanguinea.* None of the species in our list are confined to the south and east coast (categories 8 and 13), and we suspect that only *Macrocystis pyrifera* reaches neither Cape Point nor Port Nolloth (category 1).

With extended knowledge, some species may have to be referred to different components, but we believe that the components themselves will remain valid, even if with modified boundaries and "personnel."

CHANGE OF NUMBER OF SPECIES
FROM WEST TO EAST

It is often supposed that, in colder waters, species are fewer and individuals more numerous, and that, in warm waters, there is a greater variety of species and fewer individuals of any one species. The latter part of this thesis certainly needs qualification for intertidal species.

On the cold west coast of South Africa, for example, there are various organisms that occur in great quantity or even in mass formation: such animals as *Gunnarea capensis, Bunodactis reynaudi, Littorina knysnaënsis, Mytilus crenatus, M. meridionalis, Patella argenvillei, P. cochlear,* and such seaweeds as *Ecklonia buccinalis, Laminaria pallida, Champia lumbricalis,* species of *Gigartina,* and various others. There are other species on the west coast that are common without being abundant, and still others that are quite rare. It is not at all a question of small numbers of species that are all abundant, but of many species (even though fewer than in Natal) in varying quantities.

The same thing applies to the other coasts. On the south coast, such animals as *Pyura stolonifera, Pomatoleios crosslandi, Octomeris angulosa, Tetraclita serrata, Chthamalus dentatus, Mytilus perna, Patella cochlear,* and such seaweeds as *Hypnea spicifera, Plocamium corallorhiza,* and *Gelidium cartilagineum* are immensely abundant; and here again other species are present but less numerous.

In Natal, a warm region, the number of species is greater and, for many species, so is the number of individuals. Some of the species already mentioned for the south coast (*Pomatoleios,* the barnacles, *Mytilus perna, Hypnea spicifera,* and *Plocamium corallorhiza*) are again amazingly plentiful, and other warm-water species are also abundant, including zoanthids, *Ostrea cucullata,* several tropical snails and crabs, certain limpets, and algae of the genera *Caulerpa, Gelidium, Galaxaura,* and *Rhodymenia.* A great many species are represented, and they exhibit, as usual, all degrees of abundance; nevertheless, forms that are about as numerous as they well could be are by no means few.

Furthermore, some species, such as *Patella granularis, P. barbara, Siphonaria capensis,* and *Asterina exigua,* are plentiful the whole way round the coast, from the Orange River to Natal. Perhaps in the present state of our knowledge all we are entitled to say is that, although our work does indicate a genuine increase in the number of species as the water becomes warmer, there is no lack of species rich in individuals in either warm or cold waters. All degrees of abundance of individuals are exhibited throughout the temperature range, and there is no sign of a fall in individual abundance offsetting the rise in number of species. Possibly the proportion of species showing low degrees of abundance is higher in Natal than on the west coast, but to demonstrate this a quantitative study on a positively staggering scale would be necessary.

It may be added that changes in temperature affect individual species in two ways: (1) the more stenothermic species have an optimal temperature range within which each reaches its greatest profusion, waning as the limits of the optimal area are reached; and (2) the eurythermic ones often undergo some change, whether in size, colour, or abundance, as the average temperature changes in different parts of their range.

Figure 3.4 shows the distribution of 318 tidal species among 36 localities on the

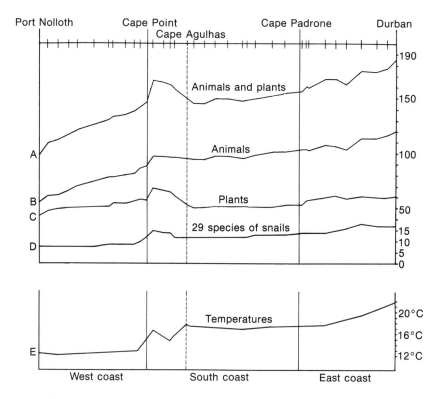

FIGURE **3.4**

The distribution of 318 tidal species on the South African coast compared with sea-surface temperatures. The numbers *0–190* along the right scale indicate the numbers of the 318 species present at various localities. Details and horizontal scale are as in Figure 3.2. Three vertical scales are included; lines *A–C* are drawn to the scale indicated on the right by the figures *50–190;* line *D* is drawn to twice this scale, as indicated by the figures *0–15* (all lines are referred to the same zero-line, but the scale has been increased for line *D* because it refers to much smaller numbers than the others, and is thus rendered more easily comparable with them); and line *E* shows sea-surface temperatures, in degrees Celsius, based upon relevant figures given by Isaac (1937) and Marchand (1932). For the most part, the temperatures given are annual means.

South African coast. The information on which the table is based was taken from our own records. The number of animal species (line *B*) present at each locality tends to rise steadily from west to east, the number at Durban being more than twice that at Port Nolloth. The number of species of algae (line *C*) shows this gradient, but less markedly, as there is a stronger western element among the algae, possibly due to a special outburst of vigour connected not only with low temperature, but also with a very favourable supply of phosphate and nitrate from the Antarctic upwelling along the west coast. When we combine the number of animal and algal species (line *A*), the west-to-east gradient is quite obvious.

All the lines show minor irregularities, partly because of genuine local variations and partly because of the nature of the information available. Undoubtedly significant is the decided peak shown by all three curves for the region between Cape Agulhas and Cape Point. This is the region where the eastern warm-water component of the population has not yet been reduced to its minimum, the western cold-water one is still strong, and the south-coast one is still at its maximum. The peak is accentuated by the addition of contributions from local components. It should be noted that the tip of this peak coincides with the one in the temperature line, E, at the bottom of the figure, which represents the warmer northwestern corner of False Bay. It is followed by a dip, indicating the return to colder conditions in the stretch of coast between Cape Hangklip and Danger Point. The irregularities present in the population curves of the other principal region of overlap (the eastern overlap from Port Elizabeth to Port Edward) are not as clear as the peak in the western one, and cannot be safely interpreted from available data.

This example of the phenomenon of an increase of species as one proceeds from colder to warmer water is unexpectedly definite.

WORLD RELATIONSHIPS OF THE SOUTH AFRICAN BIOTA

The components of intertidal shore life of South Africa must also be viewed from the standpoint of their relation to the biotas of the rest of the world. The coastal populations include members of the following groups.

Indic or Indo–West-Pacific Species

These have a distribution over part or the whole of the Indo–West-Pacific area. Many of these reach the limit of their distribution in Natal, but some continue further and may even extend around the whole coast or beyond (*Asterina exigua*, for example). It is perfectly possible that some Indic or Indo–West-Pacific species have even spread, in the course of ages, round the Cape and across the South Atlantic to South America, ending up in the West Indies region.

Southern-Ocean Species

There is probably a circumpolar shallow-water population, adapted to very cold water, round Antarctica; but although there is a cold-temperate temperature belt round the southern regions that contains the subantarctic islands and the southern ends of South America and New Zealand, there seems to be no single circumpolar cold-temperature littoral fauna. Rather, there seem to be several faunas of considerable independence, with limited interconnections, but with definite associations with the antarctic fauna. South Africa certainly possesses species derived from some of these southern-ocean populations (*Plagusia chabrus*, for example) found also in New Zealand, Australia, and Chile.

Atlantic Species

South Africa also has species that have a wide distribution in the Atlantic, even some that occur in the North Atlantic. For example, *Actinia equina* is found from Arctic shores all down the coasts of Europe and round Africa to Durban. It is replaced by similar forms in Australasia and elsewhere. *Marthasterias glacialis* occurs on Arctic shores, and on those of Europe, the Mediterranean, and East Africa. The distribution of Atlantic species may be genuinely discontinuous or it may go below tidemarks along the central parts of the eastern Atlantic margin. This type of distribution seems to be more characteristic of some animal groups than of others.

Endemic Species

South Africa undoubtedly has a high percentage of endemic forms, possibly about 60%; but because species are endemic to a country, it does not follow that they all have the same distribution along its coast. On the contrary, our work showed clearly that in South Africa there are warm-water endemics, cold-water endemics, and so on. There are endemic species in 11 out of our 13 possible categories, and they seem to form an overlapping series round the coast.

Cosmopolitan Species

There are certain more or less cosmopolitan species in South African waters, such as the algae *Centroceras clavulatum* and *Ulva lactuca*.

If we study the proportions present in the various geographical components of the South African population, we find that South Africa is an independent region rich in endemic forms. The outside relationships of the South African biota are mainly with the Indo-West-Pacific biota, with subsidiary connections with Atlantic, southern-ocean, and cosmopolitan biotas. Thus, the warm-water component of the east coast includes a large proportion of Indic and Indo-West-Pacific forms, but is not without endemic species and others, whereas the south-coast and cold-water components consist primarily of endemic forms, although they possess subsidiary numbers of species belonging to the Indic, Indo-West-Pacific, and other regions. It might be expected that species contained in small local categories must necessarily be endemic, but this does not follow: *Macrocystis*, for example, occurs primarily outside South Africa, and merely has a local outpost there because conditions happen to suit it.

It may appear at first sight that the complex series of constituents demonstrable in the South African biota does not fit in with the simpler idea of three coasts each with its own population. However, the two conceptions do agree because the underlying combinations group themselves in such a way that each of the three main components persists along a considerable stretch of coastline, although of course there are overlaps where conditions are more complex (see Plate 6).

The problems here presented were further elucidated during our later, though inevitably less detailed, surveys of the Atlantic and Pacific coasts of North America.

NORTH AMERICA

THE BIOTAS OF THE ATLANTIC
AND PACIFIC COASTS

A careful study of the text figures for this chapter will provide a clearer understanding of the distribution of the biotas along these coasts than would longer written accounts. There are two sets of graphs for each of the two coasts, one set compiled from data we collected during our field work (Figures 3.7 and 3.10), and the other from information taken from American publications (Figures 3.8 and 3.11).

The populations of each coast can be separated roughly into three main groups—northern, southern, and central. Also, there are a few species that are widespread and extend the full range from north to south, or vice versa, albeit usually with changes in size, colour, and so forth. The extent of these groups along the American coasts is illustrated in a map (Figure 3.5) and in two sets of diagrams (Figures 3.6–3.9). An analysis of these figures is given below.

General Comment

In our graphs, the identifications of the species have been as scrupulous as possible, but it is obvious that allowances have to be made for various possibilities of error, as, for instance, our taking more notice of understone forms at one place than another, and the great variations in wave action between different areas. Nevertheless, our graphs do seem to reflect real conditions. They show (1) the distribution of the biota from an ecological point of view rather than from a purely geographical one, emphasising the dispersal of common species flourishing at various centres; (2) the essential distribution of key species (ecologically permanent forms), rather than a summary of complete geographical ranges; and (3) a distribution of populations. They also demonstrate more clearly than a geographical type of figure how sharp may be the changeover between populations.

The West Coast

A particular feature of the west coast is the relatively high proportion of species with a long range from north to south, and vice versa (Figure 3.6, categories 1 and 6).

The Southern Group (*Categories 2–5*).   This is the most distinctive group. The areas showing the most definite change from southern to central species are those around Pacific Grove and Point Conception (Figure 3.7). There is, undoubtedly, a southern biota south of Point Conception that includes warm-temperate forms that die out in Lower California or in the Gulf of California. The tropical fauna en masse does not reach the United States. Some southern species have a long northern extension beyond Point Conception, and have a marked influence on the composition of the central group.

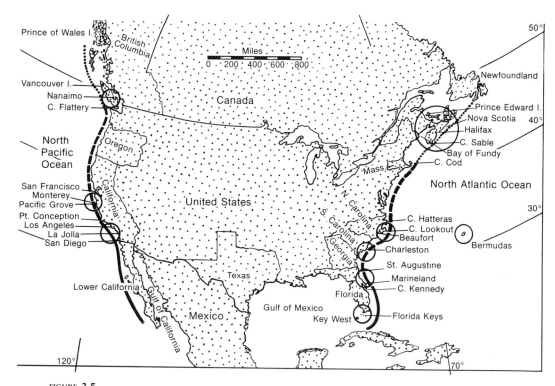

FIGURE **3.5**

A map of North America showing the locations of places named in the text. Circles indicate the areas studied. The thick lines along the east and west coasts illustrate the extent of the three main groups of species: *unbroken lines,* the southern groups; *broken lines,* the central groups; *dotted lines,* the northern groups.

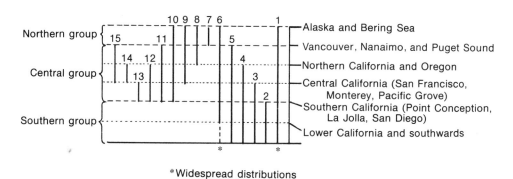

*Widespread distributions

FIGURE **3.6**

The various categories into which the three main groups of west-coast species may be subdivided according to the occurrence of the species along the coast between specified stations. The number of categories included in each group is partly dependent on the number of species studied. On the right-hand side are given the names of the stations, and on the left, braces indicate the stations defining the geographical limits of each group. An analysis of this figure is given in the text.

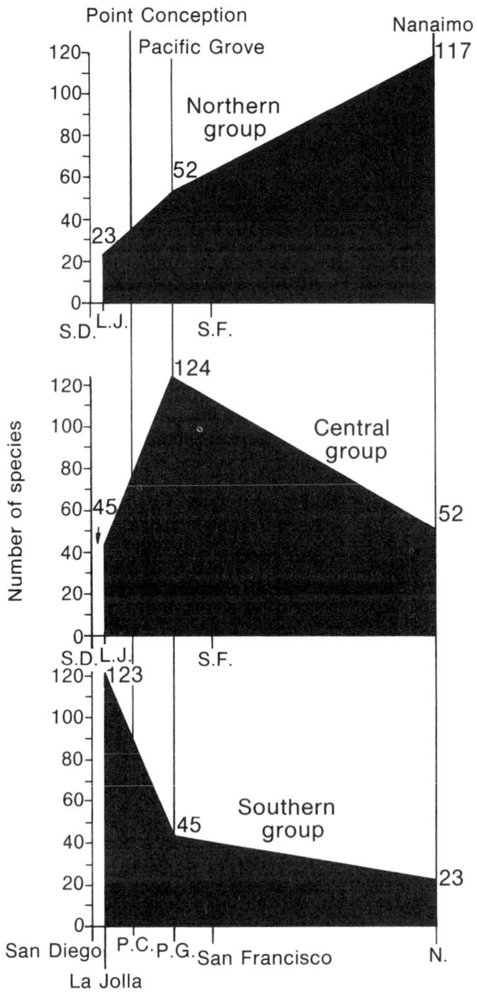

FIGURE **3.7**

The distribution of the three main groups of the biota of the west coast of North America. The graphs are based on data collected by the authors, and refer mainly to the commonest species. The graph for the northern group shows that of the 117 species collected at Nanaimo, 52 extended to Pacific Grove and 23 to La Jolla; the other graphs can be similarly interpreted.

An examination of the known southern limits of this southern group reveals that about three-quarters of the species that belong to categories 2, 3, 4, and 5 have their southern limits in Lower California or in the Gulf of California. Some may overlap into tropical regions, but they are only outliers of the main populations. Only about one-quarter seem to be genuine tropical species that extend north. This is probably significant, although the region needs more exploration. La Jolla is not really tropical, as it lacks any great number of stenothermic tropical forms; it can only be regarded as a warm-temperate area.

The Northern Group (*Categories 7–10*). This is not as clear-cut as the corresponding group on the Atlantic coast, but it certainly exists. It includes many species with long southern extensions that affect the populations of the central regions. There

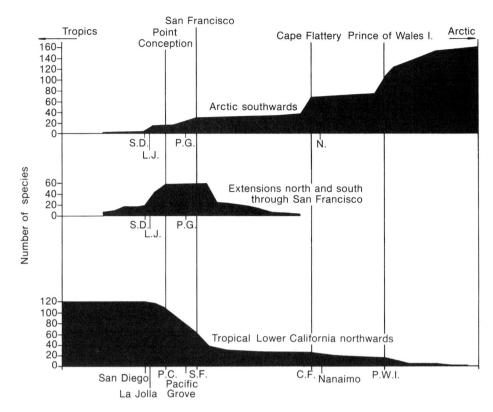

FIGURE **3.8**

Graphs similar to those in Figure 3.7 compiled from data taken from Keen (1937). These graphs, though based on greater numbers of species, indicate the same three main groupings, the central one with stronger northern affinities. The greater number of species smooths out the lines.

is no place at which northern forms are delimited as markedly as the southern ones are at Point Conception (Figure 3.8).

The Central Group (*Categories 11–15*). Many of the species of this group extend a very long distance—southwards to San Diego and northwards to Vancouver Island. Its greatest number of species are on the coast between San Francisco and San Diego.

From the data given above, it is clear that a warm-temperate biota is the dominant one on the west coast. From our own work and from other records of the ranges of species it seems that there is a definite central group of species with its extreme limits at San Diego and Vancouver. At Pacific Grove, only 1% of true southern species extend into this extensive central group, but 13% of northern species do so, which indicates that the central group has more affinity with the north than with the south.

Naturally, further work may show that there are more than three main groups, but it must be remembered that our classification was founded, on the whole, on the commonest animals.

THE EAST COAST

The species on the east coast fall roughly into the same groups, but with different affinities (Figure 3.9).

The Northern Group (*Categories 1–10*).　Most of the species belonging here are genuine northern ones that show varying degress of southward extension. A limited number are widespread (categories 10, 9, 5, and 4), and a few of these have a long northern range into the Arctic (categories 5 and 4). At least two species—*Crassostrea virginica* and the crab *Neopana texana savi*—are characteristic of what has been termed the "Transatlantic Provinces" (Cape Cod to Cape Kennedy), but have an unusually long northward extension (categories 3 and 8). Of the more northern species of this group (categories 2 and 7), more disappear north of Cape Hatteras than continue past it (Figure 3.10).

The Southern Group (*Categories 11–14*).　This group is equally distinct. It consists of species of which more than three-quarters are strictly tropical and do not occur north of St Augustine near the Florida border (category 11). Most of the rest reach South Carolina or North Carolina (categories 12 and 13), while a mere handful are known north of Cape Hatteras (category 14). There are a few widespread forms: notable are the barnacle *Chthamalus fragilis* (whose tropical form is present at Woods Hole, Massachusetts, but was found by us to be more characteristic of the Carolinas), and the gastropod *Cantarus tinctus* and the crab *Sesarma cinereum* (which seem to be Texas-Carolina types that dip south onto the Florida Keys).

The Central Group (*Categories 15–17*).　The species in this group, which our graphs show centred on the coast of the Carolinas, exhibit practically no purely northern

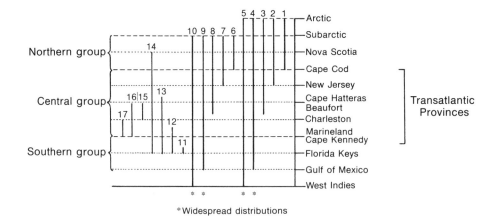

FIGURE **3.9**

A diagram similar to Figure 3.6 for the east coast of North America. An analysis of this figure is given in the text.

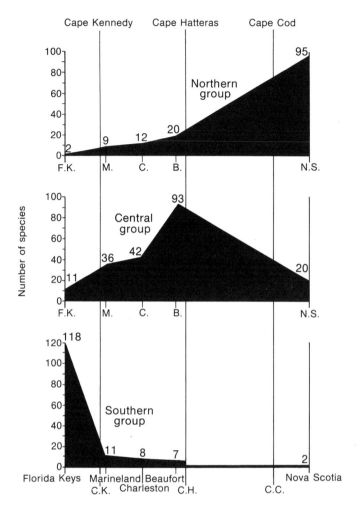

FIGURE 3.10

Graphs similar to those in Figure 3.7 compiled from data gathered by the authors on the east coast of North America.

affinity, but a few are widespread. Most of the species fall into two groups: (1) tropical species with a northern extension, or at least those that have one end of their distribution in the tropics (categories 16 and 17); and (2) species that seem characteristic of the Transatlantic Province (category 15). It might be expected that our graphs would show a higher percentage of species reaching southwards, and more species of the Florida Keys. The explanation for this is probably as follows.

First, the species designated as tropical with northern extension would perhaps not be regarded as such if more were known about them. Some may be more characteristic of the intertidal zones in the north than in the tropics. It does not follow that, in the tropics, they are all either intertidal or in the most flourishing condition; the polyzoan *Anguinella palmata*, for example, which is found from the tropics to Cape Cod, seems to achieve its best development between Cape Hatteras and Cape Kennedy. Second, there are very definite environmental differences between the shores of the Carolinas and of the Florida Keys. The former are the more oceanic. Sedimentation occurs on both, but the sand scour of Marineland and the less violent shifting of sand around the breakwaters of Charleston and Beaufort have not all the same effect as the

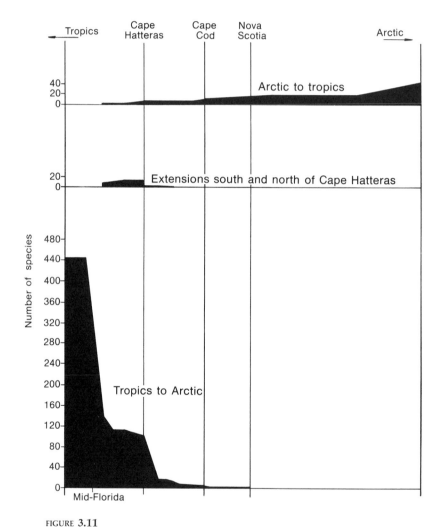

FIGURE **3.11**

Graphs similar to those in Figure 3.7 compiled from data for the east coast of North America taken from Johnson (1934). Compare with Figures 3.8 and 3.10. Note that the central group, unlike that of the west coast, has stronger southern than northern affinities.

smothering type of fine sediment to be found on the Keys. Such differences may have a selective action on possible inhabitants, and perhaps if we had worked on some West Indian beach more like Marineland, it might have provided more Carolinian forms than seem to occur on the Florida Keys. Be that as it may, Marineland completely lacks any stenothermic tropical species.

There seems to be no doubt that on the east coast of North America—intertidally, at least—there is a distinct population adapted to intermediate temperatures between Cape Kennedy and Cape Hatteras, with a fair extension between the latter and Cape Cod. As this population markedly lacks the stenothermic tropical forms of the Florida Keys, it may suitably be described as warm-temperate, although in addition to species that seem generally endemic to it (although often extending beyond it), it has several that are shared with the tropics (Figure 3.11). Thus, by contrast to the central group of the west coast, the affinities of the central group of the east coast are much more with warm-water than with cold-water species (Figure 3.5).

SELECTED REFERENCES

Isaac, W. E., 1937. South African coastal waters in relation to ocean currents. *Geogr. Rev.* 27(4): 65–664.

Johnson, G. W., 1934. Marine Mollusca of the Atlantic Coast. *Proc. Boston Soc. Nat. Hist.* 40: 1–203.

Keen, Myra, 1937. *West North American Marine Mollusca.* Stanford, Calif.: Stanford University Press.

Marchand, J. M., 1932. Hydrographic investigations during 1930. *S. Afr. Fish. Mar. Biol. Surv. Rep.* (8).

Southward, A. J., 1958. The zonation of plants and animals on rocky sea-shores. *Biol. Rev. Cambridge Phil. Soc.* 33: 137–177.

4

TROPICAL SHORES

INTRODUCTION

Typical coasts in the tropics have certain unique features that disguise the fact that their zonation follows the pattern so widespread on the world's shores. The two respects in which tropical shores differ markedly from cold and temperate ones are as follows.

First, mangroves fringe many tropical coasts, occupying both the open beach and estuaries. They can grow on rock, mud, or sand, and very commonly have mud between their exposed strut roots. They can grow so densely as to become almost impenetrable. The trees support a characteristic fauna of species that live on the trees themselves or on their roots, and in the sand and mud among the roots. Mangroves usually inhabit intertidal regions, and their lower parts and strut roots are regularly uncovered at low waters and covered at high waters (see p. 315).

The second unique feature of tropical shores is that coral reefs appear where conditions are suitable. We will give no account of the variety, origin, or types of coral reefs, on which there is an enormous literature. They will be simply considered in relation to the zonation on them.

Where temperatures are more nearly tropical—as, for instance, at Durban, South Africa (see p. 105)—a few reef corals appear in rock pools between tidemarks, but there is no general growth of them in the infralittoral. Other tropical organisms, such as zoanthids and tropical weeds, may flourish extensively. Otherwise, the arrangement of animals on the shore is very much as on any warm-temperate coast. But where the surface waters of the ocean reach sufficiently high temperatures (a mean of not less than 22°C) reef corals are no longer restricted to pools, and spread out to form infralittoral reefs that may be as dominant as the laminarian forests of colder seas.

ZONATION

The presence of coral reefs on the lower part of the shore is not associated with any marked difference in the intertidal barnacle zone and the supralittoral fringe, both of which tend to develop the same series of patterns as in other parts of the world. It is possible that this has not been realised because of the peculiar modifications of coral-bound shores. In many places, there is a wide, flat area in the lee of the coral reef more or less covered with water, beyond which there is a sandy beach. This wide flat may allow for a great, horizontal extension of the barnacle zone, not easily recognisable as such, because its character is so different from the usual substrate of a rocky shore. These reef flats can be a mile or so wide, while the supralittoral fringe is represented by periwinkles, and so forth, on outcrops of rock on the sandy beach. In some places, sand entirely suppresses the supralittoral fringe.

However, there are tropical rocky shores where the usual relations between zones are clearly visible, as at Mauritius, or where various conditions prevent the formation of reefs, as at Ghana, on the west coast of Africa (see pp. 138-144). Mangrove swamps, of course, have their own characteristic biota (see Macnae, 1968).

The greatest complexities occur where coral reefs and mangrove swamps are mingled, as at Low Isles, off Queensland. There, the mangrove swamp is situated on a wide, flattened area, which would elsewhere be occupied by barnacles, above which, on some outcrops of beach rock, there is a collection of supralittoral species.

Thus, coral reefs may be regarded as seashores that demonstrate the widespread features of zonation, but in an unusual setting.

MAURITIUS

We visited Mauritius during December 1934 and January 1935. The survey of South Africa was by then well under way, although it was some years before any theory of widespread zonation was formulated. The notes we made referred mainly to the reefs around Mauritius, which we were comparing with those we had visited on the Great Barrier Reef Expedition of 1928-29. Later, Hodgkin and Michel (1961) studied the general zonation of the island's rocky shores. Constant reference is made to their work in the following account.

Mauritius is situated in the Indian Ocean between lat. 19°50′ and 20°35′S and between long. 57°18′ and 57°48′E. It lies in the belt of the southeast trade winds and also in the path of cyclones. The wet season lasts roughly from December to May; rainfall is heavy, the mean monthly amount varying from about 12 inches in March to 2.6 inches in September and October.

The island is of volcanic origin and is composed mainly of basalt with limited outcrops of volcanic tuffs and with deposits of limestone of varying hardness (Figure 4.1). The island rears up from a depth of more than 1000 fathoms. The 100-fathom line comes within a mile of the southwest coast, and elsewhere within one and one-half miles of the fringing reefs. These reefs protect the shores of the greater part of the east coast and of the south coast from Souillac to Flic-en-Flac. North of this, reefs are intermittent, and there are stretches of exposed coast either in the form of high cliffs descending

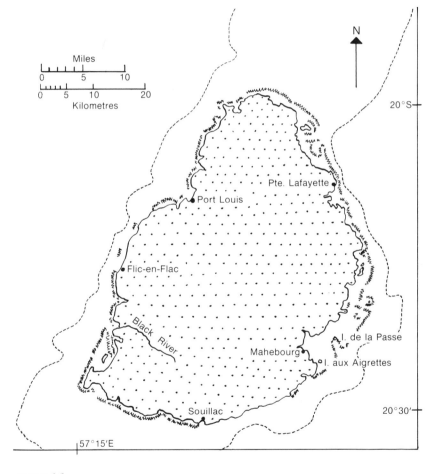

FIGURE **4.1**

A map of Mauritius marking the stations studied: *dashed line,* 200-meter depths; *jagged lines,* the approximate positions of the main coral reefs.

into deep water, or of cliffs with an irregular platform extending from the foot and rising 6–10 feet above low-water level, or of boulders cemented together by limestone. The island consists of a central plain and hills 500–2700 feet high. They rise into spectacular peaks, giving a false impression of their height. There are regions covered with tropical forest where a few indigenous plants persist.

Temperature

The surface temperatures of the sea around Mauritius range from 22°C in August and September to 27°C from January to April. Salinity is fairly constant at about 35‰, the runoff of fresh water in the rainy season having no significant effect.

Tides and Wave Action

The tides are semidiurnal and the range is small—from 2 feet (0.61 m) to 2.7 feet (0.82 m)—with, of course, exceptionally high levels on all the coastline during cyclonic disturbances.

The oceanic swell is mainly from the southeast and is almost continuous throughout the year, although more variable in the summer. The unprotected south coast is the one most subject to perpetual wave action; but almost everywhere there is considerable wave movement, except in lagoons and in some sheltered areas on the west coast.

General Ecology

The following account is a brief summary of the main features. For details of the many intertidal species, see Hodgkin and Michel (1961).

The coast can be divided into very clearly defined regions: (1) the exposed, wave-beaten oceanic coasts with wave action from moderate to violent; (2) sheltered shores bounding lagoons and reef flats; (3) the lagoons and reef flats, sheltered and normally with moderate wave action only in certain areas; and (4) islands with exposed limestone platforms, often deeply undercut, which produce an extensive splash zone.

Exposed Shores. The following information is based on the work of Hodgkin (1960) and Hodgkin and Michel (1961) at Poste Lafayette and Ile de la Passe (see Figure 4.1) and elsewhere. These localities show the typical basic zonation, strongly reminiscent of South Africa, with some very interesting modifications. Perhaps unique is the abundance of echinoderms.

In the supralittoral fringe, littorinids crept about on the myxophyceans, with a scattering of *Tetraclita squamosa rufotincta* below. Throughout the midlittoral there was the ubiquitous turf of short algae, here chiefly *Ectocarpus* and *Chnoospora*, which grew in intermittent patches, the density greater in the wetter places. Larger species, such as *Laurencia flexilis*, occurred lower down, and in gullies and crevices there were *Dictyosphaeria, Cladophora, Chaetomorpha*, and so forth. Throughout the midlittoral there was a large and varied population of crabs, chitons, gastropods, sea anemones, and so forth, as well as numerous echinoids.

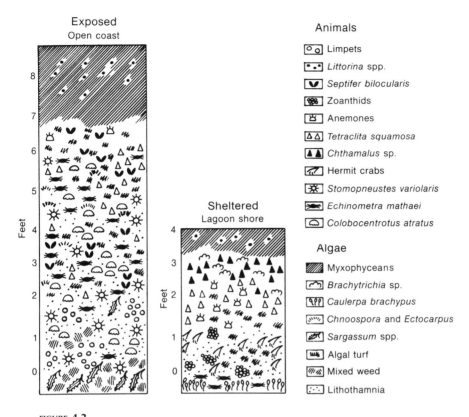

FIGURE **4.2**

The arrangement of common species on fairly steep shores on exposed and sheltered coasts of Mauritius. [After Hodgkin (1960).]

In the infralittoral fringe, such algae as *Sargassum, Padina, Turbinaria,* and *Polysiphonia* reached their maximum size and development. Here, lithothomnia formed encrustations often 1–2 centimetres thick, which gradually thinned out higher up to disappear just above the lower midlittoral.

Sheltered Shores. Shores bounding lagoons and reef platforms, as at Flic-en-Flac, had a modified biota (Figure 4.2). The supralittoral fringe had a narrower belt of myxophyceans and *Siphonaria* sp., while *Tetrachthamalus obliteratus* marked very clearly the top of the midlittoral zone, with *Tetraclita squamosa rufotincta* below. Echinoids seemed to have disappeared with the mussels (*Septifer bilocularis*), but gastropods were plentiful and there was a much mixed growth of algae. In the lower midlittoral, patches of zoanthids and alcyonarians were common, with numerous hermit crabs and occasional *Echinometra mathaei*. Where the shore merged into the lagoon there was often a dense growth of *Caulerpa racemosa*.

A number of common midlittoral animals are not mentioned above. Those common on open shores were *Acanthopleura barbonica, Chiton rusticus, Cellana cernica, Cypraea mauritania, Onithochiton maillardi, Patella chitonoides, Patelloida profunda, Septifer bilocularis,* and *Siphonaria* spp. Those species common on lagoon shores were *Cerithium morum* and *Cypraea annulus. Drupa* spp. were common on shores of both types.

Reefs and Lagoons. It was with these formations that we were mainly concerned, concentrating on reefs around Mahebourg, Ile aux Aigrettes, Ile de la Passe, Souillac, and Black River. At Black River, on the west coast, the seas are less violent than on the east. All the stations are on exposed shores, except Ile aux Aigrettes, which lies in a lagoon. Figure 4.3, based on the area around Souillac, gives a general idea of the layout of a fringing reef off Mauritius. Reefs were roughly of three types.

At Souillac, the reef platform was mainly flat, about 200 yards wide, sloping gradually and then more steeply to the breakers, and sloping gently inshore, with trickles of water flowing into a boulder zone. At low water of spring tides there were only a few inches of water covering the platform, whose surface consisted of a rough, pitted, and honey-combed pavement, with deep pools and holes, that became more broken as it merged into the boulder zone. In many places, the seaward part of the platform had a well defined zone of *Sargassum* mixed with small corallines, with dense mats of *Halimeda* and other weeds, such as species of *Turbinaria* and *Hydroclathrus.* Of the many corals, *Favia* was common, as were encrusting colonies of *Montipora, Porites,* and *Psammocera. Palythoa* and alcyonarians were seen in many places near the breakers, but they were neither abundant nor large. Lithothamnia was plentiful, but never formed a dense, pink encrustation. Farther from the sea, the profuse mosslike growths of algae comprised such genera as *Boodlea, Dictyota, Galaxaura, Halimeda, Padina,* and *Valonia,* associated with a varied fauna of sponges, starfish, fan-worms, and so forth, with *Echinometra mathaei* in burrows in the rock, and a belt of holothurians. Of the many molluscs (see Hodgkin, 1960), the most conspicuous were the clams *Tridacna gigas* and *T. maxima,* widely scattered on the reef and usually with mantles of a dull color.

To landward stretched the sandy reef flat with coral rock underlying it and emerging here and there as scattered outcrops. The flat was partly exposed at low water, and near the beach flourished thick fields of sea grass, mainly *Diplanthera uninervis.* Common

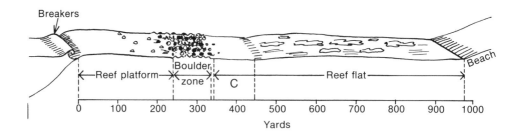

FIGURE **4.3**

The topography of a Mauritian reef at Souillac. *C,* concentration of *Holothuria vagabunda.* Modifications of this general scheme at other localities are described in the text.

FIGURE **4.4**
Southerly shores of Mauritius showing waves breaking on a fringing reef.

animals were holothurians, the fan mussel *Pinna*, and myriads of *Echinometra mathaei*.

At Mahebourg, the reefs appeared to be variants of the ones at Souillac. We studied a number of places in the area, and each had its own specialities. The following account is a summary of the most common features.

The reef platform rose to a crest near the sea and tilted slightly shorewards; the water cascaded directly over it onto the reef flat, as the boulders there did not form a zone but were scattered about over the reef. The platform was usually awash, but emerged a foot or more at spring tides. It was far more broken and pitted than at Souillac, and water welled up through the interstices. Algae were abundant on the seaward side, with extensive patches of *Boodlea*, in particular. There was a comparatively rich growth of coral—although, as a whole, of encrusting and stunted types, with nothing of the quality of the best growths on exposed midtropical reefs.

The reef platform was backed by a very shallow lagoon, and its inner edge was sharply defined. The platform had a sandy floor with scattered boulders and coral fragments lying on it. There were several well-developed, isolated coral colonies (species of *Acropora* and *Pocillopora*, for example). There were tufts of such algae as *Turbinaria conoides*. In places, the inner edge of the reef platform was populated only by quantities of *Palythoa* and large numbers of *Onchidium*. Other life was similar to that at Souillac, but the coral growth and the biota in general were much richer.

At Black River, the highest part of the reef, which was above the sea at low water, consisted of a rough, irregular reef crest of loose coral boulders and coral fragments

FIGURE **4.5**
Aerial view of a coral reef near Mahebourg, Mauritius.

of medium size, with coral shingle and occasional patches of sand and gravel between them. There was no pavement. Shorewards, the whole area looked very bare, but in some places there were plentiful algae and discontinuous crusts of lithothamnia. To seaward, the reef broke up to form deep gullies of a depth of 5 feet or more, and on the emergent ridges was a rich and varied coral population. At one place, the reef crest was separated from the fissured reef platform by a distinct outer moat, knee-deep and filled with flourishing algae, of which *Padina* was the most conspicuous. Were these ridges consolidated, they would form an outer ridge reminiscent of the reefs of the outer barrier in Australia. However, the great quantity of algae in Mauritius was countered by the relative poverty of the coral growth and lithothamnia.

Ile aux Aigrettes lay in a lagoon and, therefore, was not normally subject to more than moderate wave action. It was a raised, dead reef with a high continuous rim of coral rock eroded into points and hollows on the north, east, and south sides. This rim was deeply undercut, producing a very pronounced overhang like that of Ile de la Passe. Such formations produce an unusual amount of splash from very moderate wave action. From under the overhang, a rocky slope stretched to the sea, lying mostly in the deep shade of the undercut. On the north shore, the slope terminated in shallow water with fine sand and weed; but the water deepened along the east coast, and on the south and west there was a channel into which the slope dropped abruptly. The

Sea   Shore

FIGURE **4.6**

The principal organisms on a series of boulders lying across the reef platform at Flic-
en-Flac, Mauritius. For a key to the symbols, see Figure 4.2. [After Hodgkin (1960).]

slope was covered mainly by a turf of *Laurencia* and other sand-retaining algae. Upper
regions were bare, or had only a thin cover of lithothamnia or a felt of small algae.
(For a detailed list of the fauna of the slope, see Hodgkin, 1960.) Limpets and a selection
of gastropods were common on suitable sites, as were *Echinometra mathaei* and *Stomopneustes variolaris* in crevices and hollows in the lower parts of the shore.

The zonation on the boulders lying on the reef varied with their positions and other
physical factors; Figure 4.6 illustrates the various types of animal communities typical
of a series of five boulders on different parts of Flic-en-Flac.

The abundance of echinoderms has already been noted. Besides numerous ophiuroids
and asteroids, the most plentiful and spectacular species were echinoids and holothurians.
A study of Figure 4.2 gives an idea of the quantity of the echinoids and their positions
on the shore under the extremely variable conditions. The principal species were:
Echinometra mathaei and *Stomopneustes variolaris,* both of which occupied burrows in
rocks, their holes usually deeper inshore; *Tripneustes gratilla,* which occupied shallow
depressions and wandered about in sandy pools; and, from the midlittoral downwards,
the peculiar, flat species *Colobocentrotus striatus,* which crept over rocks and boulders,
much in the manner of a limpet.

Holothurians were present in enormous quantities, and Figure 4.7 is based on counts
of them made by several people on two occasions, March 1 and 3, 1955. The general
plan was that at each of four selected places on the shore at Souillac, a central spot
was chosen, and within a radius of roughly 10 yards from that spot, all holothurians
were collected, sorted, and counted. The exceptions were that at one of the stations
on the reef flat there was a shallow pool so full of *Actinopyga echinites* that the radius
of the circle was reduced to 5 yards; and that near the sea not all the holothurians could
be collected, although it was possible to gather enough to give an adequate idea of the
population.

Synapta maculata was ignored, as it occurs more or less everywhere, though most
abundantly on parts of the reef flat covered with sea grass. In some places, *Holothuria
vagabunda* was extraordinarily abundant immediately landward of the boulder zone.
It was not absent from the reef edge, but was rarely seen there. *H. cinerascens* appeared
to be very common near the sea, but as it lives buried under the surface, its general
distribution could not be followed.

54

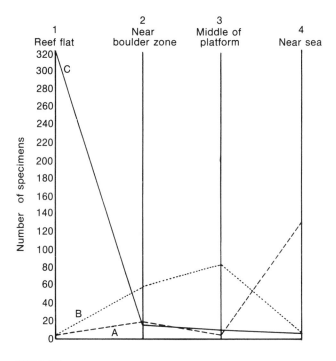

FIGURE **4.7**

A graph based on counts made by the authors of holothurians found across a stretch of reef at Souillac, Mauritius: dashed line (A), *Actinopyga mauritania;* dotted line (B), *Holothuria vagabunda;* solid line (C), *Actinopyga echinites.*

Because holothurians and ophiuroids were not listed among Hodgkin's animals, a record of those we found is included here. The holothurians were *Actinopyga echinites, A. mauritania, Bohadochia argus, Holothuria cinerascens, H. impatiens, H. monacaria, H. vagabunda, Microthele bedfordi, M. difficilis, Stichopus variegatus, S. variegatus* var. *hermanni, Synapta maculata,* and *S. maculata* var. *andrea.* The ophiuroids were *Astrobos clavata, Ophiocoma erinaceus, Ophioplocus imbricatus, Ophiothrix trilineata,* and *O. robillardi.*

This very brief account of the topography of the fringing reefs of Mauritius gives a general picture of their zonation; but there are, of course, numerous local variations, and, as usual, the zonation is dependent on a large number of different factors.

Mauritius has a thoroughly characteristic tropical shore with an abundance of algae and a moderately luxuriant coral growth, although the latter is not comparable with the fine displays of the Australian reefs and the Coral Sea.

SELECTED REFERENCES

Hodgkin, E. P., 1960. Patterns of life on rocky shores. *J. Roy. Soc. W. Aust.* 43: 35–43.
———, and C. Michel, 1961. Zonation of plants and animals on rocky shores of Mauritius. *Proc. Roy. Soc. Arts Sci. Mauritius* 2(2): 121–145.
Macnae, W., 1968. A general account of the fauna and flora of mangrove swamps and forests in the Indo-West-Pacific region. *Advan. Mar. Biol.* 6: 73–270.

5

BERMUDA

This chapter is more detailed than others because it is the first published account of work we did in Bermuda between June 25 and September 23, 1952. It is, as far as we know, the only complete account of these shores. Not all constituents of the flora and fauna have received the same attention, but we have tried to select the species best suited to ecological studies.

INTRODUCTION

The Bermudas lie in the North Atlantic Ocean about 580 miles from Cape Hatteras, east by south (lat. 32°19′N, long. 64°5′W), in the fringe of the Sargasso Sea. They are surrounded by warm water and are within reach of the Gulf Stream. In the open Atlantic, outside the West Indian area, islands are rare, and their isolation makes them particularly interesting to one concerned with the origin of the plants and animals that inhabit them. There are more than 200 islands in the Bermudas; all lie very close together, however—with only about half a dozen of any size, and these linked by bridges—making a very compact archipelago (Figure 5.1), whose greatest length is only

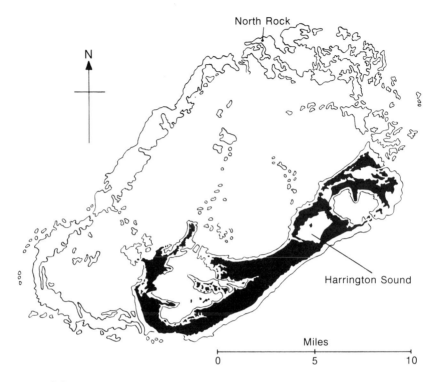

North Rock

N

Harrington Sound

Miles

0   5   10

FIGURE **5.1**

The Bermuda Islands. The land is shown black and the main sand banks from which the reefs arise are indicated in outline.

about 18 miles. The group is surrounded by shallow submarine sandbanks that rest on the summit of a rather flat-topped submarine mountain. The islands are small peaks of this mountain that emerge from the sea. Reefs lie on the sandbanks. These reefs not only have a definite arrangement in relation to the islands, but show unique peculiarities of structure.

The large islands consist of low hills, and much of the coastline is steep and precipitous. Land vegetation is typically subtropical. The rock of which Bermuda is made is a curious aeolian sandstone that is rather soft in places. It weathers along the shores into all manner of fantastic shapes, and is much pitted and honeycombed by rain, sea, and wind; the surface is often very sharp and spiky. It was formed originally from dunes built up by wind-blown calcareous sand.

Wave Action

Because the Bermuda platform lies in the open ocean, it is subject to wave action on all sides. Owing to the absence of prevailing winds, the direction from which waves come changes frequently, and all sides of the platform are exposed to roughly equal amounts of wave action. Naturally, the force of waves is modified by the configuration of the coast, and the position, number, and presence or absence of reefs on the edge of the platform. To the northwest of the crescent formed by the islands, reefs form a rough oval enclosing an area in which reefs are fewer; but this oval is hardly a lagoon, because the outer reefs are for the most part submerged and form an imperfect barrier against oceanic swells, although with the sandbanks around them they do reduce considerably the force of the waves. However, along the southeastern coast of the islands, the strip of shallow, sandy sea floor is much narrower, and reefs tend to form two lines: a fairly regular outer one, over which ocean swells are constantly breaking; and an inner one of irregular and isolated reefs. The coast is extremely precipitous, and depths of 100 metres may be attained within a couple of hundred metres of the shore. Consequently, long-crested waves do not feel bottom until within 20-30 metres of it, where they hump up and break, to dissipate their energy directly on the shore, producing very heavy wave action.

Temperature

An examination of the annual sea-temperature curve (Figure 5.2) shows that the Bermudian waters cannot be classified as tropical, and that, although the term subtropical may be suitably applied, the Bermudas have the peculiarity that the summers are almost as warm as in some marginal tropical areas, such as the Florida Keys. The winters, however, are decidedly cooler than those of any tropical region, and even cooler than in some subtropical ones, of which Durban in South Africa is an example. Consequently, many warm-water species that inhabit the Bermudian seas are living in winter at temperatures fairly close to the lower limits they can tolerate. In 1901, there was an exceptionally cold spell, which—although the temperature at its lowest was well above zero—caused the death of large quantities of warm-water fish, as well as of corals and other invertebrates.

Tides

The tides are semidiurnal, and the inequality is not marked. Their range is about 4.7 feet (1.35 m). For a diagram of the tidal amplitudes, see Figure 2.3 (p. 12).

Note on *Sargassum*

The position of Bermuda on the fringe of the Sargasso Sea gives it interest from another point of view. The Sargasso Sea is an immense eddylike area of the Atlantic in which float masses of brown sargasso weed. The weed belongs mainly to two pelagic species of *Sargassum* that are usually infertile. The origin of this weed is unknown: is it weed originally torn up from sea floors in the West Indian region where *Sargassum* grows, or is it a self-perpetuating growth that originated from beds of attached weed in the remote past? These species of *Sargassum* are not identical with any species of the genus now known to grow attached to the substrate. Although the pelagic *Sargassum* has not been produced experimentally from any attached species, it is known from experimental

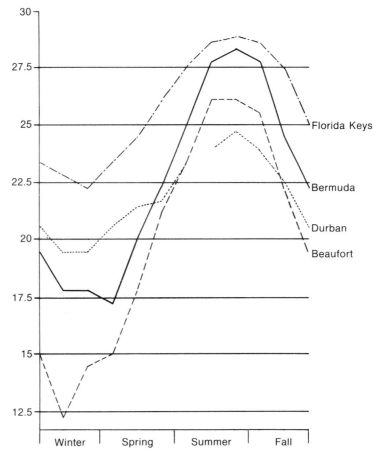

FIGURE **5.2**

Comparative sea-surface temperature curves, in degrees Celsius, for the Florida Keys, Bermuda, Beaufort, and Durban. Figures for the Florida Keys, Bermuda, and Beaufort are for December 1950 through November 1951; those for Durban are for January 1951 through December 1951. The seasons are different in the Northern and Southern hemispheres: thus, in the Florida Keys, Bermuda, and Beaufort, winter comprises the months of December, January, and February; in Durban, however, it comprises the months of June, July, and August. Because the lines on the graph are matched by season and not by month, therefore, the line for Durban has been shifted: it begins near the centre of the graph (summer: January, 1951), runs to the right, is continued again from the left side, and ends near the centre of the graph (summer: December, 1951).

evidence that the growth form of other brown weeds (*Fucus*) can be changed if they are grown tangled up in wire netting instead of attached to the sea floor, and that the Fucales in general tend to become modified if they grow unattached. An interesting view on the subject is that of Parr (1939), who surveyed the known beds of *Sargassum* from which the floating weed could have been derived, and concluded that they are altogether too small to have supplied the vast amount of weed in the Sargasso Sea, except on the assumption that it can be maintained by quite a small annual contribution, and that it lives and grows for a very long time after it has become detached.

In this connection, we should like to draw attention to a question concerning the Bermudas. In any voyage round these islands one is likely to see floating *Sargassum*. Is there any source in Bermuda from which some of it is likely derived? Although it has long been known to botanists that species of *Sargassum* occur attached in Bermuda, it has perhaps not been realized how abundant they really are. From a little above the level of low water downwards, the growth of at least two species of this genus is extremely profuse all round Bermuda and in Harrington Sound. Many of the plants near low-water level are very short (a few inches long), but they are plentiful; and, below the surface, healthy groves of plants a foot or more long are common. These plants are certainly very different from the floating ones, but no more so than plants from any other possible source. Whatever the quantity of attached *Sargassum* round Bermuda may be in relation to the quantity of floating weed, it remains substantial. As Bermuda is subject to very powerful wave action, and also to occasional hurricanes, much *Sargassum* must be torn away, and cannot but form a significant source of floating weed, considering how long this process has been going on—that is, of course, if the weed *is* capable of survival and transformation after detachment.

Note on Grazing

Throughout our studies in Bermuda, we observed evidence of intensive grazing in the supralittoral fringe and the upper midlittoral zone. The surface of the rocks bears innumerable markings of various types, showing where animals have been browsing. Which of the marks are made by which animals? Certain facts were established by close observation in the field and the aquarium. The crab *Grapsus* picks off myxophyceans with its claws, but leaves no marks on the rocks. Chitons feed very speedily over large areas, and their radulae help to make the lacy pattern so commonly noticed in the pallid zone. *Littorina*, *Siphonaria*, *Nerita*, and *Tectarius* graze quickly and extensively in the black and yellow zones, and this must contribute to the smoothing of the surface of the rock over which they wander. The larger grazing marks consist of parallel groups of sparse scorings extending to the lower part of the supralittoral fringe. Parrot fish and wrasse have beaks—3 inches across at the base—with short, hard teeth set the right distance apart to fit these big marks, and they have been observed pecking food off the rocks far up the shore at high water. However, many of these scars appear above the level that fish can reach, and it seems possible that turtles may be responsible for them, as certain species do graze seaweeds when they come ashore (Stephenson and Searles, 1960).

THE OPEN COAST

Our classification of widespread zonation is clearly demonstrated on the shores of Bermuda. The following is a summary of the information gathered at the coastal stations listed below (the reefs, Ferry Reach, and Harrington Sound are dealt with separately;

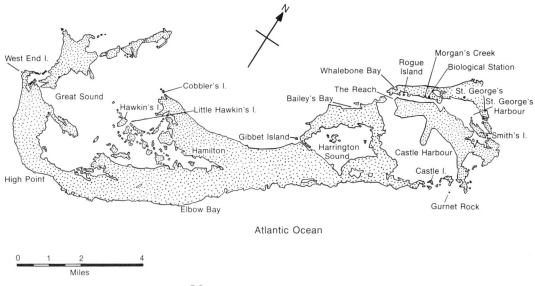

FIGURE 5.3
Map of Bermuda, showing all the stations studied.

see Figure 5.3). The stations are listed from the northwest counterclockwise round the archipelago.

Whalebone Bay	West End Island
Bailey's Bay	Elbow Bay
Gibbet Island	Castle Island and Gurnet Rock
Cobbler's Island	Smith's Island
Hawkins Island	North Rock (isolated off the
Little Hawkins Island	north coast; see Figure 5.1)

On the typical open coasts, where the land rises to a sufficient height above sea level, there is a maritime flora of coarse grasses, *Borrichia arborescens, Conocarpus erecta* (button tree), sometimes low cypress, *Lantana,* and prickly pear. The lower branches of the button trees may be wetted at high water of spring tides in such places as Ferry Reach.

The Supralittoral Fringe

The supralittoral fringe extends seaward from the line of maritime vegetation, at first, as a bare grey zone, which then merges into the usual belt of rock blackened by myxophyceans. In exposed areas with strong wave action, the black zone is often wide with very irregular boundaries; in quiet, sheltered neighbourhoods, it tends to be narrow with a clearly marked upper boundary. The variations in width and definition can be seen very clearly around the shores of Castle Island. Gradually descending, the blackening often fades into a soft grey or white—the pallid zone, which marks the division between the fringe and the midlittoral zone—and the rock becomes progressively smoother, until, in the lower part of the zone, the contours become gentle. This is not entirely due to the mechanical action of the elements, but also to the grazing of many animals (see Note on Grazing, p. 61).

In the supralittoral fringe, the rock is often very dry and the macroscopic algae are limited to specially adapted forms that can live in hollows and crevices. The commonest is *Bostrychia tenella*—sometimes mixed with *B. binderi* at the higher levels—which flourishes best as a belt that may reach down to the upper part of the midlittoral, where it competes with *Polysiphonia howei*. In certain places there are tufts of bright green *Enteromorpha flexuosa*, which are conspicuous when in season.

The commonest animal is the isopod *Ligia baudiniana*, particularly numerous at Whalebone Bay. The lively crab *Pachygrapsus transversus* clatters about in the black zone picking up food all over the place. At Castle Island, nimble little land crabs, *Gecarcinus lateralis*, scamper about quite high on the cliffs. The molluscs are not particularly interesting. The one fairly common littorinid, *Littorina ziczac*, is plentiful on Gibbet Island and in an area in Harrington Sound, but it is not common elsewhere. *L. angulifera* is recorded as single specimens from Ferry Reach and from West End Island. *Mytilus exustus* often occurs in crevices and cracks of the pallid zone. Periwinkles, limpets, and chitons that stray up from the midlittoral are listed below, at the end of the description of the upper yellow zone.

The Midlittoral Zone

The midlittoral zone shows very marked upper and lower divisions, which, because of their colouring, are referred to as the upper and lower yellow zones.

The Upper Yellow Zone. The characteristic animal is *Chthamalus stellatus thompsoni*, which can form a definite belt of varying width, as on Gibbet, Cobbler's, Hawkins, Little Hawkins, and West End islands; or the barnacles may be scattered about or clustered in little groups, as on Castle Island, Gurnet Rock, and at Elbow Bay. Always associated with the barnacles is an algal moss of mixed *Bostrychia tenella* and *Polysiphonia howei*, sometimes with subsidiary *Herposiphonia secunda*. The number of barnacles among the algal turf varies greatly, and in some places they are almost excluded by a dense algal moss. *Mytilus exustus*, as usual, is common, extending widely in suitable sites. The following species are typical of this region and as wanderers into the fringe above: *Chiton tuberculatus, Nerita peloronta, N. tesselata, N. versicolor, Siphonaria alternata, Tectarius muricatus, T. tuberculatus,* and *Onchidium* sp.

The Lower Yellow Zone. The barnacles of the upper yellow zone are here replaced by the vermetid *Dendropoma irregularis*, which can make a continuous veneer over the rock surface, or form irregular patches, or produce an extensive zone, as at places on Castle Island. Mixed with it is a sparse brown pile of small algae, of which species of *Gelidium, Laurencia,* and *Polysiphonia* are the most common. *Dendropoma* can extend downwards to overlap the *Padina*, as at Gibbet Island; or, as at Elbow Bay, can make a dense fringe above the turf in pools and runnels on small platforms. This turf is a particularly rich dark green, containing much *Laurencia papillosa* and the bright green, creeping *Cladophoropsis membranaceae*. At Little Hawkins Island, *Polysiphonia howei* makes a beautiful, dark brown, velvety pile over the vermetids.

In spite of these concentrations, much of the upper part of the lower yellow zone has a singularly bare appearance, with large areas supporting algae only in small tufts or as a scanty fur. In some places, at the bottom of the zone, there is an invasion of larger algae from the infralittoral fringe. The following are notable features: (1) *Padina sanctaecrucis,* growing thickly on shelving slopes and round the edges of pools in damp places on Gibbet Island; (2) the black *Codium intertextum,* encrusting shaded places in many areas; (3) very delicate corallines growing on Cobbler's Island; and (4) *Acetabularia sertularia,* present only in one pool in Bailey's Bay.

The fauna is plentiful and consists of an assortment of animals from the zone above and the fringe below. Crabs form a very active part of the population, and pools contain small corals and anemones (see Table 5.2).

TABLE **5.1**

Various algal species common to the reefs and open coasts of Bermuda

Common myxophyceans	Characteristic algal species that form dense carpets extending throughout the infralittoral fringe into deep water†
Calothrix pilosa	
Calothrix scopulorum	
Entophysalis crustacea	CHLOROPHYCEANS
Hydrocoleum lyngbyaceum	*Anadyomene stellata*
Microcoleus chthonoplastus	*Caulerpa racemosa*
Microcoleus tenerrimus	*Cladophora fulginosa*
	Halimeda monile
Typical turf-forming algae of the supralittoral fringe	*Halimeda scabra*
	Halimeda tridens
Bostrychia binderi	*Neomeris annulata*
Bostrychia tenella	*Penicillus capitatus*
*Enteromorpha flexuosa**	*Valonia macrophysa*
Typical turf-forming algae of the upper yellow zone	PHAEOPHYCEANS
	Dictyopteris justii
Bostrychia tenella	*Dictyota cervicornis*
Herposiphonia secunda	*Dictyota dentata*
Polysiphonia howei	*Dictyota dichotoma*
	Dictyota divaricata
	Dilophus guineensis
Typical turf-forming algae of the lower yellow zone	*Pocockiella variegata*
	Sargassum bermudense
Centrocerus clavulatum	*Sargassum polyceratium*
Cladophora fuliginosa	*Zonaria zonalis*
Cladophoropsis membranacea	
Gelidium pusillum	RHODOPHYCEANS
Lophosiphonia cristata	*Digenia simplex*
Polysiphonia ferulacea	*Eucheuma isiforme*
Taenioma perpusillum	*Galaxaura obtusata*
Laurencia papillosa	*Laurencia obtusa*
Valonia ocellata	*Laurencia papillosa*
	Wurdemannia miniata

* A seasonal species.
† In addition to the forms listed, species of coralline algae are also often present.

TABLE **5.2**
Animals representative of the reefs
and open coasts of Bermuda

Sponge
  *Cliona lampa**

Anemones
  Actinia bermudensis
  *Bartholomea annulata**
  *Condylactis gigantea**
  *Lebrunia danae**
  *Phymanthus crucifer**
  Rhodactis (Actinotryx)
    sanctithomae

Corals
  *Agaricia fragilis**
  Diploria labyrinthiformis
  Diploria strigosa
  Favia fragum
  *Isophyllia dipsacea**
  Madracis decactis
  *Millepora alcicornis**
  Montastrea annularis
  Montastrea cavernosa
  Oculina diffusa
  *Porites astreoides**
  *Porites porites**
  *Siderastrea radians**

Gorgonians†
  Muricea muricata
  Plexaura crassa
  Plexaura flexuosa
  Plexaurella didiotoma
  Pterogorgia americana
  Rhipidogorgia flabullum
  Xiphigorgia citrina

Barnacles
  Balanus amphitrite
    *hawaiiensis**
  Catophragmus imbricatus‡
  Chthamalus stellatus
    *thompsoni**
  Tetraclita squamosa
    stalactifera‡

Isopod
  *Ligia baudiniana**

Anomuran
  Clibanarius tricolor

Crabs
  Gecarcinus lateralis
  Grapsus grapsus
  Pachygrapsus transversus
  Percnon gibbesi
  Plagusia depressa

Amphineuran
  Chiton tuberculatus

Limpets
  Fissurella barbadensis
  Siphonaria alternata

Snails
  Astraea longispina
  Batillaria minima
  Charonia tritonis§
  Echinius nodulosus
  Littorina angulifera
  *Littorina ziczac**
  Nerita peloronta
  Nerita tesselata
  Nerita versicolor
  Tectarius muricatus
  Tectarius tuberculatus
  Thais deltoidea‡
  Thais rustica‡

Vermetid
  *Dendropoma irregularis**

Bivalves
  *Arca occidentalis**
  *Arca umbonata**
  *Chama macrophylla**
  Mytilus exustus

Echinoderms
  Diadema antillarum
  Echinometra lucunter
  Eucidaris tribuloides‡
  Lytechinus variegatus
    *atlanticus**
  Tripneustes esculentus

Ascidians
  *Ascidia nigra**
  *Clavelina picta**
  *Ecteinascidia turbinata**

* Recorded for Harrington Sound.
† On reefs and offshore only.
‡ On reefs only.
§ On North Rock only.

The Infralittoral Fringe

The infralittoral fringe, which is submerged except at spring tides, has a particularly abundant growth of seaweeds of many genera—among which *Sargassum* is ubiquitous—and its own characteristic fauna. This dense belt of algae often has a strip of *Padina* at its upper limit. Among the algal forest the following animals can be found: sponges; a selection of small corals; large anemones in pools; echinoderms in holes, with *Diadema antillarum* straying up from greater depths; and oysters in crevices, which can extend upwards into the lower yellow zone. Gorgonians can be seen in varying quantities in deeper water, and are particularly noticeable at Cobbler's and West End islands, which have wide areas of shoal water around them.

REEFS AND BOILERS

The water never becomes sufficiently warm in winter to support anything approaching a full growth of reef-building corals. In tropical seas, the framework of coral reefs is composed predominantly of the skeletons of corals and coralline algae. The Great Barrier Reef of Australia is a classic example. The reefs of Bermuda are not of this type, nor are they merely rocky outcrops; they are intermediate between the two. The core of the submerged reefs is made of aeolian limestone, and both the shore and the inshore and outer reefs are protected from further wave action by a veneer of the vermetid *Dendropoma irregularis* and its associates near the surface, and by other organisms at lower levels. In deeper waters, these reefs have a growth of corals, but with a smaller selection of species than on Caribbean reefs. In Bermuda, the branched colonies so plentiful in warmer water are represented only by *Millepora; Acropora* is not recorded.

Among the most spectacular features of the Bermudian reefs are the huge lilac, purple, fawn, and brown sea fans, and other gorgonians. These are not peculiar to Bermuda, but are characteristic of Atlantic warm-water reefs, as distinct from Pacific ones, where alcyonarians of different and more fleshy growth forms prevail.

The isolated reefs of Bermuda can be divided roughly into two varieties. Reefs of the first variety are wholly submerged at all tides, although the water may be very shallow over them at the lowest level of spring tides. These reefs are irregular in shape, and are extremely variable in size and in the heights to which they rise above the sea floor. They are often very sharply delimited, and have precipitous and overhanging sides. In some places, there is a direct transition from coastal rocks to reefs of this type.

Reefs of the second variety are locally called "boilers." Because their highest parts often project only a little, or are awash at low water, they can be explored only in calm weather. The simplest forms are shaped like a mushroom standing on the sea floor, expanding into a disc near the surface, the margin having a raised rim overhanging the disc (Figure 5.4). The stem is a funnel: it may be hollow right down to the sea floor, and its sides may be perforated; thus, water enters the hollow from below and roars up to the surface, hence the term "boilers." More complicated forms may consist of a cluster of irregular funnels grouped together to produce a ribbonlike formation above the surface of the sea (Figure 5.5).

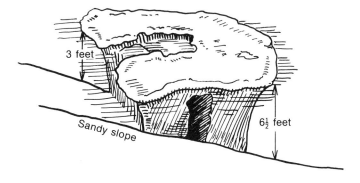

FIGURE **5.4**

An isolated funnel-shaped boiler. Note the cavern in the
near side of the boiler.

FIGURE **5.5**

An elongated "boiler," with an indication of the shape of the parts lying below the surface of the sea.
Inset, a section of a similar boiler.

Characteristically, there are raised ridges around and traversing the tops of the boilers
that are made of masses of vermetid tubes, over which grows a profusion of short
seaweeds.

A Typical Boiler

A typical boiler is one near Gurnet Rock on the south coast. It is a long, narrow, sinuous
reef that lies more or less parallel to the shore. It rises steeply from the sea floor, and
its sides are everywhere vertical or overhanging. They drop to considerable depths on

the seaward side. There is no flat or continuous top, and the only part of the reef above water level consists of the usual vermetid ridges, which provide a raised outline of the reef contours. These outline ridges cross the reef at irregular intervals, producing a general labyrinthine effect. Between the ridges are pools of all sorts and sizes: these pools vary from shallow basins a few inches deep to large, deep holes big enough to swim in. All the ridges have a dense covering of seaweed, dull brown in colour and extremely slippery.

No part of the reef is high enough to support a zone of *Padina*, but there is a rich carpet of algae (see Table 5.1). Lithothamnia is present in fair quantity, but the growths are thin and pale. The algae continue intermittently down the precipitous seaward slopes of the reef below low water.

The reef has an immense population of sea urchins: quantities of variously coloured *Echinometra lucunter* live in pools and holes in the ridges, and in larger holes and on submerged open rock are enormous numbers of *Diadema antillarum,* which sometimes can denude the rock surface of all algal cover. Other fauna is listed in Table 5.2.

Inshore Reefs on the Southeast Coast

From a little west of High Point, an extraordinary line of reefs runs to Elbow Bay: probably none of the reefs are more than a hundred feet from the shore, and some are attached to it. Along a good deal of the coast, but at no great distance, there is a second outer line of reefs. The innermost reefs make a very irregular line and rise from a sandy bottom of varying depths. The different structural forms are actually variants of that of the boiler. They can be completely or partly circular, crescent-shaped, irregularly linear, and so on, but all share certain characteristics. Their outlines are sharply marked, and their sides are steep or overhanging. Their heights are diverse, some emerging, some constantly awash, and the tops are often irregularly flat, solid, and smooth, and can be walked on with ease. The circular ones usually have funnel-shaped hollows of various shapes and depths, communicating with the water outside. For the fauna and flora, see Tables 5.1 and 5.2.

West End Reefs

Near West End Island, the bottom is a wide sand flat that deepens irregularly to its ocean edge, with patches of sea grass growing on it. From this sand flat arise reefs of many types that increase in height as the water deepens. Some reach nearly to the surface of the water but none protrude above it. Generally, they rear up very abruptly with steeply cliffed faces. All are compact masses with good growths of coral on them, and they appear to have rocky foundations.

These reefs have an overall covering of dull brown algae of the submerged type, but there is no red-brown *Dilophus,* which is characteristic of the crests of boilers and appears to be an intertidal feature of much-washed reefs that break the surface. There are numerous small to medium-sized corals, and sea fans and other gorgonians, some of which may attain a height of 5 feet. There is a sprinkling of larger corals 2–3 feet across. *Diploria* was the largest seen, and appears to be the commonest genus, with massive *Porites* next, and *Orbicella annularis* making extensive encrusting sheets. The only branching coral noted was the common red-brown *Millepora*. The corals are mostly yellow to brown; most sea fans are violet and other purplish shades, often deep and dusky with brighter lilac edges, and a few are grey. The other gorgonians show similar ranges of the same violet colour. This growth of corals and gorgonians becomes richer a short distance offshore. A simple ascidian of a beautiful translucent orange covers some gorgonian stems, and being gregarious, forms a sort of thick stocking round them. This colour scheme of purple, brown, and yellow prevails in this region.

Fishes must be mentioned here because they are so numerous: fry and large fish of all sorts—trunkfish, barracuda, garfish, angelfish, parrot fish, and herds of as many as 500 chub. There are also many turtles.

Reefs Around North Rock

These are of exactly the same type as the outermost of the West End reefs described above, but some of them do break the surface at low water. The growth of corals on the rock is abundant enough to give the appearance of a true coral reef. Colours are particularly brilliant, with the iridescent weed *Zonaria zonalis* and swarms of such fish as "Blue Fry"—which from a distance resemble a dense growth of metallic sea grass— "yellow tails," *Abudefduf*, "Blue Gregorys," and many others.

Observations Made When Diving off West End and Cobbler's Islands

A diver off West End Island can see the direct transition from shore to submerged reef. On reefs close to the coastline, the usual growth of brown sublittoral seaweed is dominated in places by corals up to 18 inches across and gorgonians up to 2 feet high. In places, the corals are sufficiently large and close together to give the appearance of a Pacific coral reef. However, on a submerged reef off Cobbler's Island, which is a continuation of the infralittoral fringe, all the steep sides and overhangs, unlike those of West End Island, are completely covered with epiphytes. Corals are incidental and usually small. There are huge anemones (*Condylactis gigantes*, for example, a foot in diameter and with tentacles a foot long), oysters, sponges, and so forth, but gorgonians are not numerous. See Tables 5.2 and 5.3.

HARRINGTON SOUND—A MARINE "LAKE"

Harrington Sound is a land-locked marine "lake" that—apart from underground ones— has only one entry, a channel roughly 40 feet across under Platt's Bridge. The lake is almost entirely surrounded by a fringe of rocks—some sloping, but mostly steeply cliffed or overhanging—in which caves are common. The limestone rock varies from the usual spikily eroded condition to considerably smoother slopes; horizontal strata

TABLE **5.3**
Algal species typical of the "boilers" of Bermuda

Species	Occurrence
Acetabularia sertularia	Bailey's Bay
Avrainvillea nigricans	encrusting pools
Caulerpa racemosa	in pools
Cladophora fuliginosa	mats in pools
Codium intertextum	encrusting pools
Corynomorpha clavata	in pools
Dictyopteris justii	reef edges and just below water
Dictyosphaeria sp.	pools, reef surfaces
Dilophus guineensis	reef edges and just below water
Eucheuma isiforme	pools
Laurencia papillosa	on surfaces and edges of reefs and just below water
Padina sanctaecrucis	pools, reef surfaces
Pocockiella variegata	pools
Polysiphonia ferulacea	pools
Rivularia bullata	a myxophycean on reef surfaces
Sargassum bermudense	pools
Sargassum polyceratium	central hollows
Zonaria zonalis	central hollows

are visible in many places. The bottom is mostly sandy with rock masses rising from the sand to various heights. The greatest depth is about 11 fathoms, and the shallowest parts are not quite shallow enough to wade.

Harrington Sound is well sheltered, but slight wave action can throw spray 6 feet high or more on rocks along some shores. It obviously has a much smaller tidal range than prevails outside—it is estimated at 12–15 inches—and the width of the intertidal zone is correspondingly restricted.

These shores closely resemble those described in the section on general zonation of the open coasts of Bermuda; the following account is not a detailed description of the biota, but emphasizes the exceptional features in Harrington Sound.

The Supralittoral Fringe

The black zone of the supralittoral fringe is narrow and well marked, with the clear-cut upper edge characteristic of this zone in sheltered positions; on sloping faces, however, it tends to broaden, and its limitations are more vague. *Bostrychia tenella* seems less in evidence, and *Enteromorpha flexuosa* more so, than on outside shores. *Littorina ziczac* can be locally numerous, and may wander throughout the midlittoral zone. *Ligia baudiniana* is as ubiquitous as usual.

The Midlittoral Zone

In many shady and damp places in the midlittoral zone, there is a very pronounced dark zone of "moss-algae" that covers the lower two-thirds of the upper yellow zone, with *Polysiphonia howei* the main constituent at lower levels, mixed, in some places, with *Gelidium pusillum.* In other places, however, the whole yellow zone carries only a film of microphytes. *Chthamalus stellatus thompsoni* is the common barnacle, and in one place it forms a very definite belt. Single specimens of *Balanus amphitrite* occur from place to place. There is no intertidal zone of *Dendropoma irregularis,* but it grows in mass formation with *Petaloconchus nigricans* in some places along and below low-water level. *Fissurella barbadensis* is also fairly common here. *Mytilus* tends to form sheets as well as to inhabit crevices.

The Infralittoral Fringe

The infralittoral fringe of Harrington Sound has a richer and far longer growth of algae than elsewhere in the Bermudas, and the epiphytes on these algae are abundant. Many animals live on the ledges and platforms of the numerous submerged rocks. (The species of algae characteristic of Harrington Sound are listed in Table 5.4, as are the notable

TABLE **5.4**
Characteristic algae and notable animals
of Harrington Sound

Algae	RHODOPHYCEANS
	Acanthophora spicifera
CHLOROPHYCEANS	*Acanthophora beauvoisii*
Anadyomena stellata	*Bostrychia binderi*
Bryopsis pennata	*Bostrychia tenella*
Caulerpa racemosa var.	*Centroceras clavulatum*
occidentalis	*Laurencia obtusa*
Caulerpa verticellata	*Polysiphonia howei*
Cladophora crystallina	*Spyridia filimentosa*
Codium pilgeri	
Enteromorpha flexuosa	Animals*
Halimeda tuna	
Neomeris annulata	SPONGES
Ulva lactuca var.	*Dysidea crawshayi*
rigida	*Ianthella ardis*
	Ircinia fasciculata
PHAEOPHYCEANS	POLYZOAN
Colpomenia sinuosa	*Schizoporella floridana*
Dictyota cervicornis	CRAB
Dictyota dichotoma	*Panopeus herbstii*
Padina sanctaecrucis	VERMETID
Padina vickersiae	*Petalocochus nigricans*
Pocockiella variegata	ECHINODERM
Sargassum polyceratium	*Stichopus badionotus*

*For further animal species, see items marked with asterisks
in Table 5.2.

animal species.) In many places, there is a very noticeable fringe of sertularian hydroids, many of them with plumes 6 inches long, growing at the water's edge. Anemones (many of them very big), ascidians, and sponges are common, but there appear to be no gorgonians. The holothurian *Sticophus badionotus* lies large and conspicuous on the sandy floor. Sea urchins are numerous, with *Lytechinus variegatus* var. *atlanticus* locally common just below low water. The crab *Panopeus herbstii* runs about at the water's edge. In some places, corallines are present in quantity. There is no vigorous growth of lithothamnia.

Colours are subdued dull browns, greens, and yellows, but the red sponges and bright green corals that occur in some areas bring vivid beauty to an otherwise drab scene. Fish, many of them brightly colored, are plentiful.

FERRY REACH

Ferry Reach is a narrow enclosed waterway—bounded by low hills on the north and by an airfield on the south—the coast of which could not be studied. Ferry Reach lies roughly northeast to southwest, with an opening into St George's Harbour at the northerly end and one into the open sea at the other. Except in hurricanes, the water is quiet, but gales can produce a choppy sea and definite wave action on windward coasts. Four areas were selected for special study: (1) the entrance to the reach, with its small headland; (2) Rogue Island; (3) the northern shore, from the entrance to the station jetty; and (4) Morgan's Creek, an opening into the north shore. Descriptions of the first three areas can be contained in one account, but Morgan's Creek requires special mention.

The main population of the first three stations, both in flora and fauna, is, with a few reservations due to differences in physical conditions, much the same as that of the open coasts (see Figure 5.6). Wave action is moderate, sedimentation is very heavy, and—in places—the reach sometimes resembles a hot, still lagoon, its maximum sea-surface temperatures being higher than those of the water outside.

The black zone shows the same restriction as in Harrington Sound. The growth of *Bostrychia tenella* under trees is exceptionally heavy. *Littorina ziczac* is replaced by occasional *L. angulifera*. On the north shore (which is particularly thick with sediment), in addition to *B. tenella*, there is a far-reaching, short, greenish grey moss consisting of *Calothrix pilosa* and *Gardnerula corymbosa* that extends into both upper and lower yellow zones and covers the vermetids in the latter. There is also a yellow-green turf of *Laurencia papillosa* and *Padina sanctaecrucis*, in which *Laurencia* seems to predominate on the landward and *Padina* on the seaward slopes of the shore.

On Rogue Island, the low, mat-forming *Cladophora fuliginosa* and other short species are found with *Laurencia*. *Valonia ocellata* is locally common in less sedimented places, and on the eastern end of the north shore it occurs above *Laurencia*, in association with the firm cushions of the slimy, cartilaginous myxophycean *Phormidium crosbyanum*. The cemented bivalve *Chama sinistrorsa* is common on Rogue Island.

There is the typical *Sargassum* in the infralittoral fringe, and other species are

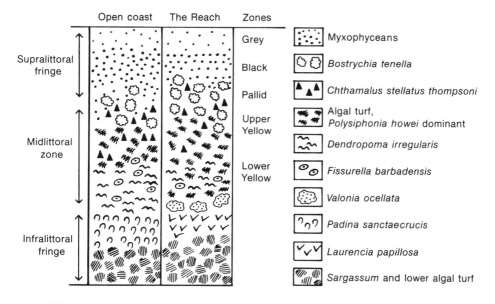

FIGURE **5.6**

The types of zonation on open and sheltered shores of Bermuda. Note the sharp line of dense myxophyceans in the Reach, the greater amount of algae and turfs, and so forth. The extent of the colour zones is also indicated.

abundant (see Table 5.1). In some places, at the upper edge of the *Sargassum*, there is enough *Padina* to give the impression of a *Padina* belt, but this is misleading. A belt of *Laurencia papillosa*, with a little *Padina* in it, is the real counterpart of the typical *Padina* belt of the outer coasts, where *L. papillosa*, if present at all, is entirely subsidiary.

Morgan's Creek

This small cove, which opens into the northern shore of Ferry Reach, is a quiet stretch of water rarely more than rippled, and is very reminiscent of the Florida Keys. It has some special features: the zonation is not as clearly defined as it is in many places, because the lower shore is covered with sediment; the black zone is narrow, pitted, and frequently browsed into lacy patterns by gastropods; and there are no parallel tooth marks, as there are at Castle Island.

In the supralittoral fringe, and spreading into the upper midlittoral, are exceptionally large numbers of *Ligia baudiana*. *Bostrychia tenella* forms a very dense green moss, but with and beyond it is another equally thick fur of *Rivularia atra* and *R. polyotes*. *Littorina angulifera* is sparse, and we did not find any *L. ziczac* or *Tectarius*. The active crab *Goniopsis cruntata* was common.

The midlittoral supports the usual growth of algae, including *Anadyomene stellata, Acrainvillea nigricans, Caulerpa crassifolia, C. sertularoides, Laurencia* spp., *Neomeris annulata, Petrosiphon adhaerans,* and *Valonia ocellata.* Animals particularly noticeable are the crab *Panopeus herbstii,* sea urchins (*Lytechinus variegatus,* for example), and hydroids in fringes at the lowest levels.

In the infralittoral fringe, despite the sediment, there is a rich and interesting—if dully coloured—flora, beginning very suddenly at a fixed level. The algae include *Caulerpa racemosa, Cymopolia barbata, Dictyota cervicornis, Halimeda monile, Padina sanctaecrucis, Penicillus capitatus,* and *Sargassum polyceratium.* There are two common sea grasses, *Thalassia testudinum* and *Syringodium filiforme.* Sea urchins and the anemone *Condylactis gigantea* are conspicuous animals.

Caves in the Reach

There are some deep, underground, sea-filled caves south of the opening to the sea. These contain, among other things, huge *Condylactis* and *Bartholomea* with incomplete tentacular bands. There are also large brittle stars (*Ophioderma cinereum*) and small holothurians under stones (*Holothuria arenicola*). At the mouths of the caves grow exceptionally long, healthy fronds of *Sargassum polyceratium.*

COMPARISON OF BERMUDA WITH THE FLORIDA KEYS

It has long been recognized that the marine population of the Bermudas is related to that of the West Indies, with the Gulf Stream acting as a highway for the conveyance of larvae and of individuals on floating objects.

The Florida Keys are subject to exceptional physical circumstances: the seas adjacent to their coasts are unusually shallow, there is a peculiarly heavy fall of sediment, and wave action is much reduced. On the other hand, the Bermudas lie isolated in mid-ocean, and their open coasts have none of these features. But there do exist, among the Bermudas, sheltered areas of enclosed sea with an environment comparable to that of the Florida Keys. Ferry Reach is the best example of these. There, the population is very reminiscent of that of the Keys, and is distinct from that of the outer wave-swept coasts of Bermuda. These may be regarded as rough-water modifications of the Keys, with some significant differences.

Some features common to the Florida Keys and Bermuda are:

1) A well established land vegetation (more luxuriant and extensive on the Keys, where mangroves, which are absent from Bermuda, may be found in many places).
2) A conspicuous black zone of myxophyceans.
3) Myriads of *Ligia baudiniana* infesting the supralittoral fringe.
4) *Littorina ziczac* is present (it is common on the Keys, but is often sparse in Bermuda; the species is plentiful in Harrington Sound, and is replaced by *L. angulifera* in Ferry Reach).
5) There is an extensive turf of *Bostrychia tenella* and *Polysiphonia howei* in many places in the midlittoral.

6) *Nerita peloronta, N. tesselata,* and *N. versicolor* are common.

7) *Chthamalus stellatus* is plentiful (*C. stellatus augustitergum* on the Keys and *C. stellatus thompsonii* in Bermuda).

8) *Mytilus exustus* and *Siphonaria alternata* are present (they are more abundant on the Keys than in Bermuda; but the characteristic midlittoral *Fissurella barbadensis* of the latter we have not found in the Keys).

9) *Batillaria minima* is present (very common in the Keys, but recorded only on Hawkins Island in Bermuda).

10) A belt of *Dendropoma irregularis* is present (on the Keys, it is associated with an extensive growth of *Valonia,* which occurs only in patches in Bermuda).

11) Lithothamnia are sparse or absent (they seem to be nonexistent on the Keys, and are never abundant or conspicuous in Bermuda).

12) A dense carpet of short *Laurencia papillosa,* with an admixture of other algae, such as *Centroceras clavulatum, Polysiphonia ferulacea,* and *Padina sanctaecrucis,* occurs in the infralittoral fringe.

13) The same species of anemones and echinoids occur in both regions.

The infralittoral fringe of the Keys is properly developed only to seaward; on sandy flats inshore it has a restricted population including much *Thalassia* and *Mytilus.* In Ferry Reach, in spite of sediment, there is a dense mixed algal carpet, as well as *Thalassia.* On open shores elsewhere, the fringe is fully developed. A more detailed description of the fauna and flora of the Florida Keys in Chapter 9 further emphasizes how much there is in common between the two areas.

SELECTED REFERENCES

Agassiz, A., 1895. A visit to the Bermudas in March, 1894. *Bull. Mus. Comp. Zool. Harvard Coll.* 26(2): 207-281.

Bernatowicz, A. J., 1950. Seasonal aspects of the Bermuda algal flora. *Pap. Mich. Acad. Sci. Arts Lett.* 36: 3-8.

Carson, R. L., 1951. *The Sea Around Us.* London: Oxford University Press.

Darwin, C., 1854. *A Monograph on the Sub-class Cirripedia, with Figures of All Species.* London: The Ray Society.

Deacon, G. E. R., 1942. The Sargasso Sea. *Geogr. J.* 99: 16-28.

Henry, Dora P., 1958. Intertidal barnacles of Bermuda. *J. Mar. Res.* 17: 215-234.

Klingel, G. C., 1944. *Inagua.* London: Readers Union.

Livingston, W., 1944. Observations on the structure of Bermuda. *Geogr. J.* 104: 40-48.

Parr, A. E., 1939. Quantitative observations on the pelagic *Sargassum* vegetation of the western North Atlantic. *Bull. Bingham Oceanogr. Collect. Yale Univ.* 6(7): 1-94.

Stephenson, W., and R. B. Searles, 1960. Experimental studies on the ecology of intertidal environments at Heron Island, 1: Exclusion of fish from Beach Rock. *Austr. J. Mar. Freshwater Res.* 11(2): 241-267.

Verrill, A. E., 1900. Notes on the geology of the Bermudas. *Amer. J. Sci.* (4)9: 313-340.

————, 1901a. Additions to the fauna of the Bermudas from the Yale Expedition of 1901, with notes on other species. *Trans. Conn. Acad. Arts Sci.* 11(1): 15-62.

————, 1901b. Comparisons of the Bermudian, West Indian and Brazilian coral faunae. *Trans. Conn. Acad. Arts Sci.* 11(1): 169-206.

————, 1902-1903. The Bermuda Islands. *Trans. Conn. Acad. Arts Sci.* 11(2): x, 413-911.

————, 1905. The Bermuda Islands, IV: Geology and paleontology; V: An account of the coral reefs. *Trans. Conn. Acad. Arts Sci.* 12: 45-348.

6

WARM-TEMPERATE AND COLD-TEMPERATE SHORES

The approximate range of water temperature for warm-temperate regions is from below 20°C (but not usually below 10°C) in the winter to above 16°C (but not usually above 25°C) in the summer. For cold-temperate regions, the range is from below 10°C (but not usually as low as 0°C) in the winter to above 12°C (but not usually above 20°C) in the summer. However, although the temperatures that limit these regions are fairly clear-cut, too little is known of the populations for these to be correspondingly well defined.

As shown in our accounts of the South African and North American coasts, the biota of temperate regions is a mixture of (1) tropical, arctic, or antarctic species that can tolerate a rather wide range of temperatures; (2) cosmopolitan species; and (3) a few species limited to areas of restricted moderate temperatures.

On cold-temperate shores at low water of spring tides, there is, characteristically, a heavy forestlike growth of large phaeophyceans, commonly laminarians, many of them of very great size. It is interesting to note that both the corals of the tropics and the phaeophyceans of the colder seas are bulky growths of sedentary organisms that liberate photosynthetic oxygen into the surrounding water—although in the case of corals, of course, this oxygen is produced by the contained zooxanthellae.

Any generalisation about warm-temperate coasts is necessarily limited; but on such of them as are known, the shore lacks both the heavy growth of huge phaeophyceans common to cold-temperate shores and the growth of corals possessed by tropical shores. Fair-sized phaeophyceans can be quite common, but the giants never are. Instead, there may be a short turf of algae among which rhodophyceans of various species are prominent, or perhaps a continuous sheet of some particular organism, such as simple ascidians (for example, *Pyura* in South Africa and parts of Australia).

In the extensive areas where biotas of different regions overlap or blend, a number of tropical species on the fringe of their range may be mixed with warm-temperate ones; the abundant zoanthids on the Natal coast provide a good example of this. However, the limits of the stenothermic inhabitants of icy and tropical localities usually are fairly definite, as evidenced by the distribution of reef corals and of the profuse growths of laminarians.

The world distribution of various populations is particularly complicated for those species that are temperate, or transitional, or both. Sea-temperature belts on coasts are, naturally, not wholly dependent on degrees of latitude. Temperatures are also determined by the courses and nature of the great ocean currents. Thus, the arrangement of water masses of various temperatures along any given coast depends both on the relation of that coast to the oceanic systems nearest to it, and on the interaction of these with the inshore topography.

To recapitulate briefly: In South Africa, where the warm Agulhas Current extends tropical conditions southward on the east coast, and the cold upwelling and the Benguela Current extends cold-temperate conditions northward on the west coast, three water masses with different ranges of temperature support three distinct populations: a subtropical population on the Natal coast, a warm-temperate one on the south coast, and a cold-temperate one on the west coast (see Stephenson, 1939, 1943). Between the populations are areas of overlap. Similarly, on the North American Atlantic coast— where the Gulf Stream lies very close inshore off southern Florida, farther off from Cape Kennedy and Cape Hatteras, and then diverges eastwards into the Atlantic, and where further north the coast comes under the influence of the cold Labrador Current—a tropical population exists in southern Florida, a warm-temperate one between Cape Kennedy and Cape Hatteras, and a cold-temperate one, tinged with Arctic species, in Nova Scotia. There is an extensive region of transition between Cape Hatteras and Cape Cod, but a very short region of transition between subtropical southern Florida and the warm-temperate region to the north.

On the Pacific coast, there are again tropical, warm-temperate, and cold-temperate regions, but the last is long and drawn out because of a cold current, and the overlapping is so extensive that the distinctiveness of the regions becomes blurred.

If the distribution of the faunas in the Southern Hemisphere is examined (see Plate 1), their relation to the Antarctic Convergence and the Subtropical Convergence can be seen, as can their relation to other oceanic features, such as the Agulhas Current. It will be noted that the antarctic littoral faunas all lie south of the Antarctic Convergence, and that the subantarctic and the cold-temperate littoral faunas (largely on islands) are on this convergence, or between it and the Subtropical Convergence—except for those of South Africa and south Australia, where there are cold upwellings north of the Subtropical Convergence. The warm-temperate areas appear north of the Subtropical Convergence, and the tropical areas still farther north.

The rocky shores of the British coasts represent an interesting variant of the usual pattern of zonation found on temperate shores (see Figure 2.2). Until comparatively recently, these shores were little studied on a broad scale; but this deficiency has now been remedied to a considerable extent, so that we know enough about British coasts to compare them with others. It has become clear that the widespread features of zonation that we described earlier do apply to Britain, but that their application here is unusually difficult. We feel that this difficulty need not be taken too seriously, because when Britain is considered in relation to the rest of the world, it does not stand out as an area of particular importance. In fact, there is some truth in a jesting comment by an Australian biologist, to the effect that Britain is "a small untypical area in the Northern Hemisphere of little importance to anybody."

The difficulties connected with these islands are due partly to their complicated coastlines, partly to their relation to a rather confused system of ocean currents, and partly to their being situated off the European coast in a geographical region with but vaguely marked limits in which there are extensive overlappings of northern and southern species. The most notable difficulty, as far as zonation is concerned, is that of defining the upper limit of the midlittoral in Britain. As we have already noted, the barnacles characteristic of this zone do not belong to the same species and genera everywhere. For example, there are shores where a species of *Balanus* is the dominant barnacle of the midlittoral, others where this position is occupied by a species of *Chthamalus.* On a *Balanus* shore, the upper limit of the occurrence of *Balanus* in quantity marks the boundary of the midlittoral; on a *Chthamalus* shore, it is the upper limit of the *Chthamalus* that forms this boundary. In Britain, the distribution of barnacles is fairly complicated, but there are *Balanus* shores, *Chthamalus* shores, and intermediates where both these barnacles are present in varying proportions. If the two kinds reached their upper limit at the same level, there would be no problem; but it so happens that *Chthamalus* extends to a higher level than *Balanus* under the same conditions, though the difference between the two levels is not great. The problems of zonation presented by the British coasts are not insoluble. The upper limit of the midlittoral on *Balanus* shores is at a slightly lower level than the same limit on *Chthamalus* shores; but the variation in level occurs over a narrow belt situated very approximately between high water of neap tides and high water of spring tides, as these appear in sheltered places, or between the comparable levels of less sheltered places. Where the two barnacles occur together, the upper limit of the midlittoral may be taken, for descriptive purposes, as the level at which the species of barnacle occurring in quantity highest up ceases to be abundant. In the case of intermediate shores, there will sometimes be complications, but each case can be treated on its own merits without too much difficulty. In other words, we look for the upper boundary of the midlittoral between certain limits. It will never lie below a certain level (represented by the *Balanus* line in Britain) nor above a certain other level (represented in Britain by the *Chthamalus* line). That a real and significant boundary of this nature should vary in level from one shore to another, within a belt limited by the lines along which different species of barnacles cease to be abundant, seems to us eminently natural and to be expected.

Figures 8.5 (p. 104), 10.9 (p. 217), 10.34 (p. 266), 11.5 (p. 286), and 11.44 (p. 341) illustrate some of the variations in selected common species and in their zonation on the cold-temperate and warm-temperate coasts of different countries. A more detailed account of these biotas is provided in the text of the relevant chapters. See Chapters 2 and 13 for further information on temperate shores.

SELECTED REFERENCES

Stephenson, T. A., 1939. The constitution of the intertidal fauna and flora of South Africa, I. *J. Linnean Soc. London Zool.* 40: 487–536.

———, 1943. The causes of the vertical and horizontal distribution of organisms between tide-marks in South Africa. *Proc. Linnean Soc. London* 154: 219–232.

7

SHORES AFFECTED BY ICE

THE GENERAL EFFECTS OF COLD
AND ICE ON SHORE POPULATIONS

It is important to realise that ice and cold can kill or damage organisms independently, although they often act together. There is no doubt that spells of unusual cold, altogether apart from ice, can cause mortality on a large scale in shallow-water populations. The exact effect of a cold spell depends upon how quickly the temperature falls and then rises, upon the extremity of the low temperature and how long it persists, upon the state of the tides and of the prevailing weather, and upon whether the fall in temperature occurs in a warm or a cold region. Cold spells in warm areas may kill vast numbers of fish and certain invertebrates and algae, even when the water temperature remains well above 0°C. For instance, in a region such as that just south of Cape Cod, where southern species extend rather far north, some of those may be greatly reduced in numbers during severe winters, though they may recover subsequently. Many other organisms may succumb to temperatures between 0° and −10°C. However, there are some marked exceptions, and species that can dry out may survive very low temperatures. Many species can sustain a certain amount of ice in their tissues—their chances for survival thus being a function of their water content—and many are unharmed by low temperatures so long as their tissues do not freeze. Some animals migrate to below the low-water level in cold winters (*Ilyanassa obsoleta*, for example, and perhaps species of *Acmaea* and *Littorina*); others gain some protection by retiring under stones, to crevices, to the undergrowth of *Fucus*, to low-level pools, and so forth.

The effect of ice on the biota of seashores is far more drastic. Obviously, in cold-temperate regions where winters are not always cold enough for the formation of ice on the shore, a succession of mild winters may permit a good growth of *Fucus*, barnacles, and mussels, and then the return of ice may bring about their wholesale destruction. This is the result not only of cold but of mechanical action by ice.

Ice behaves very differently in different areas. In a quiet, landlocked inlet, ice may remain static for long periods, gently rising and falling with the tide. In addition, ice may form a sheet on the rocks, which for a time remain stationary. This sheet may be formed in more than one way—for instance, by the flooding of the rocks by the water of a rising tide, which freezes to their surface. If the thickness of the sea ice is greater than the tidal amplitude, as is common in many parts of the Arctic, the entire shore may become completely covered by ice. However, when the tidal amplitude is great—say 20-30 feet—only the upper reaches of the shore freeze solidly and form an ice foot (see Figure 9.23; see also Ellis, 1955, p. 233, fig. 7). These conditions will cause damage by lowering the temperature of organisms beneath the ice or frozen into it; but, on the other hand, the ice sheet protects these survivors from adjacent moving ice, and prevents their being scraped off.

On more open shores, the ice is more restless, breaking up and moving in mild and windy weather, and subsequently causing greater damage to the fauna and flora. On exposed shores, where ice is rarely still, maximum injury occurs, particularly in times of thaw, when ice can exert immense grinding pressure and scraping action along the coast. Where this action is greatest, there is usually a definite belt around sea level scraped bare of all organisms save those in crevices, hollows, and other protected positions. Rocks that have been denuded by ice may be repopulated in summer. Algal species that typically live under these circumstances are *Chordaria flagelliformis*, *Scytosiphon lomentaris*, *Chorda filum*, and *Fucus vesiculosus* (see p. 86). But fucoids, barnacles, and mussels cannot reach anything like their full development in one summer; to develop fully, they must be undisturbed for several seasons. However, there is often a brief growth of annual algae. On soft bottoms, as well as on rocky ones, species may be wiped out, for the time being, after particularly extreme winters.

Tidal pools are features of importance on these shores. Apart from providing a certain degree of protection from the elements, their variations in temperature and salinity afford suitable conditions for certain plants and animals. In the uppermost zones, the salinity of these pools depends on the amount of thaw water draining into them and on the spray carried up into them by rough seas. Their characteristic flora comprises species of chlorophycean annuals. Midlittoral pools, in many places, have dense growths of small brown annual weeds, such as *Fucus vesiculosus* and *F. distichus*, while the pools that are frequently washed by waves support large communities of small kelps and leafy rhodophyceans, with other infralittoral species.

Madsen (1936) suggested that the amount of intertidal fauna on the shores of these cold seas depends not only on such special littoral factors as mentioned above, but may

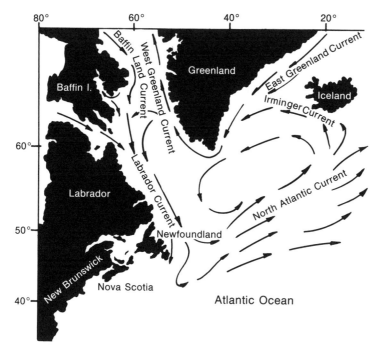

FIGURE 7.1
The general direction of the main currents around eastern Canada.
[After Admiralty Chart 5310.]

be linked with the behaviour of sea currents. Subarctic surface waters have a very high nutrient content, which is increased where warm and cold currents meet. Such conditions prevail at the mingling of the less cold Irminger Current and the East Greenland Polar Current off Angmagssalik on the east coast of Greenland, and there is biological evidence—the presence of numerous cod (*Gadus callarias*)—of an Atlantic admixture on the west coast of Greenland and the extreme southeast of Baffin Island (Figure 7.1). It has been noted already that certain usually intertidal species retreat to the infralittoral fringe under extremely cold conditions. Some of these, particularly those in shallow water, require a rich supply of phytoplankton in the early stages of their life cycles. Currents may assist in distribution also by carrying larvae from place to place.

THE ICE REGIONS OF NORTHEAST CANADA

The marine arctic has been defined as comprising those areas in which unmixed water of polar origin is found in the surface layers to a depth of at least 200–300 metres. The marine subarctic has upper layers of water of mixed polar and nonpolar origin (Dunbar, 1953).

The greater part of the marine subarctic lies on the Atlantic side of the world, from

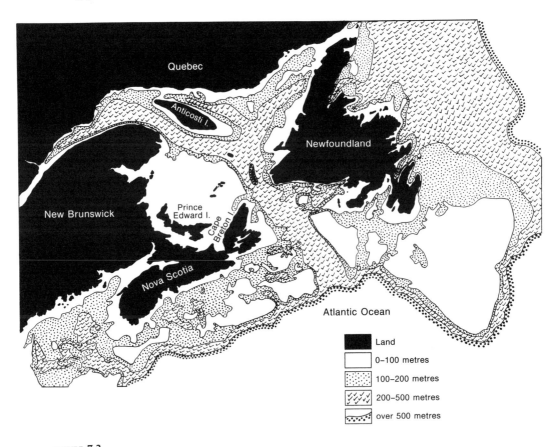

Quebec

Anticosti I.

Newfoundland

New Brunswick

Prince
Edward I.

Cape
Breton I.

Nova Scotia

Atlantic Ocean

■ Land

□ 0–100 metres

100–200 metres

200–500 metres

over 500 metres

FIGURE **7.2**

The submarine topography of the Gulf of St Lawrence and neighbouring areas. Scale: 1.23 inches = approximately 100 nautical miles. [After Sandstrom (1919).]

the Scotian and Hudson Bay shelves to the Barents and Kara seas off Russia. It includes almost the whole coast of west Greenland and the waters around Newfoundland and Iceland, the boundaries depending on how far the arctic waters extend during a winter season (Figure 7.2).

The Atlantic Coast of Nova Scotia

From Halifax westward to Cape Sable there is usually no ice in the open sea other than broken slushy material, which has little effect on seashores. Halifax and Peggy Cove are virtually untroubled by ice (Figure 7.3).

In smaller inlets and harbours, conditions are very different. For example, inlets leading off St Margaret Bay may be frozen over, in a hard winter, with a continuous sheet of ice 2–3 feet thick. But the growths of *Fucus* and *Ascophyllum* seem permanent,

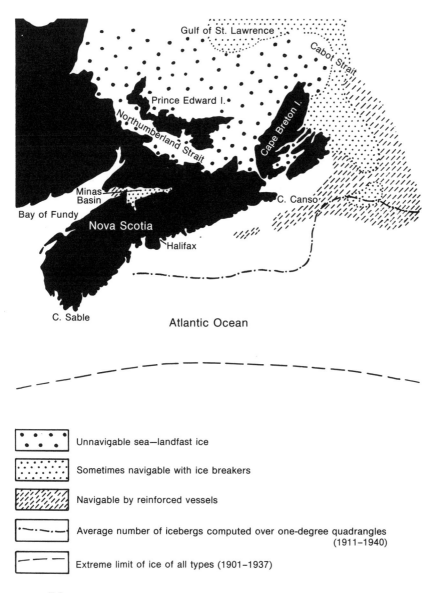

Unnavigable sea—landfast ice

Sometimes navigable with ice breakers

Navigable by reinforced vessels

Average number of icebergs computed over one-degree quadrangles (1911–1940)

Extreme limit of ice of all types (1901–1937)

FIGURE **7.3**

The extent of winter ice in Nova Scotia and Prince Edward Island. [After U.S. Navy Hydrographic Office (1952).]

and the floating rafts of fucoids and mussels that are to be found may represent what has been torn away from the shore. The lack of barnacles may be due to the distance from the sea, and not to the presence of ice. This area seems to provide an example of the relatively slight damage caused by prolonged exposure to gently moving ice.

The Gulf of St Lawrence

The Northern Coast of Prince Edward Island. This is not a typical subarctic area. The summer sea temperatures are comparatively warm (see Figure 9.18), so that, despite the winter cold, a number of southern animals are able to breed and survive among the northern majority. Southern animals are not able to survive to any extent on the Atlantic coast or in the Bay of Fundy, because, in these places, sea temperatures are always too low. Ice forms in the bays and harbours in late December, and when the Gulf ice arrives in January some parts of the shore are already protected by an ice foot, which forms where the tidal range is small and the sea ice is thicker than the height of the intertidal area. The pressure in northerly gales is tremendous, and the ice piles in great pressure ridges that may fall flat when the wind changes. When offshore winds blow, the ice may move off a few miles and loosen up, returning to the shore as the wind reverses direction. This continues until late April, by which time the ice has moved east. Along the stretch of open coast between Cavendish and North Rustico, broken ice movable by wind and tide may pile up nearly as high as the cliffs, and may even form a succession of high ramparts in the course of a winter (see Figures 9.19 and 9.20). Here, the effects visible during the summer of 1948, which were probably due to the ice of the preceding winter, were extremely marked. The soft rock showed a planed appearance—the whole surface, in some places, apparently having been recently removed by the action of moving ice. The barnacles were abnormally scarce, and adult specimens were restricted almost always to crevices and hollows where they had been afforded some protection from the action of ice. Nearly all the individuals on open rock were young ones from the current summer's settlement, suggesting that during the preceding winter any others living on the open rock had been scraped away. Algal growth between tidemarks was at a minimum, much of the seaweed was confined to cracks and angles, as were the relatively few mussels. Some small plants of *Fucus,* and some of the more transient algae, were apparently that season's growth—apart from plants that may have been regenerated from holdfasts that may have survived the winter. Below low water, there were no laminarians; and, although here the seaweed growth was extensive, there was even then a concentration of some species round potholes and other protected surfaces. The dominant beardlike algae of open rock (*Chordaria, Delmarea, Chorda, Dictyosiphon,* and *Scytosiphon*) are all species defined as annuals; but, from the appearance of the growth, it seemed possible that some of their basal parts had survived the winter and had regenerated vigorously in the spring. The absence of whelks from the shore might in part be connected with the paucity of barnacles, but the lack of limpets would seem more likely to be due to ice.

For further discussion of the Gulf of St Lawrence, see pages 189 and 199.

Northumberland Strait

In the winter, Northumberland Strait becomes filled with ice, and strong tidal currents keep the ice moving east or west—the easterly motion predominating—according to the state of the tides. By mid-April, the largest amount has passed through to the east.

Local conditions determine the effect of ice on this coast, which follows the pattern described earlier in this chapter.

Souris, on the south coast of Prince Edward Island, lies in a small, well-marked bay more subject to static ice than to moving ice. It has rocks similar to those of the northern shore, and these support a plentiful normal type of intertidal growth. The only direct suggestion of ice action is the absence of laminarians. Souris would seem to illustrate the possible biota of the northern coast were that coast not subject to strong ice action.

The Bay of Fundy

At Hall's Harbour, Nova Scotia, on the Bay of Fundy, there is usually some ice in February, but it is gone before April. It occurs as pieces of various sizes, including thick cakes that pile up on the shore, moving and eddying about. This ice can exert great pressure, and can damage or sweep away the large fishtraps used locally, foundation boulders and all. The action of ice is combined with the scouring of stones and boulders, and it is difficult to determine which has the more harmful effect. Possibly, the moving stones and boulders are more destructive than the ice in this area. Cliffs and the hard basaltic rocks of higher zones form angular profiles; but lower down, the rocks are much smoothed and rounded, giving the appearance of great elephant's feet (see Figure 9.22). Barnacles were so plentiful that ice action was not immediately suspected, but close examination revealed that nearly all the large species likely to have survived the winter were in crevices and under overhangs, and that those on open rock were almost all from the current year's settlement. The best *Fucus* and *Ascophyllum* showed the same restriction to protected positions. Places further south in the Bay of Fundy showed that ice is not a regular or extreme hazard to intertidal life there.

GREENLAND AND BAFFIN ISLAND

The following is a very brief summary of the type of shore life to be found on subarctic coasts that extend into arctic regions, specifically those of Greenland and Baffin Island (see Figure 7.4). The information has been obtained mainly from work by the following authors: Bertelsen (1937); Dunbar (1947, 1951, 1953); Ellis (1955, 1956); Huntsman (1930); Madsen (1936, 1940); Taylor (1954); and Wilce (1959).

Tides on the coasts of Greenland and Baffin Island are semidiurnal, and tidal amplitudes are usually small, ranging from 4 to 12 feet (1.2–3.6 m), although they may be much greater on the southeast coast of Baffin Island, reaching as much as 36 feet (11 m) in Frobisher Bay.

The boundaries between arctic and subarctic shores can be determined by the sudden disappearance of three common animals, *Mytilus edulis*, *Littorina saxatilis* var. *groenlandica*, and *Balanus balanoides*. The last two are the most decisive indicators, as their disappearance is more abrupt and *M. edulis* may continue further north in the infralittoral fringe. *B. balanoides* and *M. edulis* are able to survive under ice in the subarctic in west Greenland; the intertidal zone, therefore, may be recolonised from this source each year.

In west Greenland, these three species survive to a little north of Upernavik (lat. 74°N), but the limit is not very well defined (Figure 7.5). However, at Angmagssalik on the east coast the change of regions is very clear, with a sharp drop in the number of shore animals both above and below the low-water level.

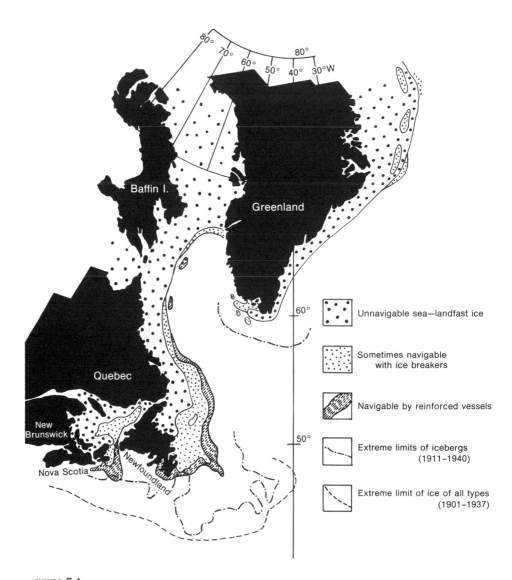

FIGURE 7.4

The extent of winter ice in the East Arctic of Canada. [After U.S. Navy Hydrographic Office (1952).]

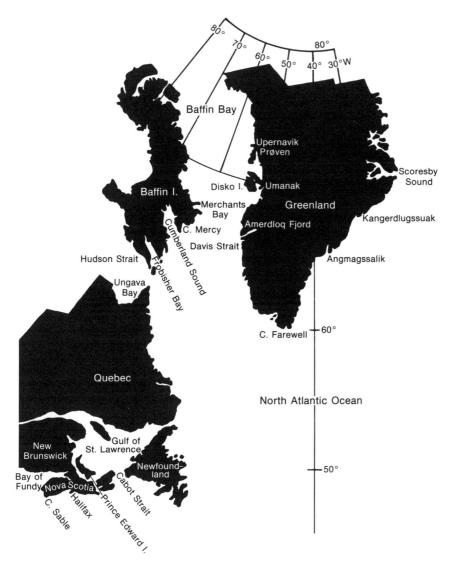

FIGURE **7.5**
A map of Greenland and eastern Canada, showing the localities mentioned in this chapter.

West Greenland

Umanak (*Lat. 72°42′N*). Here, the sea-water temperatures for June range from 0.2° to 1.0°C. The littoral fauna of this area is comparatively rich. *Littorina saxatilis* is described as present, often as adult specimens, and most notably in crevices. *Fucus* occurs with the *Littorina* and densely in the infralittoral fringe. *Mytilus edulis* is fairly plentiful, extending from the midlittoral into the infralittoral fringe throughout the shore. Large *Actinia* occur attached to the mussels in the upper part of their range.

The molluscs *Buccinum groenlandicum, Margarita helcina,* and *Modiolaria faba* are characteristic of these shores, as are the crustaceans *Gammarus locusta, Calliopus laeviusculus, Caprella septentrionalis, Ischyrocerus anguipes,* and *Pontogeneia inermis.* The hydroid *Laomedea longissima* occurs, as do numerous large copepods and nematodes.

A number of species that live both on *Fucus* floating in shallow water and between tidemarks are common along the entire coast of Greenland. These include *Margarita helcina, Modiolaria discors* var. *laevigata, M. faba,* the mussel *Cyamium minutum,* and the polychaet *Spirorbis spirium.*

When the surface-water temperature rose to 4°C in July, *Littorina* occurred in great numbers. Particularly interesting were the isopod *Jaera marina* and the hydroid *Campanularia lacerata,* which were found at that time in tidal regions.

Prøven (*Lat. 72°23'N*).   Prøven lies a little farther south than Umanak, and the harbour is a creek opening into a narrow strait between two islands. Ice conditions are not severe, and surface-water temperatures reached 6.9°C in July, a temperature never attained during the work at Umanak. The intertidal life is much the same as at Umanak, but it is much more abundant. Among the more notable animals, *Littorina* occurs in great numbers with large specimens present on flats about 0.5 metre above low-water level. *Mytilus edulis* is very abundant, with many recently settled individuals, and *Acmaea testudinalis* is conspicuous, large, and plentiful. *Buccinum groenladicum* is present about 10 cm above low-water level, and the sea urchin *Strongylocentrotus droebachiensis* occurs in a line at the water's edge. *Asterias glacialis* and various actinians are present in the infralittoral fringe. Large but scattered specimens of *Balanus balanoides* live up to 1 metre above low-water level, but only in one place—a slope with a southern exposure.

Southwest Greenland

Amerdloq Fjord (*Lat. 67°N*).   The amplitude of the tide here varies from about 8 feet (2.5 m) to 12 feet (3.7 m) in the spring. Surface-water temperatures range from −1.2° to 5.5°C. This part of the coast is less affected by ice than any other in Greenland. As would be expected, the biota of this coast is as rich in variety and numbers as that of any subarctic coast already listed; it also has a number of species of circumpolar distribution. There is also a definite balanoid zone with *Fucus.*

East Greenland

Angmagssalik (*Lat. 65°30'N*) and Kangerdlugssuaq (*Lat. 68°N*).   In the neighbourhood of Angmagssalik during August, the surface-water temperatures in most localities lie between 5° and 8°C; at Kangerdlugssuaq, however, the surface water is much colder, more like that of Scoresby Sound and Franz Josef Fjord, and the average surface-water temperature is near 2°C, varying from 1.52° to 2.38°C.

The outer shores of these areas descend steeply into the sea, and three zones of vegetation can be distinguished: a *Fucus* zone, which can extend up to high water mark, but is most profuse at low-water level; a laminarian belt, with a substantial mixture of *Alaria pylai;* and, finally, a zone of *Desmarestia.* At Kangerdlugssuaq, the laminarians may be replaced by a fairly dense growth of red algae.

The littoral fauna at both places is poor—strikingly so at Kangerdlugssuaq, which is within the Arctic, and where the few animals that occur live at depths less than 4-5 metres, and the shallower the water is, the more extreme are the variations in temperature and salinity. The only large littoral animals are *Mysis oculata* and *Gammarus wilkitski.*

At Angmagssalik, which is subarctic, characteristic animals are *Balanus balanoides, Littorina saxatilis,* and *Mytilus edulis;* there are various crustaceans common to arctic regions, such as the two found at Kangerdlugssuag and a number of burrowing forms.

Baffin Island

The climate of Baffin Island is colder and drier than that of Greenland, and the east coast has the same arctic and subarctic divisions (see the paragraph on currents on p. 82). The coast is broken by long fjords and estuaries; typical are Frobisher Bay and Cumberland Sound, where sloping sand and mud flats, often permanently covered by water, occupy the upper reaches. These areas support a special community of shallow-water species, restricted by the low salinity of surface water during summer thaws and the depth of freezing in winter. These species do not extend into arctic regions.

The intertidal fauna and flora is to be found on the steep rocky shores at the entrances to sounds, and these are more continuous further north at Merchants Bay. Here, the ice and its abrasive action are the dominant limiting factors. The biota of the area, as given in a detailed list by Ellis (1955), is almost identical with that of Prøven and Umanak.

In arctic regions, intertidal life on rocky shores is almost entirely absent; in subarctic areas, it exists in a much modified state. Apart from the usual restricting factors, it has to contend with a number of others, including reductions in salinity, deep freezing, and the abrasive movements of ice. In most places, there is no supralittoral fringe, any life at that level being confined to temporary pools inhabited by annual algae. In many places, there is a midlittoral zone, perhaps with an upper belt of characteristic *Littorina, Balanus,* and *Fucus,* and a growth of *Ascophyllum* and *Mytilus edulis* below. Algae constitute the bulk of the population of the infralittoral fringe, and the main animal life is associated with these plants, whereas in temperate climates this would occupy the lower midlittoral and the infralittoral fringe. In arctic conditions, laminarians are replaced by a turf of small red algae. Algae flourish best in areas of moderate exposure, and in the infralittoral zone they are dense; some can reach an immense size, but there are relatively few species, *Desmarestia* and *Phyllogigas* being the dominant ones.

THE SUBANTARCTIC REGION

The following summary is made from an account of the work done by J. H. Price and P. Redfearn on the marine ecology of Signy Island, South Orkneys, near Antarctica.

In the antarctic regions, all the effects of ice on intertidal shores described earlier in this chapter exist in their more extreme forms. The South Orkneys lie well within the limits of winter fast ice, and the conditions are chiefly influenced by the West Wind Drift and, to a lesser extent, by the northerly component of the cold East Wind Drift from the Weddel Sea. Sea-surface temperatures range from $-2°$ to $0°C$ in winter, and from $0°$ to $12°C$ in summer. The tides are more or less regularly diurnal, with a range of about 7.5 feet (2.286 m) at extreme spring tides.

Signy Island, South Orkneys

The biota of the intertidal region of Signey Island is meagre, even during summer months. Owing to the rigorous conditions, most of the biota is confined to lower levels of the shore, and there is no supralittoral fringe. The coast is broken by many bays and inlets with boulder and gravel beaches. The south coastline is covered by the snout of a glacier.

The Midlittoral Zone. Highest on the shore is the alga *Porphyra*, from 5 to 7 feet (1.83-2.13 m) above extreme low water of spring tides in suitable damp, shaded, or wave-washed places. Where it is not present, there can be a more or less continuous band of *Ulothrix* and *Urospora* along the centre of the zone, varying in width from a few inches to 6 feet (1.83 m) according to circumstances. This belt of filamentous algae is often the only growth visible on the shore.

The remaining intertidal flora exists only in crevices and sheltered pools at lower levels, therefore there is often a bare strip of more than half the vertical height of the intertidal region between the bottom of the *Ulothrix* belt and the algal growth of the infralittoral fringe.

The fauna of the shore is sparse. The migratory limpet *Patinigera polaris* occurs in the summer in groups in crevices and on sheltered surfaces as far up as the *Ulothrix* belt. Apparently, it travels upwards to graze on the algae, but its exact habits are not yet known. Amphipods, nemertines, and flatworms are found clustered under stones and rocks in damp places in the lower midlittoral. Pools in this area support a flora comparatively rich by antarctic standards. For example, short growths of such algae as *Leptosomia simplex*, *Iridaea* spp., and *Adenocystis utricularis* are common—to which can be added, from the deeper and lowest pools and crevices extending into the infralittoral fringe, *Desmarestia menziesii*, *Curdiea racoviteae*, *Monostroma hariotii*, and various species of *Plocamium* and *Myriogramme*.

The Infralittoral Fringe. Many of the algae here are extensions from the infralittoral zone, where alone the main bulk of the marine biota can survive. A number of species occur here that appear elsewhere in intertidal zones.

Desmarestia menziesii, and another large brown alga, *Ascosiera mirabilis*, may extend 4-5 feet (1.22-1.52 m) below extreme low water of spring tides, sometimes with *A. mirabilis* predominating. Another *Desmarestia* (*D. anceps*) overlaps the lowest growth of *D. menziesii* and reaches far down into the infralittoral zone, attaining an enormous weight and size—up to 12 pounds (5.44 kg), with lengths to 8 feet (2.44 m). It is the prevailing plant down to 25 feet (7.62 m). Deepest of all, from 23 to 28 feet (7.01-

8.53 m) below extreme low water of spring tides, grows the huge *Phyllogigas grandifolius*, which, with a length of up to 20 feet (6.10 m), is the longest of all these great seaweeds, and the dominant one at these depths.

Apart from the fauna mentioned above that inhabit sheltered pools and crevices, most of the marine animals live in the holdfasts and under the fronds of these mighty plants, which support an immensely rich and varied small fauna, representing most animal groups—worms, sponges, hydroids, molluscs, and so forth. For a detailed account of these, see Price and Redfearn (1968).

As in the Arctic, there is no supralittoral fringe in many places on Signy Island. There is a midlittoral zone of very restricted content, with an increasing population beginning in the fringe of the infralittoral zone and reaching its greatest numbers below the tide levels among the swaying plants of the great seaweeds.

Melchior Peninsula, South Shetlands

The following notes are based on an unpublished report on the marine vegetation of the Antarctic Peninsula (Graham Land) made by Drs Rene Delepine, I. Mackenzie Lamb, and Martin H. Zimmerman, who visited the neighbourhood in 1964-65. The names they use are Argentinian ones, and the area lies off the southeast coast of South America (see Figure 3.3).

The flora of the intertidal region of the Melchior Archipelago (South Shetlands) shows some variations from that described of Signy Island of the South Orkneys group.

The Supralittoral Fringe. This does exist, but in a very impoverished form. The shore is free from ice and snow only in early February. It contains a number of pools, many of which are temporary. They are subject to great variations in salinity—depending on the amount of fresh thaw water and sea spray—to extremes of temperature, and to contamination by great quantities of nitrogenous and phosphatic compounds from the guano of nesting sea birds. In many places, there are such species of lichens as *Verrucaria centhocarpa* and *Caloplaca cirrochroides*. The algae of the pools are mainly unicellular chlorophyceans.

The Midlittoral Zone. This has a vertical depth of approximately 1.5-2.0 metres, and here also life is seasonal. It begins with a growth of the filamentous chlorophyceans *Urospora* and *Ulothrix* in pools and on smooth surfaces; these algae form a dark green band that indicates the upper edge of high water level. Individual plants of *Porphyra endiviifolium* appear in cracks. The lower limit of this zone is marked by the upper limit of the encrusting corallines (*Lithophyllum acquabile*, for example) on sloping seaward surfaces. The lichen *Verrucaria serpuloides* is also present in some places in this zone. Algae are confined to pools and crevices, and consist of scraps from those growing attached at lower levels. During February, these infralittoral species make a rapid invasion upwards, and can conceal the encrusting corallines. For a list of these algae, see page 92 in the description of Signy Island in this chapter.

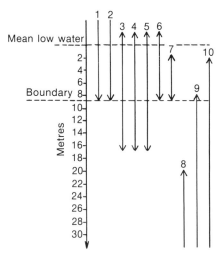

FIGURE **7.6**

The distribution of some important algae in the infra-littoral regions in the South Shetlands (Melchior Archipelago), Antarctica: 1, *Leptosomia simplex;* 2, *Verrucaria serpuloides;* 3, *Curdiea racoviteae;* 4, *Iridaea* sp.; 5, *Gigartina* sp.; 6, encrusting corallines (e.g., *Lithophyllum acquabile*); 7, *Ascoseira mirabilis;* 8, *Picconiella* sp.; 9, *Phyllogigas grandifolius;* 10, *Desmarestia anceps.* Species *1* and *2* extend into the lower midlittoral; *3, 4, 5, 6,* and *7* are inhabitants mainly of the infralittoral fringe; *8, 9,* and *10* are deep-water forms, two of which can reach into the infralittoral fringe. *Boundary,* the division between the infralittoral fringe and the infralittoral zone, which is always submerged.

The Infralittoral Fringe. This appears to be practically identical with that of Signy Island. Figure 7.6 illustrates the approximate levels at which the larger algae occur. It was noted that, in January, there was a week during which an excessive growth of the diatom population affected the clarity of the water and coated the larger algae. A somewhat similar phenomenon has been observed off the north coast of Prince Edward Island.

SELECTED REFERENCES

Bertelsen, E., 1937. Contributions to the animal ecology of the fjords of Angmagssalik and Kangerdlugssuaq in east Greenland. *Medd. Grønland* 108(3): 1-58.

Broch, H., 1933. Einige Problema der biogeographischen Abgrenzung der arktischen Region. *Sond. Mitt. Zool. Mus. Berlin* 19: 1-20.

Dunbar, M. J., 1947. Note on the delimitation of the arctic and subarctic zones. *Can. Field Natur.* 61(1): 12-14.

———, 1951. Eastern arctic waters. *Bull. Fish. Res. Board Can.* 88: 1-131.

———, 1953. Arctic and subarctic marine ecology: immediate problems. *Arctic* 6(2): 75-90.

Ellis, D. V., 1955. Some observations on the shore fauna of Baffin Island. *Arctic* 8(4): 224-236.

———, 1956. Some temperature and salinity records from the Canadian Arctic during 1954 and 1955. *J. Fish. Res. Board Can.* 13(4): 591-598.

Ellis, D. V., and R. T. Wilce, 1961. Arctic and subarctic examples of intertidal zonation. *Arctic* 14(4): 224-235.

Fisch, H. J. L., 1961. Recherche de mollusques antarctiques. *J. Conchyliol.* 150: 81-86.

Fleming, R. H., 1955. Review of the oceanography of the northern Pacific. *Int. N. Pac. Fish. Comm. Bull.* (2): 1-43.

Huntsman, A. G., 1930. Arctic ice on our eastern coast. *Biol. Board Can. Bull.* (13): 3-12.

MacKintosh, N. A., 1960. The pattern of distribution of the antarctic fauna. *Proc. Roy. Soc. London* (B)152: 624-631.

Madsen, H., 1936. Investigations on the shore fauna of east Greenland with a survey of the shores of other arctic regions. *Medd. Grønland* 100(8): 1-79.

———, 1940. A study of the littoral fauna of northwest Greenland. *Medd. Grønland* 124(3): 1-24.

Mieghem, J. van, and P. van Oye, 1965. *Biogeography and Ecology in Antarctica.* The Hague: Junk.

Neushul, M., 1959. Biological collecting in antarctic waters. *Veliger* 2(1): 15-17.

Parr, A. E., 1939. Quantitative observations on the pelagic *Sargassum* vegetation of the western North Atlantic. *Bull. Bingham Oceanogr. Collect. Yale Univ.* 6: 1-94.

Price, J. J., and P. Redfearn, 1968. The marine ecology of Signy Island, South Orkeney Islands. In *Symposium on Antarctic Oceanography.* Cambridge: Scott Research Institute.

Sandstrom, W. J., 1919. The hydrodynamics of Canadian Atlantic waters. *Rep. Can. Fish. Exped. 1914-1915.*

Skottsberg, C., 1941. Communities of marine algae. Addition. *Kgl. Svenska Veten-skapsakad. Handl.* 19(4): 1–92.

Taylor, W. R., 1954. Algae: non-planktonic. *Bot. Rev.* 20(6, 7): 363–399.

U.S. Navy Hydrographic Office, 1952. *Ice Atlas of the Northern Hemisphere.* Washington, D.C.: U.S. Government Printing Office.

Wilce, R. T., 1959. The marine algae of the Labrador Peninsula and northwest New-foundland (ecology and distribution). *Nat. Mus. Can. Bull.* (158): 1–103.

8

AFRICAN COASTS

SOUTH AFRICA

INTRODUCTORY NOTE ON THE CAPE PENINSULA

Our original intention was to study only the Cape Peninsula of South Africa. This narrow tongue of land, on which Cape Town is situated, lies near a meeting place of warm and cold water masses, though these vary in position with the season and the weather. Nevertheless, it is approximately correct to say that, during the summer, the northern part of the False Bay (or eastern) coast of the peninsula is chiefly affected by warm water, and the northern part of its Atlantic (or western) coast by cold; on the same day at the same latitude, the difference between the temperature of the sea on opposite sides of the peninsula may exceed 8°C. In winter, the difference is much less marked. There is possibly no small area in the world where water of such different temperatures is separated by so little land, and it seemed to us that such conditions must naturally lead to interesting effects in the ecology and the distribution of the biota.

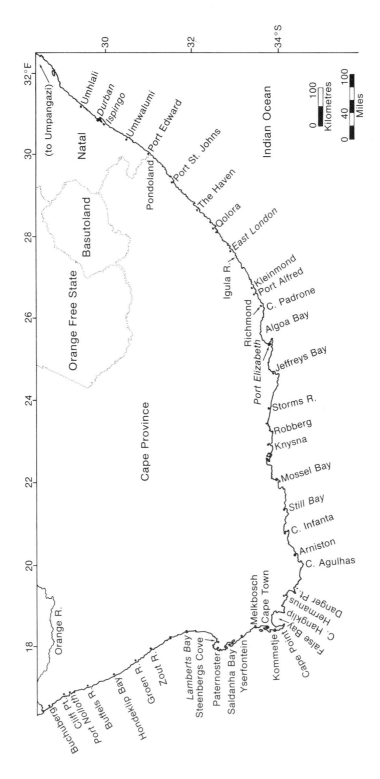

FIGURE **8.1**

A map of South Africa, showing the approximate positions of the chief districts visited during the survey with the main stations in *italics*. For other localities on the Cape Peninsula and False Bay, see Figure 8.28.

However, when we began to make a systematic study of the area, we found that conditions, both hydrographic and in respect to shore populations, were so complicated that they could not be understood without first obtaining some knowledge of the distribution of the biota round the entire South African coast. Ten years later, armed with this knowledge, we returned to the Cape Peninsula and found it much easier to understand many of its problems. (See p. 131 for further notes on the Cape Peninsula.)

GENERAL ACCOUNT OF THE SURVEY, 1932–1940

The coastline of the Union of South Africa is well over 1800 miles long (Figure 8.1). In order to obtain a preliminary picture of the occurrence of plant and animal species round such a lengthy coast, eight stations were chosen, and the shore at each was studied in detail. An account of each of these stations was published, and by the time the series was complete, the general picture was clearly visible, and it was possible to select ecologically important species for further attention. The next step was to visit the coast at a great many other stations collecting further information on the chief species only. About 39 such stations were visited. The section of coast between Durban on the east and Port Nolloth on the west was examined at stations that averaged about 44 miles apart. Most of these stations were visited more than once and, as a rule, by two or more workers at a time. Throughout the investigation, the work was confined to rocky

FIGURE 8.2
Typical wave action on the intertidal zone of South African coasts. [Photograph by A. Blahovsky.]

intertidal areas. Zonation, as well as horizontal and numerical distribution, was always taken into account. Apart from the main survey, certain more detailed studies were carried out; and during the main survey itself, special attention was paid to the distribution and habits of limpets, of which South Africa has a unique assortment.

Table 8.1 lists the stations visited in the course of the survey, and gives the distance between them. Two or more localities were visited in the vicinity of each of the stations mentioned (especially in the Cape Peninsula), but it does not seem necessary to mention all their names. The number given after each name represents the distance in miles between that station and the one next below it. The distances are given to the nearest mile, and were measured with an opisometer on a large-scale sectional map of the Union. They represent approximate distances following the coastline, not as the crow flies.

For purposes of description and reference, with due consideration to the distribution of topographical features, currents and organisms, we originally divided the coast into three sections (see Figure 8.3): (1) the East Coast (from Natal to Cape Padrone, the eastern limit of Algoa Bay; (2) the South Coast (from Cape Padrone to Cape Point,

TABLE **8.1**

Stations visited in the course of the survey of South African coasts, and the distances between them. (Stations are listed in order round the coast from east to west, and distances are calculated as miles of coastline. The eight main stations are in boldface type.)

Station	Distance to next station (miles)	Station	Distance to next station (miles)
1. Umpangazi*	131	21. Arniston	23
2. Umhlali	41	22. Cape Agulhas	56
3. **Durban (Isipingo)**	39	23. Danger Point	22
4. Umtwalumi	47	24. Hermanus	31
5. Port Edward	61	25. Cape Hangklip	45
6. Port St. Johns	63	26. **St. James**	58
7. The Haven	41	27. **Oudekraal**	24
8. Qolora River	45	28. Melkbosch	36
9. **East London**	69	29. Yserfontein	53
10. Kleinmond	10	30. Paternoster	17
11. Port Alfred	21	31. Steenbergs Cove	61
12. Richmond	71	32. **Lambert's Bay**	68
13. **Port Elizabeth**	54	33. Zout River	34
14. Jeffreys Bay	72	34. Groen River	42
15. Storms River	38	35. Hondeklip Bay	48
16. Robberg	21	36. Buffels River	32
17. Knysna	66	37. **Port Nolloth**	11
18. Mossel Bay	54	38. Cliff Point	31
19. **Still Bay**	42	39. Buchuberg†	
20. Cape Infanta	43		

*Umpangazi is 91 miles south of the border of Mozambique.
†Buchuberg is 11 miles south of the mouth of the Orange River.

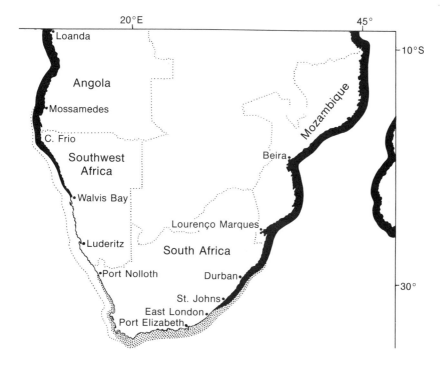

FIGURE **8.3**

A map of South Africa, showing the approximate distribution of its principal coastal faunas. The two tropical faunas involved are shown in *black*—the southern part of the tropical West African fauna on the left, the southeastern termination of the Indo-West-Pacific fauna on the right. The "Cape Fauna"—in the strictest sense, a warm-temperate fauna confined to the south coast of the Union of South Africa—is *stippled*. The cold-water fauna of the southern part of the west coast (the "Namaqua" fauna of many authors) is shown in *white bounded by a dotted line*. The overlapping of these four faunas is shown conventionally and should not be interpreted too literally, though it is shown as accurately as present knowledge and the simple nature of the figure will allow.

the tip of the Cape Peninsula); and (3) the West Coast (from Cape Point to the mouth of the Orange River). In these three sections there are three principal populations:

1) A population of tropical species, which is strongly represented in the Durban district (mixed with south-coast forms), but which disappears, except for a few of its constituent species, between Durban and East London. The population of the southern part of the east coast is similar to that of the south coast.

2) A population of organisms that flourishes in warm temperate conditions, which is characteristic of the south coast. Many members of this series continue, for varying distances, on one or both of the other coasts.

3) A population of cold-loving organisms, which inhabits the west coast and extends in full force not farther south than the Cape of Good Hope, although many of its constituent species pass that point on to the south coast, along which they are to be found for varying distances.

This rough division into three chief populations remains valid, although it became obvious, as work progressed, that the true picture was much more complex. It must be emphasized that the three populations overlap markedly; and although the general nature of each biota is maintained quite distinctly along a considerable stretch of coastline, there is no significant strip within which no progressive change takes place—considering the distribution of the basic components, it is clear that such a static situation would be impossible.

Therefore, for descriptive purposes, we have subdivided the coast not into three, but into five sections, in order that the regions where each population is most fully developed may be distinguished from those of most marked overlap. They are as follows: (1) the east-coast (or Natal) population (from Umpangazi to Port Edward); (2) the south-coast population (from Port Elizabeth to Cape Agulhas); (3) the west-coast population (from Port Nolloth to Kommetje); (4) the eastern overlap (from Port Edward to Port Elizabeth); and (5) the western overlap (from Cape Agulhas to Kommetje). There is nothing dogmatic about these subdivisions. It would be quite legitimate to assign other limits to them; but whatever limits are established, much the same results ensue.

DISTINCTIVE FEATURES OF THE EAST COAST

In this region the warm-water component is paramount; only a limited number of species from other components appear, though these increase in number southward.

FIGURE 8.4

The coast at Isipingo, Natal, looking north. In the foreground is a wide, flat-topped, lower platform, with low rocks belonging to the seaward belt at its foot. An upper platform can be seen on the left against the background of one of the coastal hills.

The Supralittoral Fringe

Littorina africana is the most abundant animal of the supralittoral fringe, accompanied throughout by smaller numbers of *L. obesa* and *Tectarius natalensis,* the first of which typically inhabits the highest part of the zone. A fourth species, *L. scabra,* appears locally. The periwinkle prevalent in the other geographical regions, *L. knysnaënsis,* seems rare or absent in northern Natal; however, it becomes common at the southern end of the region, though still overshadowed by *L. africana.* Other snails frequently found are species of *Nerita.* Algae, when they occur (for in sunny, open places the rock may be extremely bare), usually form mosslike patches, and include species of *Bostrychia, Rhizoclonium, Gelidium,* and *Herposiphonia.* In steep and more shaded places, these turflike growths may be extensive, continuing into the upper part of the midlittoral zone.

The Midlittoral Zone

Figure 8.5, which presents the maximum complexity on a fairly long slope at Umhlali with direct access to the sea, will elucidate the general account of the midlittoral zone.

The upper part of the midlittoral zone is populated primarily by barnacles and limpets, the former often occurring in dense sheets. The leading species are *Chthamalus dentatus, Tetraclita serrata,* and *Octomeris angulosa,* but the bulk of the *Octomeris* occupies the more seaward positions. The limpets are *Patella granularis, P. variabilis, Cellana capensis,* and species of *Siphonaria.* Another characteristic species is the Indo-Pacific oyster *Crassostrea cucullata,* which occurs in close-set abundance, mainly in the middle and upper parts of the zone, and extends in quantity farther inshore than barnacles on upward-facing surfaces (Figure 8.6). A very common snail is *Oxystele tabularis;* in Natal, it largely replaces the *O. variegata* of the rest of the coast. The algae present are mostly small and of relatively few species, primarily *Gelidium reptans* and *Caulacanthus ustulatus.*

The lower part has many areas dominated by algae instead of animals. Many of the algae are short forms that make a more or less mosslike coating or a short turf on the rock, which varies in composition and density from place to place and from one level to another. This algal covering may contain short corallines, or may be dominated by them; where it consists of noncalcareous species, it sometimes looks very much like velvet. Typical constituents of the turfs are *Gelidium reptans* and *Caulacanthus ustulatus* at higher levels, and *Gigartina minima, Hypnea arenaria, Centroceras clavulatum,* and *Herposiphonia heringii* at lower levels. Barnacles compete with the algal turf, as do the gregarious polychaet *Pomatoleios crosslandi* (whose tubes form one of the most characteristic features of this area) and invading zoanthids.

Still further seaward, in the lowest part of the zone, the barnacles and *Pomatoleios* are much reduced and often absent, being unable to compete with the zoanthids (which are at their maximum here), with the dense low-region turf mentioned above, with the corallines, and in some places with larger algae. In exposed positions with strong wave action, *Hypnea spicifera* is well developed, displacing the zoanthids; with and above it is the brown mussel *Mytilus perna,* forming groups or sheets. Zones of turf, *Hypnea,* and *Pomatoleios* change their positions according to the degree of exposure and the disposition of the rock in the vicinity (see Plate 4).

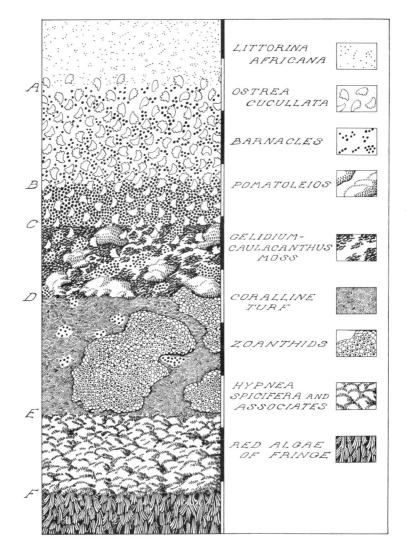

FIGURE **8.5**

The zonation on a fairly exposed slope in Natal, based on a measured traverse near Umhlali. Each division on the scale represents 10 feet. *A*, the bottom of the *Littorina* zone; *A-B*, the upper balanoid zone; *B-C*, the middle balanoid zone; *C-D*, the first lower balanoid zone; *D-E*, the second lower balanoid zone; *E-F*, the *Hypnea* zone; *F*, the top of the infralittoral fringe. Only some of the overlapping can be shown: to show the penetration of *Littorina africana* into the balanoid zone, for example, would only create confusion.

The *Hypnea* Zone. A conspicuous growth of bright green *Hypnea spicifera* is a common feature of exposed parts of the Natal coast. At its densest, it is transitional in position between the lowest part of the midlittoral zone and the infralittoral fringe. With its accompanying corallines and other short algae, it is the belt most nearly comparable to the *Patella cochlear* belt so marked on the south coast, though its

FIGURE **8.6**

Crassostrea cucullata and barnacles colonising the rock surface in the balanoid zone at Isipingo, Natal.

maximum growth is somewhat lower. Its noteworthy animals are *Patella barbara*, which strays into higher regions on the shore; *P. sanguinans*; *P. longicosta*; and *Diedora natalensis*, which, in some areas, in association with lithothomnia, may become dominant.

Corals.   There is a fair representation of healthy reef corals of such genera as *Anomastrea*, *Favia*, *Goniastrea*, *Pocillipora*, *Psammocora*, *Stylophora*, and *Symphyllia*. These form large colonies in some places, but they are at their best in rock pools. The number of species is limited, and few seem common; the most plentiful is probably the curious *Anomastrea irregularis*. In the infralittoral fringe, corals are usually dominated by algae. Reef-building corals, such as *Pocillopora* and *Acropora*, are rare.

Zoanthids.   The zoanthid community is an important and striking feature of Natal coasts. In more sheltered and landward positions, it occupies the level occupied by *Hypnea* on exposed shores, and can extend down to a little below the level of low water. On gradual slopes open to waves, however, its typical habitat is on open rock immediately above the *Hypnea* in the belt of corallines; it may extend into the *Gelidium-Caulacanthus-Pomatoleios* belt and, in pools, to even higher levels. These sheets of zoanthids are often brightly coloured and amazingly extensive. Some forms require damp places, but others can stand considerable exposure to air and sun. *Palythoa nelliae* is probably the most generally abundant; *Zoanthus natalensis, Z. sansibaricus,* and *Z. eyrei* are quite common but rather erratically distributed; and *Palythoa natalensis, P. margaritae,* and *Isaurus spongiosus* are of more scattered occurrence.

FIGURE **8.7**

Masses of calcareous tubes formed by the polychaet *Pomatoleios crosslandi* in the lower balanoid zone at St James, Cape Peninsula. The barnacles are *Octomeris angulosa* and *Tetraclita serrata,* the mussels *Mytilus perna.*

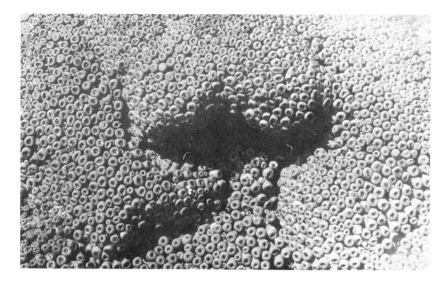

FIGURE **8.8**

A carpet of the zoanthid *Palythoa nelliae* in a sheltered position in the infralittoral fringe at Isipingo, Natal.

The Infralittoral Fringe

The population of the infralittoral fringe varies greatly from place to place, though certain features tend to recur repeatedly. In some places, the infralittoral fringe is occupied by the ascidian *Pyura stolonifera* in dense formation, with associated forms; in some places, by limpets and lithothamnia; but in most places, by a community in which algae are dominant. This growth usually takes the form of a dense sward of rather small species exhibiting various shades of red, pink, and purple, which contrast sharply with the vivid green of *Hypnea spicifera*. The composition of this carpet varies. In some places, it is dominated by the reddish-pink *H. rosea* and the deep red *Rhodymenia natalensis*, and elsewhere, by the dark, wiry *Gelidium rigidum*, perhaps with the handsome *Galaxaura natalensis* below it; there are many other combinations. A marked feature of the zone is a luxuriant growth of long corallines of several species, either alone or mixed with noncalcareous forms. Larger species may be present, and these become important in some places. At Port Edward, for instance, the large, brown *Sargassum longifolium* and *Ecklonia radiata* are plentiful; and at Umpangazi, the straplike *Dictyopteris dichotoma* and the small *Sargassum elegans* form an important part of the infralittoral flora. The green, grasslike *Caulerpa ligulata* is frequently seen.

Other Features of the East Coast

The general occurrence of reef corals in Natal has already been mentioned, but a more graphic idea of them may be conveyed by the following brief description of a deep pool at Umhlali. This contained colonies of *Favia abdita* three feet in diameter; of *Goniopora lobata*, with expanded polyps, one foot across; of *Symphyllia simplex*, with

FIGURE **8.9**

The infralittoral fringe at Shelley Beach, East London: *centre,* plants of *Ecklonia radiata; above and below,* growth of the bright green *Hypnea spicifera.* [Photograph by G. J. Broekhuysen.]

green polyps; and of another species of *Favia.* Close by, in a similar pool, were colonies of *Favia magnistellata, Acropora pectinata,* and *Anomastrea irregularis.* Other notable and common features of pools are the alcyonarian *Xenia coerulea* and the dark blue ascidian *Sigillina coerulea.* There is nothing striking about the sea anemones of the east coast, apart from occasional fields of *Actinoides sultana, Anemonia natalensis,* and *Anthopleura annae;* the other species present are common to all South African coasts.

The limpet population has been described on page 103. There are five species of *Siphonaria,* and all are abundant. The large snail population includes over nineteen species distributed among the genera *Cominella, Drupa, Littorina, Tectarius, Monodonta, Nerita, Oxystele, Thais,* and *Turbo,* but apparently only five of these species continue onto the south coast. *Littorina knysnaënsis* is exceptional: it is not a small, warm-water snail diminishing westwards, but a cold-water species extending into Natal. The crypto-fauna is rich in bivalves, and the barnacle *Balanus amphitrite* is common and continues along the south coast. A very common chiton, which seems to be confined to this coast, is *Plaxiphora wahlbergi.*

Crabs are plentiful and extremely nimble. *Ocypode ceratophthalmus* and *O. kuhlii* make burrows in sand, and various others (species of *Plagusia* and *Grapsus,* for example) may be seen scampering over the rocks. Few of the species here, as far as we know, extend to the south coast.

There seems to be a curious lack of intertidal asteroids, apart from *Asterina exigua.* There are numerous holothurians. Sea urchins are the most notable group of echino-derms, and form one of the most conspicuous elements in the Natal fauna. The tropical forms in Natal, of which the commonest are *Stomopneustes variolaris* and *Echinometra matthaei*, quite overshadow the endemic species, such as *Parachinus angulosus*, which replaces them completely on the south and west coasts.

The bright green alga *Caulerpa* (there are at least four varieties of the grapelike species *C. racemosa*) is characteristic of these shores. At low levels, in some places, it forms extensive carpets, as do other interesting algae. Of sixteen species of seaweeds known from the east coast, we can certify only four as continuing to the south coast, where one of them, *Zonaria interrupta*, is still important.

DISTINCTIVE FEATURES OF THE SOUTH COAST

Some warm-water groups extend into the south-coast component peculiar to this region; these groups are still important but are never dominant. Certain cold water species are also strongly represented (see Figure 8.10).

The Supralittoral Fringe

Littorina knysnaënsis is incredibly abundant in the supralittoral fringe, and far out-numbers *L. africana* everywhere; other Natal representatives of the genus *Littorina* seem to have disappeared altogether. *L. africana*, although fairly common locally in the east, is a rarity at Cape Agulhas. Snails that frequently invade the lower parts of this zone are *Oxystele variegata* and *Thais dubia*, together with the limpet *Helcion pectunculus*. *Ligia natalensis* is often found. The zone is very bare and algae are few, except for the usual patches of *Bostrychia* and a variable amount of *Porphyra*. The latter is irregular in occurrence, but is present on some part of every beach examined. At its maximum, as at Cape Agulhas, it forms dense sheets; and at its minimum, it forms occasional pallid tufts, which, on this coast, are yellowish and brownish in colour. As far as we have seen, the *Porphyra* here never approaches the black luxuriance that it exhibits in colder waters. Its usual habitat is the lower part of the supralittoral fringe, from which it extends downwards into the upper part of the midlittoral.

The Midlittoral Zone

Characteristically, this zone is divisible into easily discernible upper and lower subzones. As in Natal, the upper part supports a vigorous population of the barnacles *Chthamalus dentatus*, *Tetraclita serrata*, and *Octomeris angulosa*, which exhibit their usual peculi-arities of distribution, *O. angulosa* usually favouring the more seaward positions. These barnacles grow in dense sheets—which may become intermittent, leaving bare areas— and to landward on a wide shore, both barnacles and limpets may disappear. *Patella granularis* is common everywhere, and in many places is associated with *Helcion pectunc-tulus* and species of *Siphonaria*. The Natal periwinkle *Oxystele tabularis* is replaced by *O. variegata*. Algae are scanty, but there may be *Porphyra*, *Colpomenia capensis*, and the usual mosslike *Caulacanthus ustulatus* and *Bostrychia mixta*. The last species has a very wide range, and may extend down to the *Patella cochlear* belt.

BARNACLES		MYTILUS PERNA	
CAULERPA LIGULATA		PATELLA COCHLEAR	
CORALLINES		PATELLA GRANULARIS	
GELIDIUM PRISTOIDES		PLOCAMIUM CORALLORHIZA	
HYPNEA SPICIFERA		POMATOLEIOS CROSSLANDI	
LITTORINA KNYSNAËNSIS		PYURA STOLONIFERA	
SHORT ALGAL TURF			

FIGURE **8.10**

The zonation on the south coast of Africa, of the type represented in the Port Elizabeth district. *A*, the upper edge of the *Littorina* zone; *B*, the upper margin of the balanoid zone; *C*, the upper limit of the *Patella cochlear* zone; *D*, the upper edge of the infralittoral fringe. The shore is shown at low water of an exceptionally low spring tide on an unusually calm day. The zones are represented in a very diagrammatic and somewhat telescoped manner, the *Littorina* zone being particularly reduced.

The lower midlittoral, as in Natal, has enormous stretches of the gregarious tube-building polychaet *Pomatoleios crosslandi*, almost comparable in quantity with *Gunnarea capensis*—which in some parts of this region extends downwards to become a dominant form and a more important feature than it is in Natal, but it still falls far short of the position of overwhelming prominence that it occupies farther west. A third poly-chaet, *Dodecaceria fistulicola*, is a common feature in this zone and in the *Patella cochlear* belt at its foot. Both *Dodecaceria* and *Pomatoleios* are associated with lithothamnia. As

FIGURE **8.11**
A general view of Reef Bay, Port Elizabeth. [Photograph by G. J. Broekhuysen.]

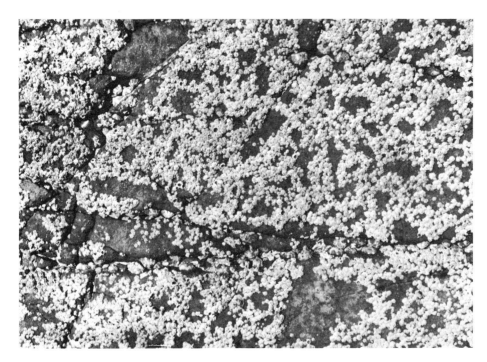

FIGURE **8.12**
Zone of *Chthamalus dentatus* on the South African coast.

in Natal, the mussel *Mytilus perna* and the barnacle *Octomeris angulosa* are numerous in seaward positions in many places; but of the west coast mussels, *M. meridionalis* was never found, and *M. crenatus* was limited to occasional specimens at the western end of this stretch of coast.

Two species of algae are very striking features of this shore. The first is the conspicuous *Gelidium pristoides*, which forms a belt of dark, often luxuriant tufts in the lower midlittoral zone, especially in the *Patella cochlear* belt, and persists through a wider range of exposure and shelter than does the *Patella cochlear* mosaic. It tends to grow in narrow belts in sheltered places, with a wider, sparser concentration in more exposed ones. The second is *Colpomenia capensis*, a species with a peculiar brainlike appearance, yellowish-brown in colour, which can become locally plentiful, extending into the upper part of the midlittoral.

Bifurcaria brassicaeformis appears abundantly at Cape Agulhas, and extends into the *Patella cochlear* belt below.

The *Patella cochlear* Belt. A well-developed typical mosaic of *Patella cochlear*, with lithothamnia, covers the rock surfaces at the bottom of the midlittoral zone in suitable positions—that is, in positions where, in ordinary weather, the wave action in the most exposed situations is moderate. Where the wave action is perpetually violent, however, the mosaic is replaced by different communities of animals, mainly *Mytilus* and barnacles, in many places associated with such algae as *Gelidium cartilagineum*. Also, in positions of increasing shelter, the *P. cochlear* mosaic becomes intermittent and finally disappears, being replaced there not by an animal community, but by a short turf of relatively mosslike algae, among which larger plants occur in varying proportions (see Figures 8.13 and 8.14).

Where the *Patella cochlear* mosaic is well-developed, the constant, intense browsing of innumerable closely set limpets prevents any large settlement of barnacle and mussel larvae,* so that only odd patches of mussels and barnacles occur among the mosaic. *P. cochlear* is less successful as an inhibitor of algae: In some places, mosaic areas are much overgrown with algae, which even flourish on the backs of limpets. There appears to be a perpetual struggle between the algae and the limpets for space.

In most sheltered places where the *Patella cochlear* belt is replaced by a turf, the short species are corallines with epiphytes, which are mixed with other algae and with sand, and the larger forms include such species as *Sargassum heterophyllum*, *Caulerpa ligulata*, *Dictyota dichotoma*, with *Laurencia flexuosa*, *L. glomerata*, and *L. natalensis*, which may grow either as a turf or in small tufts.

The one animal that seems able to intrude upon and destroy a *Patella cochlear* belt is *Gunnarea capensis*.

Algae of the *Patella cochlear* belt, other than the turf species, are numerous, and may vary in quantity and "personnel" from place to place, according to the degrees of available shelter and other factors. *Hypnea spicifera* occurs throughout this area, at its maximum forming a green carpet. *Gigartina radula* and *G. papillata* are distributed

---

*The stomach contents of limpets are varied. Although limpets eat mainly algae, they also consume small animals, which may include the larvae of barnacles and mussels.

FIGURE **8.13**

Part of the seaward face of a boulder at Oudekraal, South Africa. At the top, a few *Patella granularis* are visible. Below them is the sharp upper limit of lithothamnia incrustation, which includes two species, a coralliform species above and a finely nodular one below. Numerous *Patella cochlear* form a zone that extends from the upper limit of the lithothamnia to beyond the lower edge of the picture.

rather erratically in certain localities, but in quantities in no way comparable to those of the same species on the west coast. On certain slopes, there are sheets of the black-green fleshy lobes of *Codium stephensiae* and *C. lucasii* that may continue below low-water level.

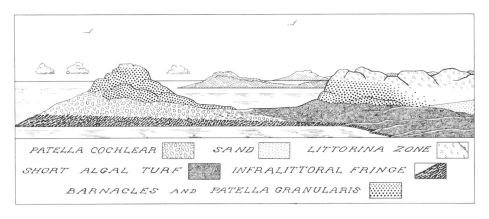

FIGURE **8.14**

The type of change that takes place in the zonation from an exposed promontory to a sheltered bay or creek between two such promontories on the south coast in the Port Elizabeth district. On the seaward face of the left-hand promontory, the infralittoral fringe, *Patella cochlear* zone, and balanoid zone are well represented, with a small area of *Littorina* zone above them; in the bay on the right, the *Patella cochlear* mosaic has disappeared to be replaced by short algal turf, and the infralittoral fringe has narrowed; inshore, the barnacles and *Patella granularis* of the balanoid zone die out as the sandy slope is approached.

The Infralittoral Fringe

In most places, the *Patella cochlear* belt has a well-marked lower limit, so that the beginning of the infralittoral fringe can easily be distinguished; in other places, however, the change is gradual, especially where the limpets are overgrown with algae and the *Pyura* of the fringe is scanty. On the south coast, *Pyura stolonifera* achieves its maximum development, and these leathery ascidians, tightly packed and in incredible numbers, must cover many miles of rock. The fauna associated with them is varied and may be highly coloured—anemones, compound ascidians, and, notably, the vivid magenta *Alcyonium fallax*. *Mytilus perna* sometimes descends freely among the *Pyura*. A certain number of algae may also grow on the *Pyura* tests (for example, *Gelidium cartilagineum* and *Plocamium cornutum*). Where the fringe is not dominated by a carpet of ascidians, algae may grow, though these are not usually very large. Prominent are the larger corallines—a richly mixed and varied growth, which may exclude, or mix with, noncalcareous species; many of the corallines extend upwards somewhat into the bottom of the *Patella cochlear* belt. Tufts of rich, deep red *Gelidium cartilagineum* may make a dense growth, in many places accompanied by the similar, but even more wiry, *G. rigidum*, especially to eastward. *Caulerpa ligulata*, occurring in sheets or patches or as a scattered growth (at its best in relatively sheltered creeks), is as common as in Natal. *Plocamium corallorhiza*, in various dull shades of purplish red, is found in every district and, at its best, forms dense carpets. *Hypnea spicifera* may extend downwards into this part of the shore. Among the larger species are the yellowish *Sargassum heterophyllum*, which flourishes especially in inlets, and, locally, *S. longifolium*. There

FIGURE **8.15**

An area in the infralittoral fringe of False Bay densely populated by *Pyura stolonifera.* The photograph was taken on a calm day at extreme low water of a spring tide. [Photograph by G. J. Broekhuysen.]

is no true laminarian zone, and *Ecklonia radiata*—though it is fairly common in pools and channels in the Port Elizabeth district and forms a belt in some places in the eastern overlaps—is not really a feature of the south coast. The brown *Zonaria interrupta* is characteristic of both south and east coasts, and is variable in abundance. There is a similar variation in the fringe population from exposed to sheltered places, as there is in the *Patella cochlear* belt.

Other Features of the South Coast

No reef corals or xeniids appear on the south coast; but sometimes, under stones and overhangs, and in pools, *Balanophyllia* may be found, *B. capensis* being common at Cape Agulhas where *Alcyonium fallax* is also present. In marked contrast to the coast of Natal, this coast has (so far as we know) only one species of zoanthid, the small, common *Isozoanthus capensis,* which is not a tropical form, and belongs primarily to the cryptofauna. There are the usual sea anemones of the whole South African coast, including *Anthopleura michaelseni, Anthothoe stimpsoni, Bunodactis reynaudi,* and *Bunodosoma capensis.*

For limpet distribution, see pages 109–113.

Of the common south-coast snails, only four out of eleven species are also found in Natal. There are no species of *Nerita. Thais capensis,* important in Natal, dwindles away and vanishes after Cape Agulhas.

The cryptofauna is plentiful, with numbers of bivalves. Particularly striking is the giant chiton *Dinoplax gigas,* a very prickly species characteristic of this coast. *Acanthochiton garnoti* occurs both on and under stones. The genus *Siphonaria* is nearly as well represented here as in Natal, though *S. oculus* is rather less in evidence.

None of the common crabs of Natal seem to extend to the south coast. The most notable in this area are *Plagusia chabrus,* at lower levels, and *Cyclograpsus punctatus,* at higher ones. Both are common along the west and south coasts, and on the southern end of the east coast. *C. punctatus* extends, but less plentifully, into Natal.

Among barnacles present, there are four species of *Balanus: B. amphitrite, B. trigonus, B. algicola,* and the giant *B. maxillaris.* The two former occur in Natal, and the two latter become more important to westward.

On the south coast, unlike the coast of Natal, there are four common asteroids, including *Asterina exigna,* and but only one abundant sea urchin, *Parechinus angulosus.* Holothurians are common.

The algae on the south coast are less invaded by species from Natal than by species from the west. An interesting eastern species that occurs on the south coast is the small red *Desmia tripinnata,* which lives at an unusually high level for so delicate a form; at Port Elizabeth, it can form a distinct belt in the lower midlittoral. Many other algae are characteristic of this coast, but are not limited to it.

DISTINCTIVE FEATURES OF THE WEST COAST

On the west coast, the cold-water component outweighs all others. The south-coast element is present, but falls off rapidly to northward, and few of the warm-water species remain.

There is an environmental factor in this region that seems peculiar to it. This is the occurrence of a persistent, dense, and often prolonged morning mist, which, in more northern areas, seems to be frequent throughout the year. It is presumably a function of the very cold inshore water of this coast. As low water of spring tides occurs in the morning, it means that during much of the period when rocks are uncovered, they will be kept cool and moist, and animals may move about and feed. In a country of such hot sunshine as South Africa, this materially affects the life of the fauna and flora (see Plate 5).

The Supralittoral Fringe

The width of the supralittoral fringe on the west coast is particularly noticeable; in exceptionally sheltered places, it extends little, if at all, above the level of high water of spring tides, but in places where wave action is strong and the spray is far-flung, it may reach far above this level. The zone is inhabited by myriads of *Littorina knysnaënsis,* but apparently by no other species of that genus. The lower section is commonly invaded by the snails *Oxystele variegata* and *Thais dubia,* by the limpet *Helcion pectunculus,* and by outliers of *Patella granularis* from the zone below. *Porphyra capensis*

FIGURE **8.16**
The shore at Lambert's Bay on the west coast of South Africa. Note the kelp on the edge of the surf.

FIGURE **8.17**
A view of the shore at Oudekraal with Lion's Head in the background. Note the smooth boulders and the dense kelp beds showing above the water. [Photograph by G. J. Broekhuysen.]

is present in many places. This alga has markedly seasonal variations in growth, particularly noticeable to the south of the west coast. It is a prominent feature of the west coast, growing in bands or as a series of separate patches, either in the upper midlittoral or in the lower supralittoral fringe. In many places, it makes a dense, blackish, healthy growth, much more conspicuous than the somewhat anaemic growth usual on the south and east coasts. Other macroscopic algae are mostly confined to pools. High-level pools of the fringe and upper midlittoral may contain *Chordaria capensis, Grateloupia filicina, Enteromorpha bulbosa* var. *africana,* and species of *Ulva* and *Cladophora,* among others; in some places, one or two of these forms will fill a pool to the exclusion of all the others.

The Midlittoral Zone

The upper part of the midlittoral zone supports so few barnacles that there is no balanoid zone as described for other coasts; and where algae are few, *Patella granularis* becomes the most conspicuous inhabitant of otherwise bare-looking rock. In such circumstances, on many parts of the shore (on exposed slopes, or broken rocks, for instance), there may be a marked gap in algal succession below the belt of high-growing *Porphyra.* On the other hand, on gentle slopes subject to reduced wave action, particularly in sandy places, the rock, instead of being bare, may be covered throughout by an unbroken succession of algae, and the *Porphyra* above can extend in reduced quantities to the

FIGURE **8.18**
A healthy growth of *Porphyra capensis* at Kommetje, Cape Peninsula.

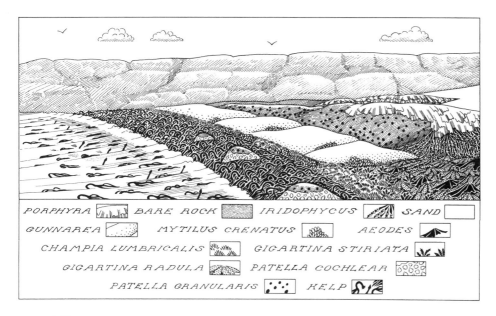

FIGURE **8.19**

The zonation in a small sheltered bay between two rocky headlands on the west coast of South Africa. The laminarian zone is well developed, and part of it is exposed to the air at low water of low spring tides. The *Patella cochlear* mosaic is reduced to isolated patches on boulders that project above the kelp. Immediately to landward of the laminarian zone is a narrow belt dominated by *Champia lumbricalis* and *Mytilus crenatus.* In the depression shown in the right-hand corner, *C. lumbricalis* is followed by a continuous growth of algae, including *Gigartina stiriata, G. radula,* and *Aeodes orbitosa,* which give place higher up to *Iridophycus capensis* and, finally, to *Porphyra.* Along the remainder of the bay, the strip of *C. lumbricalis* is followed first by fields of *Gunnarea capensis* (concealing the underlying rock completely) with some *Mytilus crenatus* interspersed, then by bare rock with *Patella granularis,* and finally by *Porphyra,* above which the sandy beach begins.

bottom of the shore. Below the main concentration of *Porphyra* are *Chaetangium saccatum, C. ornatum,* and *Ulva lactuca;* these algae, in their turn, become mixed with, and largely replaced by, *Iridophycus capensis, Aeodes orbitosa,* and two brown algae (the clublike *Splachnidium rugosum* and the filiform *Chordaria capensis*) that extend down to the kelp. Exceptionally, *Chylocladia capensis* becomes dominant between *Porphyra* and *Gigartina* (see Figures 8.19 and 8.20).

The barnacle *Chthamalus dentatus,* which is so common elsewhere, occurs sparsely, or is locally common on the south part of the coast, but disappears proceeding northwards. Similarly, two large species, *Tetraclita serrata* and *Octomeris angulosa* (as usual living farther down the shore), persist throughout the length of the west coast; but while to southward they may be fairly plentiful, occurring on open rock as well as in clefts and crevices, they become scarcer in the north until, at Port Nolloth, they are rare and are confined to specially suitable habitats.

The lower part of the midlittoral is typically indicated by the rather sudden appearance of species of lithothamnia, which cover the rocks in continuous sheets below the bare

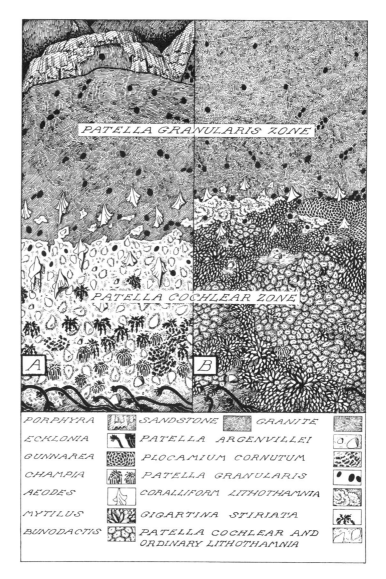

PATELLA GRANULARIS ZONE

PATELLA COCHLEAR ZONE

A   B

PORPHYRA		SANDSTONE		GRANITE	
ECKLONIA		PATELLA ARGENVILLEI			
GUNNAREA		PLOCAMIUM CORNUTUM			
CHAMPIA		PATELLA GRANULARIS			
AEODES		CORALLIFORM LITHOTHAMNIA			
MYTILUS		GIGARTINA STIRIATA			
BUNODACTIS		PATELLA COCHLEAR AND ORDINARY LITHOTHAMNIA			

FIGURE **8.20**

The zonation of two fairly exposed seaward-facing slopes of roughly 45°. *A*, an area from which lithothamnia is absent and in which the *Patella cochlear* zone is carpeted by a typical mosaic of this limpet; *B*, a granite slope on which the *Patella cochlear* mosaic is replaced by dense sheets of *Mytilus*, *Bunodactis*, and *Gunnarea*, with a little lithothamnia; the mussels in such places are often encrusted with *Balanus algicola*, not shown in the figure.

zone. This line of lithothamnia is also present on the south and east coasts, but is so obscured as not to be conspicuous. On the south and east coasts, turfs of short algae have been recorded in lower midlittoral areas; on the west coast, it is possible to find areas of short corallines, or of *Caulacanthus ustulatus* (especially among the *Aeodes*), which might be regarded as turfs, but the general tendency is either towards bare rock,

FIGURE **8.21**

A sheet of *Gunnarea capensis* tubes, in a layer 15 inches thick, completely covering the rock at a locality on the west coast of South Africa.

a sheet of lithothamnia and limpets with few algae, or a varyingly dense growth of larger species.

The commonest algae at the upper limit of continuous lithothamnia (or, in some places, just above it) are *Aeodes,* rather sparsely distributed and small, and the *Splachnidium* and *Chordaria* mentioned above. The upper limits of these three lie close together, and their width and density of growth depend on local conditions. *Chordaria,* in the south, is limited to pools and damp places, but occurs freely on open rock in the north. Often on the lithothamnia itself there are considerable quantities of *Cladophora,* notably *C. flagelliformis.*

The coarse sandy tubes of *Gunnarea capensis* form a much more important item on the west coast than elsewhere. At its greatest development, *G. capensis* can form a layer of tubes more than a foot thick, completely concealing the rock over large areas. Its distribution seems affected by three things: the amount of exposure to wave action, the angle of the rocky slope, and the proximity of sand. The reaction of this worm to these factors is demonstrated very clearly on certain promontories that are exposed to strong wave action at the tip and run shorewards towards a sandy beach and increasing shelter at the other end. As shelter increases, the tubes form a narrow band or ridge between the *Patella cochlear* belt and the main *Patella granularis* population; this band then gradually broadens, and, in the most protected sites, the tubes fuse into a continuous mass, which occupies a wide area between the midlittoral and the kelp. This belt of *P. granularis* is very common, and is at its maximum in inlets and bays and on protected reefs. On occasion, it may partly or completely replace the limpet mosaic. It is frequently colonised by other organisms, including mussels and various algae.

Pomatoleios crosslandi, the tubicolous polychaet so common on other coasts, seems absent on the greater part of the west coast of South Africa, though it is present at the southern end of the Cape Peninsula. The other common polychaet, the small *Dodecaceria fistulicola,* a species common on the south coast, continues throughout the west coast in variable quantity in the lower midlittoral, burrowing in lithothamnia on the open rock as well as in pools.

The *Patella cochlear-P. argenvillei* Belt. This is a continuation of the *Patella cochlear* belt of the south coast. *P. cochlear* and *P. argenvillei* make an interesting study of interchanging populations along the west coast. The former is abundant on the southern part of the west coast, but gradually becomes scarcer to the north until it disappears entirely. Its numbers depend on the same factors that control the *P. cochlear* belt, as given above, and its attendant fauna changes from south to north. *P. argenvillei,* however, although it becomes paramount to northward, is important along the whole coast. There is a tendency in regions where both are abundant for *P. cochlear,* in association with scattered individuals of *P. argenvillei,* to occupy the upper part of the belt, and for large, closely set individuals of *P. argenvillei* to inhabit the lower regions, often extending among the fringe of kelp. Where there is no concentration of *P. argenvillei, P. cochlear* may descend to these low levels. On the other hand, outliers of *P. argenvillei* may occur in the upper midlittoral among the *P. granularis,* where no *P. cochlear* exists, except occasionally in pools or on isolated growths of lithothamnia. It seems that the *Patella cochlear* mosaic prefers areas of submaximal exposure, but that *P. argenvillei* persists under conditions of extreme exposure. Where shelter increases, both species diminish in number, and are absent from completely protected places.

As stated earlier, masses of *Gunnarea capensis* may replace the limpet mosaic entirely where conditions are favorable to its growth. In other places, it may form large patches that eliminate all but small groups—or even single individuals—of the limpets. So also, on certain types of exposed slopes, the mussel *Mytilus crenatus* may eliminate the mosaic (see Figure 8.20). Where there is at least some reduction in wave action, and where rocks run down to the sand, the multicolored anemone *Bunodactis reynaudi* may completely cover the area, or a mixture of mussels (often encrusted with *Balanus algicola*) and anemones may occupy all available space. The algal growth of such beds is variable in species and in quantity; a notable alga is *Gymnogongrus capensis,* a species of wide occurrence on western coasts.

Four important algae of the *Patella cochlear-P. argenvillei* belt are *Champia lumbricalis, Plocamium cornutum, Gigartina stiriata,* and *G. radula. C. lumbricalis,* a plant with tubelike branches, is a constant feature all along the west coast. It is plentiful in places of moderate wave action, and is more tolerant of sheltered places than limpets. Its denser formations may replace limpets. To southward, in areas where *P. cochlear* is abundant, it tends to avoid the higher levels of the belt; farther north, where *P. argenvillei* prevails, its maximal growth is just above the limpets. *P. cornutum,* a short, often dark brown rhodophycean, appears to replace *C. lumbricalis* in areas of increased wave action—although it often mingles with it or alternates with it in patches—reaching its greatest abundance at lower levels. *G. stiriata* has brown digitiform fronds, and *G. radula* forms flat blades that are often large. They both occupy the *P. cochlear-P. argenvillei* belt,

FIGURE **8.22**
A dense growth of *Patella argenvillei* at a locality on the west coast of South Africa.

with a tendency to occur most densely just above the *Champia.* They may also extend up into the bare upper part of the zone, and are common in pools and cracks. *G. stiriata* inhabits places too exposed for *G. radula;* both live in sheltered places, but *G. radula* more luxuriantly.

A few species of corallines may grow profusely in the *P. cochlear–P. argenvillei* belt, and are particularly conspicuous in the north among the large *P. argenvillei* at the bottom of the belt.

Other Inhabitants of the Lower Midlittoral. The dominant brown *Mytilus perna* of the south and east coasts soon disappears on the west coast, and is replaced in more northerly localities by the blue-black *M. meridionalis* and the ribbed *M. crenatus.* The latter may replace the limpets in exposed positions in the *Patella cochlear–P. argenvillei* belt. In some places, it grows down among the kelp, but it can also extend upwards into the lower midlittoral zone, particularly in pools and gullies, and prefers rocky seaward positions clear of sand. On the other hand, *M. meridionalis* is at its best in sandy, quiet places, even on sand itself. In pools, it may occur fairly high on the shore. However, in spite of this general broad contrast, the two species can intermingle, and, if one may judge from the accumulation in certain places of quantities of large shells too great to have been derived entirely from the intertidal zones, both species would appear to live in abundance below tidemarks. Mussel beds may be heavily overgrown and almost concealed by algae.

Another locally important gastropod is *Vermetus corallinaceus,* which constructs extensive sheets of polychaetlike calcareous tubes. It occurs throughout the west coast in varying quantities. In most places, its tubes are embedded in lithothamnia; they may form a marked, regular, prominent ridge along the rock immediately below the upper limit of lithothamnia, or just above the limpet mosaic, into which they may freely penetrate, in some places forming ridges between the limpets or even replacing them. This vermetid may extend as far down as the infralittoral fringe, and farther upshore as well, and into more sheltered places than *P. cochlear* can colonise. Many tubes have the dwarf *Gelidium micropterum* growing on them.

Special Algae of the Southern Part of the West Coast. The most important of these is *Bifurcaria brassicaeformis.* It is characteristic of the western overlap, and has been found as far north as Paternoster. Where it is abundant, it grows large in pools and tends, on open rock, to make a belt just above *Champia* that may reach into the *Patella granularis* area. A similar, more shrubby species, *Pycnophycus laevigatus,* extends at least as far north as Groen River; it is usually found in pools among *Patella granularis,* and often farther down; it may be important locally. *Gelidium pristoides,* so much a feature of the south coast, reaches round the west coast of the Cape Peninsula to Kommetje, where it occurs throughout the lower midlittoral zone. At Slangkop, a distinct belt of *G. pristoides* and *Vermetus* is distinguishable. Northwards from Kommetje, *G. pristoides* is replaced by the allied but dwarf species *G. micropterum,* with which it coexists at Slangkop and Kommetje. *G. micropterum* can be found as far east as Cape Agulhas; it is distinct from the little *G. reptans* of the east coast, and as previously mentioned, is frequently associated with vermetid tubes.

The Infralittoral Fringe

On the west coast, the infralittoral fringe is less discernible than on the other South African coasts visited. The sublittoral region proper is occupied by giant laminarians, the three common species being *Ecklonia buccinalis, Laminaria pallida,* and *Macrocystis pyrifera* (see Figure 8.23). The transition from the fringe of this belt of kelp* to the *Patella cochlear–P. argenvillei* belt marks the lowest intertidal level, but it may not appear as a distinct zone. The kelp varies much in quantity. From some steep exposed slopes it may not be seen at all, but where the configuration of the sublittoral rocks suits it, it forms large beds offshore. These are never exposed, only wet fronds being visible above the water, and in most wave-washed positions, kelp remains permanently damp. In some sheltered bays and inlets, however, the kelp plants nearest the shore may be exposed for some hours at low water. The intertidal growth of kelp in a shelving bay may be as much as 50 feet (15 m) wide. *Macrocystis* appears confined to relatively sheltered positions on the southern part of this coast, and is therefore only locally important. *Ecklonia,* on the whole, is the dominant form to southward, as *Laminaria* is to the northward. The changeover is not strictly progressive (at Yserfontein, for example, *Laminaria* seems to be unduly abundant for the latitude), but, generally speaking, the statement holds good.

---

*By "kelp" we mean the giant algae only—*Ecklonia, Laminaria,* and *Macrocystis.*

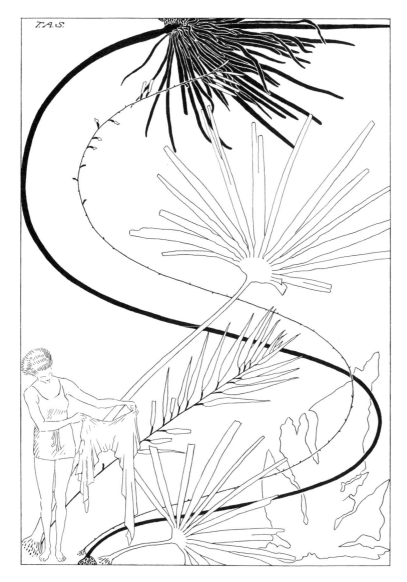

FIGURE **8.23**

The relative sizes of a human figure 5 feet 6 inches tall and specimens of some of the large algae of the Cape region of South Africa. The black figure represents a large individual of *Ecklonia buccinalis,* and following the line of its stipe is a plant of *Macrocystis pyrifera* that has lost most of its fronds. The fan-shaped white figures represent large and small plants of *Laminaria pallida.* The human figure is holding a plant of *Aeodes orbitosa,* another specimen of which is spread out in the lower right-hand corner. Below the stipe of the large *Laminaria* is a young plant of *Macrocystis.*

The fringe of kelp is inhabited by species characteristic of the lower part of the *Patella cochlear-P. argenvillei* belt (*P. argenvillei, Bunodactis reynaudi, Gunnarea capensis, Champia lumbricalis, Plocamium cornutum,* species of *Gigartina,* corallines, and lithothamnia, for example). The kelp may have smaller algae epiphytic upon it (including

FIGURE **8.24**
The laminarian zone near Slangkop Lighthouse, Cape Peninsula. [Photograph by G. J. Broekhuysen.]

Carpoblepharis flaccida, Polysiphonia virgata, Suhria vittata, and species of *Porphyra* and *Cladophora*), as well as small encrusting forms. *Patella compressa* is very common on *Ecklonia* stipes. Unfortunately, we were unable to undertake what would have been an interesting study of the inhabitants of the kelp holdfasts and the rock between them. Other red algae of this region deserving special mention are the foliose forms *Botryoglossum platycarpum, Botryocarpa prolifera, Neuroglossum binderianum,* and *Epymenia obtusa,* and the large, leathery, straplike *Pachymenia carnosa.*

On the south coast, a dense community of *Pyura stolonifera* dominates many areas; on the west coast, this species may be plentiful at low levels (especially among kelp) as isolated individuals, in clumps, or as a narrow belt. In few places does it make the dense formations of the south coast. Often, it is found in pools and creeks, and may even live in deep pools cut off from the sea at low water.

Other Features of the West Coast

Alcyonium fallax, as on the south coast, may achieve great profusion. *Balanophyllia bonaespei* is common at Oudekraal, where, in pools and clefts, there is an extraordinarily rich and plentiful population of coelenterates. The commonest sea anemones are those of the south coast, but forms peculiar to the west coast also occur, including *Isanthus capensis* and *Halianthella annularis. H. annularis* is of particular interest, as it carries its young in brood pouches. The corallimorphid *Corynactis annulata* is locally common all along the coast. A striking feature of this region, and one which must rank as one of the most beautiful natural spectacles of the world, is *Bunodactis reynaudi*. This gregarious anemone covers its column with gravel and exhibits almost endless variations of colour, including bright pink and blue; in many places, it forms a sea of colour that fills the pools and covers the open rock almost to the exclusion of everything else (see Plate 9).

On this coast, *Patella argenvillei* attains its greatest profusion and size, and the cold water species *P. granatina* and *P. compressa* are also important. Species of *Siphonaria* are common, with *S. capensis* extending throughout the coast and *S. aspera* waning northwards. *S. deflexa*, however, apparently occurs only at the southern end of the Cape Peninsula.

The most interesting fact about the chiton population of the west coast is the apparent absence of *Acanthochiton garnoti*. *Callochiton castaneous* may be found all along the coast, and other forms characteristic of much of the area are the large *Chaetopleura papilio* and the small *Chiton nigrovirens*, which may be locally abundant.

Snails are numerous, including species of *Cominella*, *Gibbula*, *Oxystele*, and *Thais*. Particularly notable is *Argobuccinum argus*.

The most frequently found crabs are the same as those of the south coast, with the addition of the sluggish *Pilumnoides perlatus* among worm tubes and laminarian hold-fasts. The "crayfish" *Jasus lalandii* is common among kelp in suitable conditions, and did not seem to have been over-fished at the time these notes were made.

Asterina exigua and *Parechinus angulosus* continue plentifully all along the west coast, as on the south, but their colours on the west coast are much duller.

Of the algae, *Caulerpa* is apparently absent, as are practically all species of *Dictyota* and *Sargassum*. *Gelidium cartilagineum* is much less important here than it is on the south coast, although it does persist all along the west coast, and may be common locally. *Hypnea spicifera* also diminishes westwards. Many other algae are common, but few, if any, are confined to the west coast.

DISTINCTIVE FEATURES OF THE EASTERN AND WESTERN OVERLAPS

Now that the three main regions have been dealt with, consideration must be given to those stretches of coast along which their biotas markedly overlap (see Figure 8.25).

The Eastern Overlap

It will be evident from Plate 6 that, on the east coast, there is a marked overlap between the biota of the warm-water component and those of the cold-water and the south-coast components. In many respects, this transition is quite gradual; but at certain points on the coast (for example, on the strip of coast between Port St Johns and Qolora*),

---

*If we had more details of the distribution of the Natal species, the fall-off between Durban and Port St Johns might prove to be more gradual than the graphs show, and that between St Johns and Qolora might prove to be even steeper.

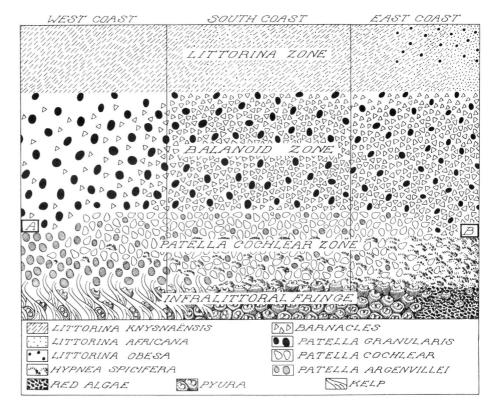

LITTORINA ZONE

BALANOID ZONE

PATELLA COCHLEAR ZONE

INFRALITTORAL FRINGE

LITTORINA KNYSNAËNSIS		BARNACLES	
LITTORINA AFRICANA		PATELLA GRANULARIS	
LITTORINA OBESA		PATELLA COCHLEAR	
HYPNEA SPICIFERA		PATELLA ARGENVILLEI	
RED ALGAE	PYURA	KELP	

FIGURE **8.25**

The principal features of the zonation on the west, south, and east coasts of South Africa. Only a few of the species that characterize the zones are included, and overlaps in vertical distribution are ignored for the sake of simplicity. The diagram represents the coast between the Orange River in the west and Umhlali in the east, and the columns corresponding to the three coasts are drawn to scale. The vertical scale has little significance, as the actual widths of zones are very variable from place to place. The word "barnacles," as used here, includes the species *Octomeria angulosa*, *Tetraclita serrata*, and *Chthamalus dentatus*; "kelp" includes *Ecklonia buccinalis* and *Laminaria pallida*; and "red algae" refers to the selection of reddish rhodophyceans common in Natal below the green rhodophyceaen *Hypnea spicifera*. *A*, the region to westward where *Patella argenvillei* is paramount; *B*, the region in Natal where the *Patella cochlear* zone is partly replaced by a *Hypnea* zone.

it appears to take place more rapidly. From our records, the number of warm-water species at Port St Johns is more than seven times the number of those belonging to the cold-water and south-coast components (17); at Qolora, the number of warm-water species (85) is twice that of the others (47). This change is much more pronounced for the animals than it is for the algae (see Figures 8.26 and 8.27).

According to Isaac (1937), the mean annual temperature for East London (45 miles from Qolora) is 17.6°C, and that for Port St Johns is 19.3°C. It would therefore appear that, where the mean annual temperature falls below 20°C, many warm-water animals cannot survive. Only one species of reef coral is common at Port St Johns, and we found no reef corals at all to the south: they probably die out altogether within 50 miles of this locality.

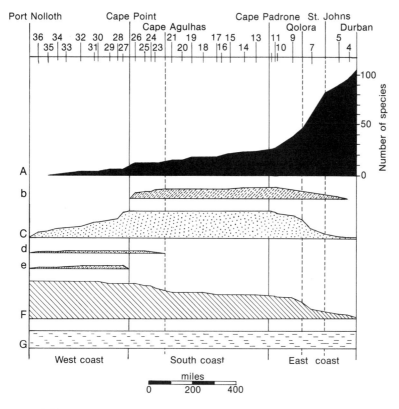

FIGURE **8.26**

The geographical components of the intertidal fauna along the South African coasts, based upon the known distribution of 202 species. The vertical scale, which indicates the number of species belonging to each component, is the same for all components: components *b–G* should be referred to the scale provided for component *A*. The approximate positions of the principal stations visited between Port Nolloth and Durban are indicated along the upper edge of the diagram; for localities indicated only by a number, see Table 8.1. The four main components are *A*, the warm-water one; *C*, the south-coast one; *F*, the cold-water one; and *G*, the ubiquitous one. Components *b, d*, and *e* are subsidiary groups of limited local range.

West of East London, the south- and west-coast components begin to outweigh the warm-water ones (see Table 8.2).

Of the warm-water species that extend farther west, three quarters of them (15 animal and 23 algal species) continue for varying distances along the south coast but, as far as we know, do not pass Cape Point. Species worthy of special mention are *Balanus amphitrite;* the periwinkle *Nerita albicella,* the only one of its genus found beyond the east coast; the minute snail *Littorina africana,* dominant in Natal, but still notable on the south coast; the plentiful limpet *Siphonaria oculus;* the common whelks of Natal, *Cominella lagenaria* and *Thais capensis;* and the important green alga *Caulerpa ligulata*

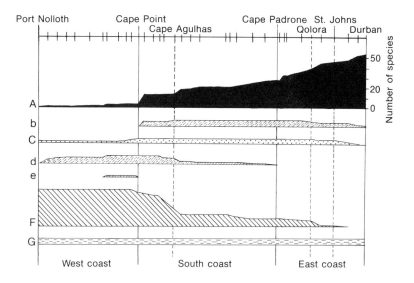

FIGURE **8.27**

The geographical components of the intertidal flora along the South African coasts, based upon the known distribution of 116 species. The diagram is constructed in the same way as Figure 8.26 (which see).

of the infralittoral fringe. The larger seaweeds include *Sargassum elegans,* common to both south and east coasts; *Halimeda cuneata,* which extends farther afield than most coral-reef algae; and members of the genera *Zonaria, Peyssonelia,* and *Dictyota.*

Warm-water species that pass Cape Point and are able to colonise the southern part of the colder west coast are few (12), but they include forms of great ecological importance. Noteworthy in waters of intermediate temperature, and in the warmer waters of the east coast, are the mussel *Mytilus perna* and the barnacle *Chthamalus dentatus* (two of the most important species). Also noteworthy are the tube-building polychaet *Pomatoleios crosslandii,* and the algae *Hypnea spicifera* and *Plocamium corallorhiza*—all three dominant south- and east-coast forms. Scarcely less important are *Balanus trigonus, Siphonaria aspera,* and *S. deflexa.*

The Western Overlap

The "personnel" of the cold-water component is very unlike that of the warm-water one. If we review the species that are largely limited to the west coast, together with those that continue along the south coast for varying distances without, so far as we know, reaching the east coast (categories 2 and 3; see Figure 3.1), we find a group of 42 species (28 algae and 14 animals). Most of these are typical west-coast algae, including not only the giant species *Ecklonia buccinalis* and *Laminaria pallida,* but a number of other more or less large forms, such as *Aeodes orbitosa, Schizymenia obovata, Pachymenia carnosa, Grateloupia longifolia, Iridophycus capensis,* and *Gigartina stiriata.* The ecologically important *Champia lumbricalis* belongs to this group, as do the principal epiphytes of the kelp, *Polysiphonia virgata, Carpoblepharis flaccida,* and *Suhria vittata.*

TABLE **8.2**
Numbers of eastern species and southern
and western species at East London
and at four stations to the west

Station	Eastern species	Southern and western species
9. East London	74	56
10. Kleinmond	60	60
11. Port Alfred	59	62
12. Richmond	54	63
13. Port Elizabeth	50	64

Most of these forms seem to die out westward of, or near, Cape Agulhas. The animals, although they include fewer species than the plants, are represented by some very important ones, such as the mussels *Mytilus crenatus* and *M. meridionalis,* the limpets *Patella compressa* and *P. granatina,* and the whelks *Argobuccinum argus* and *Cominella delalandii.*

Thirty-one species that are characteristic of the west coast continue throughout the south and into the southern part of the east coast, but now the animals (23) far outnumber the algae (8); the diminution of algae is most marked between Qolora and Port Alfred. It is difficult to single out examples for special mention in a group containing so many common species; but among the most outstanding are algae of the higher levels, *Porphyra capensis* and the smaller *Plocamium cornutum,* the former growing far more luxuriantly in the western than in the eastern part of its range. Other noteworthy algae are *Chaetangium ornatum, Gigartina radula,* and *Splachnidium rugosum.* Some of the animals that are conspicuous throughout the greater part of their ranges are the periwinkle *Oxystele variegata;* the whelk *Thais dubia;* the limpets *Helcion pectunculus* and *Patella miniata;* the barnacle *Balanus algicola* and the limpet *Patella argenvillei,* primarily west-coast forms; the gregarious polychaet *Dodecaceria fistulicola,* a species specially important to westward; and the large, multicoloured anemone *Bunodactis reynaudi,* which reaches its greatest development on the west coast. We have included the abundant periwinkle *Littorina knysnaënsis* in this group because, although it does extend into northern Natal, it is apparently so scarce there that it must be approaching the end of its range, and has ceased to be of great ecological importance.

DISTINCTIVE FEATURES
OF THE CAPE PENINSULA

The old view that the sea is cold on one side of the Cape Peninsula and warm on the other, with different faunas and floras on the two sides, although basically true, has long been seen to be too crude an expression of the facts, and probably depends largely upon the obvious contrast between the shore population at localities such as Sea Point on the western side and St James on the eastern side (see Figure 8.28).

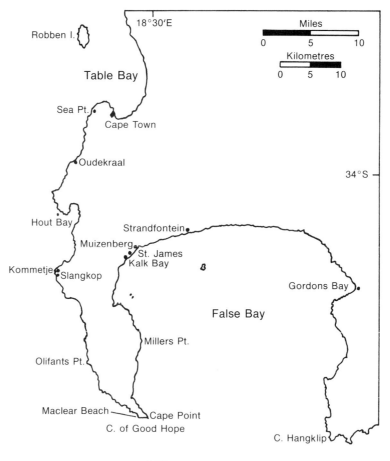

FIGURE **8.28**
The Cape Peninsula and False Bay.

As far as temperature is concerned, we cannot at present go beyond the statement that the summer sea temperature of the northern part of the Atlantic coast of the Peninsula is, on the average, well below that of the northern part of the False Bay coast; in the winter, there is much less difference. Temperatures along the southern parts of both coasts need much fuller investigation, and are probably less markedly distinct from one another than those of the northern parts, though they are distinct. We have unpublished records that show that on some days there may be extraordinary deviations from general expectation, at given localities, that are apparently determined by local variations in the weather and the physical environment. The whole subject is highly complex, and there are, undoubtedly, effects due to topography, rock formation, distribution of sand, incidence of wave action, and other such factors, as well as those due to temperature.

The Fauna and Flora of the False Bay Region

The stretch of coast from Kalk Bay Pier through St James to Strandfontein is apparently the warmest. There is no doubt that certain cold-water species, at home on the west coast, receive a severe check here. The population of this warm pocket is comparable to the warm-temperate biota of the south coast in general, although it is, of course, very different from the really warm (subtropical) coast of Natal. To the east (Cape Hangklip to Danger Point) and to the west, the populations are characteristic of colder water.

The northern part of the Atlantic coast of the Cape Peninsula gives a fair picture of the typical cold-water biota of the west coast of South Africa, but the north-to-south gradient in its population must be borne in mind. This biota provides the strongest contrast with that of the St James region.

The southern parts of both coasts of the Cape Peninsula exhibit a complex mixture of eastern and western elements. The appearance of *Gelidium pristoides* in the Kommetje region is an interesting example of the addition of south-coast species (which appear as one proceeds southwards) to a west-coast biota, and the appreciable amount of *Ecklonia buccinalis* south of Kalk Bay Pier on the False Bay coast is an example of the persistence of a cold-water form. *E. buccinalis* is more or less suppressed in the St James region, but it reappears farther east in the colder water of the Hangklip–Danger Point coast.

Both coasts of the Cape Peninsula alike are affected by the presence of some of the species belonging to our southwestern local category (category 7; see Figure 3.1), whose distribution centres on the southwestern part of the South African coast. An ecologically important example is the alga *Bifurcaria brassicaeformis*.

In spite of the fact that the southern parts of the two peninsular coasts seem to be less dissimilar than the northern ones, it is probable that the tip of the Cape Peninsula does form a significant region of change. The true kelp beds of the west coast, for example, begin at the Cape of Good Hope. It appears that many warm-water species that have persisted in False Bay fade out before Cape Point is reached, and that others extend beyond it in varying degrees, some of them perhaps not much beyond Maclear Beach.

The progressive west-to-east change of population in the southwestern region of Cape Province as a whole is more comprehensible if the exceptional features of False Bay are ignored. The west-coast biota, with an increasingly greater representation of the southern element, persists down the western side of the Peninsula, after which (jumping False Bay), in the region from Cape Hangklip to Danger Point, west-coast forms (including kelp fields) are still fairly well developed, although the south-coast admixture is now strong. South-coast elements come fully into their own in the Agulhas district, where there is a definite fall-off of various west-coast species. False Bay is really an interruption in an otherwise fairly orderly sequence of events (see Figure 8.29).

Thus, we may provisionally divide the coasts of the Cape Peninsula roughly into four regions, although no suggestion is made that these divisions are sharply marked off one from the other: (1) Cape Town to Kommetje (fairly typical west-coast population); (2) Kommetje to the Cape of Good Hope (west-coast population, with an in-

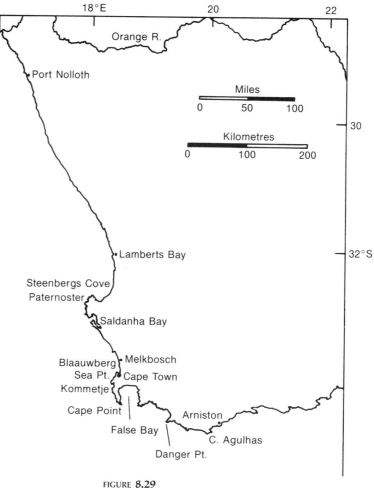

FIGURE **8.29**

The western part of Cape Province.

creasing south-coast admixture); (3) Cape Point to Kalk Bay Pier (modified south-coast population with an appreciable west-coast influence); and (4) Kalk Bay Pier to Muizenberg (population comparable to that of the south coast east of Cape Agulhas, with a recognizable, but reduced, west-coast element). This last region extends at least as far as Strandfontein.

Much more detailed study around Strandfontein and Gordon's Bay could be carried out most profitably, since Strandfontein is evidently as warm as, or warmer than, the St James area, and has very peculiar rock formations isolated amid much sand. It is hoped that this account may serve as a basis for more extensive work in the neighbourhood.

HYDROGRAPHIC CONSIDERATIONS (*Revised by Sir George Deacon*)

Because the aim of the book is biological, it does not deal with the hydrography of the South African area in as much detail as a physical oceanographer might need.

The hydrography of the South African area has been dealt with on a broad scale by a number of authors, notably Dietrich (1935a, 1935b), Sverdrup, Johnson, and Fleming (1942), Clowes (1950–51), and Hart and Currie (1960). As far as possible, they have dealt with inshore waters as well as the more general conditions offshore, but have not made a special study of the water movements, temperature, and salinity, which are likely to affect life between the tidemarks, and indeed they have been prevented from doing so by the scarcity of inshore observations. The most useful account is that of Isaac (1937), written from the point of view of a biologist interested in seaweeds; but there are also several later studies of currents off the Natal coast, such as those of Harris (1961) and Whillier (1962), and these are likely to be supplemented very soon by the results of the International Indian Ocean Expedition and subsequent work (See Figure 8.30 and the report of the Oceanography in South Africa Symposium, August 1970, published by the new South African Committee for Oceanographic Research).

To quote from Clowes (1950):

> The Indian Ocean subtropical surface water which affects the east and south coasts of the Union of South Africa, has its origin in the south equatorial current which is itself caused by the great Trade Wind Drift flowing westwards across this Ocean. At about 11°S the northern part of the south equatorial current, which has divided at Madagascar, impinges on the African coast and divides into two streams, one of which flows southwards, the Mozambique current, whilst the other flows northwards as the East African Coast current. The southward flowing Mozambique current is joined in about 26°S by a part of the south equatorial current which has passed round the southern tip of Madagascar, and the combined streams continue towards the Natal coast to flow southwards down the east coast of South Africa, and the current is now known as the Agulhas current. It is the water of this current which affects the fisheries of the east and south coasts of the Union of South Africa. The Agulhas current follows the trend of the coast of South Africa until it reaches the longitude of the Agulhas Bank, which is a wide extent of shallow depth projecting southwards. The edge of this bank serves as an obstruction to the flow of the Agulhas current which mainly follows the eastern edge of the bank. This obstruction thus causes a deflection to the south-west of the greater part of the Agulhas current, but a further obstruction is encountered in the presence of the eastward flowing water in the south of the subtropical zone of the South Atlantic Ocean. This causes the main part of the Agulhas current to be bent back on itself and to flow towards the east, being now known as the return Agulhas current. Although the main part of the Agulhas current is deflected as stated and eventually returns eastwards to the South Indian Ocean, it is also a fact that a small part of the current flows both over and round the Agulhas Bank and tends to round Cape Point into the South Atlantic, whilst according to Deacon (1937, p. 75) a portion also flows westwards south of the Agulhas Bank into the Atlantic Ocean.

Clowes gives evidence of the westward flow, and adds: "As can be well imagined, the junction of subtropical surface water from both the South Atlantic and South Indian

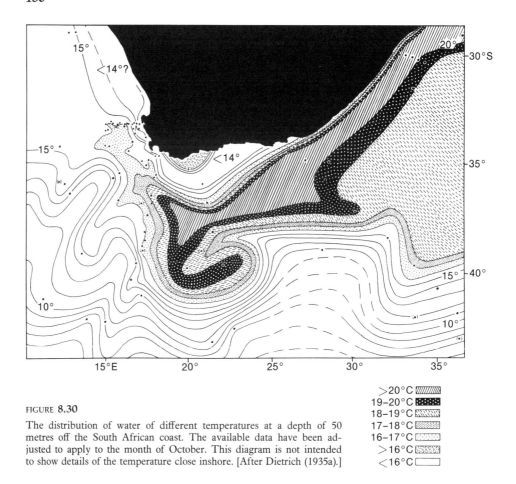

>20°C
19–20°C
18–19°C
17–18°C
16–17°C
>16°C
<16°C

FIGURE **8.30**

The distribution of water of different temperatures at a depth of 50 metres off the South African coast. The available data have been adjusted to apply to the month of October. This diagram is not intended to show details of the temperature close inshore. [After Dietrich (1935a).]

Oceans gives rise to considerable intermingling of these waters. In the area between 15°E and 20°E many sudden changes of temperature and salinity are encountered."

About the west coast, Clowes writes:

> The Benguela current which flows up the west coast of southern Africa is the current which has such a large influence on the fisheries off the west coast of the Union. In addition to its main northerly movement it has also a westerly component and takes water away from the coastal regions. It is conspicuously prominent between Cape Point and 17°–18°S. This current is very intimately connected with the circulation of air over the South Atlantic Ocean where an area of permanent high pressure, or anticyclone, lies in approximately 30°S. The current off the west coast of Africa, south of the Equator, comprises the circulation round the eastern and northern sides of this permanent anticyclone. The Benguela current forms the eastern side of this circulation, flowing northwards up and away from the coast to near the Equator. Here the bulk of the current passes into the south equatorial current which flows westwards along the northern side of the high pressure area. In agreement with the dynamics of currents in the southern hemisphere, denser water of lower temperature and salinity is found on the right-hand side of the

Benguela current so that close to the African coast a belt of cold and relatively low saline water is usually found. The current tends to have a westerly component due to the effect of the earth's rotation and surface water is taken away from the coastal region. To replace this water, upwelling takes place from moderate depths and it is this upwelling which gives to the west coast of South Africa its mixed sub-Antarctic and sub-tropical nature. Due to this upwelling a belt of cold water of relatively low salinity is found along the coastal region extending seawards for approximately 50–120 miles. The consequence to the fisheries off the west coast of South Africa of this upwelling is that an area of sea is provided in which nutrient is present in large amounts for the plankton on which fish live. The upwelled water brings large amounts of such nutrients as phosphate, nitrate and silicate into the euphotic zone where plant life and zooplankton can utilize them to grow and multiply and so form an essential part of the food of fish.

This general arrangement of currents makes it immediately intelligible that the Natal coast, where the Agulhas Stream lies close inshore, should support a sub-tropical population, partly borne southwestward from warmer regions, and partly evolved *in situ;* and also, that the west coast should support a population adapted to much colder conditions. What is less clear, at first sight, is why the south coast should support a third, warm-temperate population, distinct from either of the others. However, this is not difficult to explain if we remember that off the south coast lies the wide submarine Agulhas Bank, and that the main Agulhas Stream begins to leave the coast near Algoa Bay and flows much farther off the south coast than it does on the east, tending to follow the edge of the Agulhas Bank, upon meeting which much of it is deflected back towards the Indian Ocean, the rest flowing round and over the bank. Therefore, the effect of the current inshore will be weaker along the south coast than along the east, not only because it is farther away, but also because it is cooling down somewhat. Moreover, influences from the colder, western water play a part between the Agulhas Stream and the shoreline along at least the western part of the south coast, and there is upwelling also to be taken into account.

How does all this affect the tidal region—the coast itself—with which we are here concerned? If the arrangement of populations is any guide, there is every sign that, in a broad sense, the main currents do affect the coast directly, as one would expect. However, there is also evidence that when we come down to details there are complicating inshore factors, which—although they do not neutralize the broad effects—modify them and, locally, may even affect them quite decidedly.

Upwelling is one of these factors. It is not very easy to get a clear picture of this for the coast as a whole. Upwelling depends partly upon the dynamics of the currents, partly on prevailing winds; and it can be extremely variable from time to time, even from one day to another. That it is important and consistent is evident, in spite of its variability on the west coast; and it is also clear that, on that coast, the upwelled water brings with it plant nutrients, which encourage the development not only of phytoplankton, but also of a rich bottom flora. For the south and east coasts, the picture is less clear because, although the dynamics of the Agulhas Current imply inshore upwelling, there seems to be a great deal of local and periodic variation in the amount of it on the south coast; and because the temperature of the Natal coastal water is markedly warm, whatever upwelling occurs there must be of warmer water than that affecting the west coast. Moreover, the amount of bottom-living algae along the south and east coasts suggests that there is a less favourable supply of plant nutrients along them than along the west coast, and there is even some direct evidence of this. Probably, the average amount of upwelling, as well as the quality of the upwelled water, varies on different parts of the coast.

Another complicating factor is the existence of surface currents that are often variable, follow different directions from the main ones, and sometimes flow in the opposite direction. That these exist, in accordance with our expectations, is supported by evidence (see, for example, Gilchrist, 1902; Cleve, 1904; Stechow, 1925; Harris, 1961; and Whillier, 1962). Again, local configurations, both of coastline and sea floor, taken together with the direction of prevailing winds, account for a good deal of local variation between tidemarks—since, naturally, water in a sheltered shallow bay may be directly warmed by the sun, even if it is cool to begin with. Local flooding and the normal outflow of rivers may also be factors at inshore stations.

As yet, we have insufficient knowledge of detailed inshore conditions to enable us to describe them fully or to correlate them exactly with the observed distributions of organisms. Nonetheless, the variations in populations do seem to be closely connected with corresponding local variations in inshore sea temperatures. This is pointed out in the preceding section on the Cape Peninsula, wherein it is noted that the populations of the stretch of coast between Cape Hangklip and Danger Point, and those of the northwestern corner of False Bay, reflect immediately the local temperature differences known to exist between the two areas, and also that the populations correspond with the temperature changes known in the general region from Table Bay to Cape Agulhas (see Figures 8.28 and 8.29).

CENTRAL WEST AFRICA

INTRODUCTION

Recent studies of the marine ecology of the central West African coast have been carried out on the shores around the Gulf of Guinea, both in the neighbourhood of Victoria in Cameroun, and along the shores of Ghana, which lies between the Ivory Coast and Togo. Both areas are tropical: the equator passes through the Gulf, and the warm, east-flowing Guinea Current washes along its shores. Sea-surface temperatures range approximately from 24° to 29°C, but sometimes fall below this at Accra.

Humidity is high, and the rainfall very heavy in the rainy seasons of May–July and September–November. At these times, fresh water overflowing from lagoons and rivers produces marked variations in the salinity of inshore waters.

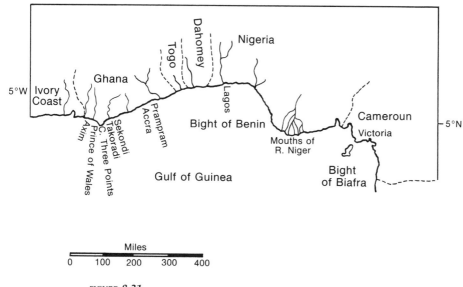

FIGURE **8.31**

A map of the Gulf of Guinea, showing places named in the text.

The coasts of the Gulf of Guinea consist of long sandy beaches, many of which have lagoons behind them, that are interrupted in many places by the mouths of rivers. There are outcrops of rock of many different types; some of them form headlands, others platforms that protrude from the sand. At Victoria, there is a beach of basalt boulders backed by low cliffs.

Tides are semidiurnal. From April to September, the lowest low waters are at night. From October to March, they are in the morning, and the hot sun has the usual desiccating effect on the organisms growing in the more seaward zones. The tidal range of spring tides is about 4–6 feet, and that of neap tides, about 2–3 feet.

From Figure 8.31, it is obvious that there will be many variations in wave action in the Gulf of Guinea. The Cameroun coast faces west on the Bight of Biafra, while the Ghana shore has a southerly aspect and is situated some distance to the west on the north coast of the Bight of Benin. The coast of Ghana stretches from the rather less exposed areas of its eastern end, near Prampram, to a western extremity near Princes Town and Sekondi, where the water is never calm. There can be a considerable surf breaking on the rocky headlands of Cameroun, but wave action is not very great on the rest of the Cameroun coast. The biota at stations in this region reflects very clearly the variation in wave exposure.

The biotas of Ghana and Cameroun show many similarities and many interesting differences. These are illustrated in Figures 8.32 and 8.33, which demonstrate the vertical distribution of a selection of their common species.

The following descriptions are only brief summaries of the biotas of the Ghana and Cameroun shores taken from papers by Lawson (1955, 1956, 1957a, 1957b, 1966), in which detailed accounts can be found.

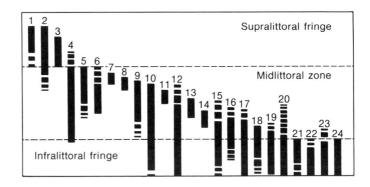

FIGURE **8.32**

The approximate vertical positions of a selection of common species on the shores of Ghana: 1, *Tectarius granosus;* 2, *Littorina punctata;* 3, *Bostrychia tenella;* 4, *Nerita senegalensis;* 5, *Chthamalus dentatus;* 6, *Siphonaria pectinata;* 7, *Enteromorpha* spp.; 8, *Pylaiella fulvescens;* 9, *Centroceras clavulatum;* 10, *Ulva fasciata;* 11, *Mytilus perna;* 12, lithothamnia; 13, *Chaetomorpha antennina;* 14, *Laurencia* turf; 15, *Balanus tintinnabulum;* 16, *Gigartina acicularis;* 17, *Hypnea musciformis;* 18, *Bryocladia thrysigera;* 19, *Gracilaria henriquesiana;* 20, *Padina tetrastromatica;* 21, *Sargassum vulgare;* 22, *Dictyopteris delicatula;* 23, *Caulerpa crassifolia;* 24, *Echinometra lucunter.* The following plants and animals are also present in Cameroun (see Figure 8.33): 1, 3–6, 9, 12–17, and 21. [After Lawson (1954).]

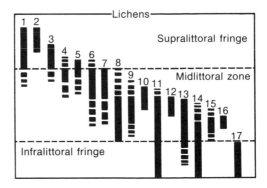

FIGURE **8.33**

The approximate vertical positions of a selection of common species on the shores of Cameroun: 1, *Tectarius granosus;* 2, *Rhizoclonium riparium;* 3, *Bostrychia tenella;* 4, *Cladophora camerunica;* 5, *Centroceras clavulatum;* 6, *Nerita senegalensis;* 7, *Chthamalus dentatus;* 8, *Gelidium pusillium;* 9, *Siphonaria pectinata;* 10, *Ostrea* sp.; 11, lithothamnia; 12, *Gymnogongrus nigricans;* 13, *Pterocladia pinnata;* 14, *Balanus tintinnabulum;* 15, *Caloglossa leprieurii;* 16, *Chaetomorpha antennina;* 17, *Sargassum vulgare.* The following plants and animals are also present in Ghana (see Figure 8.32): 1–3, 5–7, 9–11, 14, 16, and 17. [After Lawson (1955).]

GHANA

The Supralittoral Fringe

In exposed places, *Littorina punctata* is the most abundant animal; in excessively stormy places, it may reach 30 feet (9.14 m) above the high water level, and be joined by *Tectarius granosus,* a species sometimes equally common and often extending even higher still. *Nerita senegalensis* may stray in fron the midlittoral; algae, such as *Bostrychia tenella* and *B. binderi,* are rare and confined to crevices and fissures.

In moderately exposed places, the biota is very similar but the range is less. Sheltered areas are usually populated solely by *Littorina punctata,* although its range is small in very quiet places. *Bostrychia* and *Rhizoclonium* occur where there is shade.

The Midlittoral Zone

In exposed conditions, *Chthamalus dentatus* is dominant and spreads downwards until finally it gives way to the lithothamnia that coats the rock of the lower part of the zone. Other animals associated with *C. dentatus* are *Siphonaria pectinata, Nerita senegalensis,* and *Littorina punctata,* which strays in from the fringe above. In this belt, animals form the main populations, algae being sparse and rare. Where algae occur in sandy areas, lithothamnia may be covered with a matted felt of various algae about one inch high, of which a common constituent is *Hypnea musciformis.* Where the barnacles die out and the lithothamnia begins, there may be a distinct belt of short-growing *Centroceras clavulatum* and, below that, another one of *Chaetomorpha antennina.* Between these two bands of algae, a collection of characteristic but unusual species occurs, notably *Balanus tintinnabulum, Fissurella nubecola, Patella safiana,* and *Thais haemostoma.* Below the *Chaetomorpha antennina* is a strip of comparatively pure lithothamnia, which is followed by a rather poorly developed growth of *Ulva*—sometimes containing *Mytilus perna*—which merges into a short, red algal turf, of which *Laurencia* is the most characteristic constituent.

In less exposed situations, the arrangement of the biota is very similar, but there are some displacements in the positions of various algal belts—many of them related to the angle of slope of the rock faces—and algal species are more plentiful. For instance, in addition to *Ulva, Taenioma perpusillum* and *Ectocarpus breviarticulatus* grow as small tufts on barnacles, and there may be a blue-green belt of *Hydrocoleum lyngbyaceum* between barnacles and lithothamnia. The turf of algae contains additional species, including *Herposiphonia densa, Sphacellaria elliptica, Pocockiella variegata, Ralfsia expansa,* and *Bryocladia thyrsigera.*

In areas of moderate shelter, where the shore is lined by outer reefs running parallel to it, two different types of habitat may be recognized. Lawson describes the sloping seaward sides of these narrow reefs as "dip," and they are naturally more wave-washed than the steeper, landward sides, which he calls "scarp."

On "dip" slopes, *Chthamalus* forms a narrow belt covered with *Enteromorpha.* In some places, blue-green algae may be mixed with the *Enteromorpha,* or it may form

FIGURE **8.34**

An exposed shore at Winneba, Ghana, some thirty miles west of Accra, showing three main zones: an upper splashed zone of barnacles (*Chthamalus dentatus*), a midlittoral zone consisting of clusters of *Mytilus* and tufts of green *Enteromorpha*, and lithothamnia encrustations at sea level. [Photograph by R. Bassindale.]

a separate, slimy coating. In many places, there is a wide belt of *Pylaiella fulvescens*, which forms a thick, brown, sandy mat, always with tufts of *Ulva* in it, and sometimes with other algae. What are probably several species of *Ulva*, mostly small and stunted, may extend down to the infralittoral fringe. There may be a rather bare strip below the belt of *P. fulvescens*. The upper part of the lower midlittoral is predominantly green in colour, due to numerous patches of weed, but gives way to red species in the lower zones (see Table 8.3). There is often an undergrowth of corallines (*Jania natalensis*, for example). Common brown algae in the zone are species of *Padina, Dictyota*, and, locally, *Colpomenia. Padina* is seasonal and, when flourishing, may become dominant.

At the top of "scarp" slopes, lithothamnia occurs with extensive patches of *Ralfsia* and *Pocockiella* and occasional tufts of *Ulva. Hypnea cervicornis* may curl over the top edge of the reef and creep down the landward slope. There is a rich growth of *Gracilaria* at the bottom, in which *Sargassum* plants can be found.

In the most sheltered regions, the midlittoral typically has a very lifeless appearance. *Chthamalus* and *Siphonaria* are ubiquitous in fair numbers very low on the shore. In many places, a dark, narrow band of algal turf marks the upper limit of the infralittoral fringe.

The Infralittoral Fringe

A peculiar type of turf is the only form in which algae occur in very exposed places in the infralittoral fringe. It closely resembles a *Laurencia* turf, in that it is about one inch tall, and occurs in irregular patches, but its main constituent is *Dictyopteris delicatula*. Scattered about in holes among the lithothamnia is the sea urchin *Echinometra lucunter*.

Less exposed areas are invaded by *Sargassum vulgare*, which, under suitable conditions (such as those afforded by smooth slopes), becomes mixed with *Dictyopteris*. In the absence of algae, animals may dominate: the commonest are encrusting bryozoans, sponges, and sea anemones. In some places, one particular animal, such as the compound ascidian *Eudistoma ramosum*, may become completely dominant.

With moderate shelter, the number of algal species increases, although *Sargassum*, with an admixture of *Dictyopteris*, is still dominant. *Dictyopteris*, however, may become an almost pure subzone in deeper water. Table 8.3 lists the most common species growing with *Sargassum*. Sea urchins are to be found, but *Echinometra* is infrequent, being replaced here by the starfish *Arbacia lixula*.

On truly sheltered shores, lithothamnia disappears into the infralittoral zone; only a nodulated form of a different lithothamnian species appears in pools. *Sargassum* is not common, and only occasional clumps of it and small tufts of *Dictyopteris* may be found. In muddy places, other algae occur, notably *Caulerpa sertulariodes*. Tube worms and other low-level animals are common, but sea urchins are rare.

FIGURE **8.35**

Ningo Reef, about two miles east of Prampram, Ghana. This is a sheltered area of low rocks; there are patches of *Sargassum*, at low levels, with fronds up to 12 inches long. [Photograph by R. Bassindale.]

TABLE **8.3**

Some algal species common to areas of moderate shelter
on the coast of Ghana

Algae growing with *Sargassum* in the infralittoral fringe	Turf-forming algae of the lower midlittoral (contd.)
Bryopsis spp.	RED ALGAE
Cryptonemia seminervis	*Bryocladia thrysigera*
Gracilaria henriquesiana	*Gelidiopsis variabilis*
Pterosiphonia pennata	*Gigartina acicularis*
Rhodymenia palmetta	*Gracilaria henriquesiana*
	Hypnea musciformis
Turf-forming algae of the lower midlittoral	*Laurencia papillosa*
	Polysiphonia ferulacea
GREEN ALGAE	*Wrangelia argus*
Boodlea composita	
Cladophora trichotoma	
Hydrocoleum lyngbyaceum	
Struvea anastomosans	

Differences Between the Western and Eastern Shores

The length of the coastline of Ghana, from Axim on the west to Prampram on the east, is roughly 400 miles; and the changes in the biota, due to the differences in exposure to wave action, are quite definite.

There are four obvious changes: (1) *Tectarius granosus,* so plentiful high on the exposed western shores, disappears completely in the east; (2) *Ulva,* flourishing in the east, makes only a very poor showing in the west; (3) the infralittoral-fringe weed, *Sargassum,* cannot be found in the rough water of the west; and (4) *Echinometra lucunter* becomes less and less frequent toward the east, where calm weather prevails. The biota of this coast needs further study. Lawson believes that there may be further significant differences between its western and eastern extremities.

CAMEROUN

The Supralittoral Fringe

Despite differences in wave action along the coast of Cameroun, the supralittoral fringe has a fairly uniform appearance throughout. *Tectarius granosus* replaces *Littorina punctata,* the common species of Ghana, Sierra Leone, and Senegal. There is usually a bare strip above *Tectarius,* and then a well developed line of lichens. Unlike the typical supralittoral fringe, this one has an unusually good growth of algae, mainly of *Bostrychia tenella* and *Rhizoclonium riparium.* The former can make a distinct belt in the lower part of the fringe, and is at its best on vertical faces in moderately exposed places; in shelter, it tends to keep to cracks and crevices, sometimes mixed with *R. riparium,* which can extend well above *Tectarius.* Straying up from the zone below are certain turf-forming algae. (*Cladophora camerunica,* for example) and the snail *Nerita senegalensis.*

The Midlittoral Zone

In exposed places, the midlittoral zone can be divided into upper and lower subzones. The upper supports *Chthamalus dentatus,* very concentrated at highest levels, a scattering of various species of *Nerita* and *Siphonaria,* and species of *Ostrea* or *Crassostrea,* which may form a belt in the middle of the barnacles. *Centroceras clavulatum* is the only algae that occurs in any quantity; it forms a narrow belt at the top of the zone, and may extend patchily up or down. The lower midlittoral is dominated by lithothamnia, in many places with quantities of *Balanus tintinnabulum* and a short algal turf (see Table 8.4).

There is little change in the fauna on less exposed stretches of the coast, but the alga *Centroceras clavulatum* disappears and *Gymnogongrus nigricans* and *Pterocladia pinnata,* sparse under exposure, become more abundant. At lower levels, a short growth of algae appears (see Table 8.4).

In moderately sheltered places, animals that inhabit the upper levels are less frequently found. A small mussel, *Brachyodontes puniceus,* and the algae *Cladophora camerunica* occur in crevices, the latter being common. On open rock surfaces, *Gelidium pusillum* may form an extensive greyish-brown turf. An interesting collection of red algae grows in shaded places and under deep overhangs (see Table 8.4). The lower midlittoral is narrow and poorly developed. In some places, there is an encrusting belt of blackish *Ralfsia* sp. containing small patches of various algae (including *Brachytrichia maculans, Gracilaria ferox, Laurencia* sp., and *Sirocoleum guyanense*).

TABLE **8.4**
Some algal species common to the coast of Cameroun

Algae of the midlittoral zone	Algae of the infralittoral fringe
TURF-FORMING SPECIES	SPECIES FORMING FELTS
Gymnogongrus nigricans	ON BOULDERS
Herposiphonia densa	*Ceramium tenuissimum*
Lyngbya majuscula	*Gigartina acicularis*
Pterocladia pinnata	*Herposiphonia secunda*
	Polysiphonia sp.
LOW-LEVEL SPECIES	
Bryopsis sp.	SPECIES OF SHELTERED
Chaetomorpha sp.	POSITIONS
Champia parvula	*Acanthophora spicifera**
Dictyota sp.	*Callithamnion* spp.†
Padina sp.	*Dictyota* sp.
Pocockiella variegata	*Gigartina acicularis*
Struvea anastomosans	*Hypnea musciformis*
	Padina sp.
DEEP-SHADE SPECIES	*Pocockiella variegata*‡
Bostrychia radicans	*Sargassum vulgare* §
Bostrychia tenella	*Sphacellaria* spp.†
Lophosiphonia obscura	
Murrayella periclados	

*This species forms an undergrowth.   † These species form a short surface layer.
‡ A prostrate form of the species.   § The dominant species.

In sheltered areas, a large part of the midlittoral is covered by *Gelidium* turf, with an admixture of *Caloglossa leprieurii* and *Sirocoleum guyanense* at its lowest levels. Above this turf are scattered *Chthamalus dentatus,* individual oysters, a few *Nerita senegalensis, Littorina punctata,* and *L. cingulifera.* But most of the available space is covered with a growth of *Cladophora camerunica,* particularly flourishing in the shade of trees in quiet backwaters.

The Infralittoral Fringe

In exposed places, lithothamnia and lower midlittoral turfs merge into the fringe growth, and *Balanus tintinnabulum,* in association with some *Fissurella nubecula,* becomes common. There are also many species of other animals, including sponges, hydrozoans, and hydroids.

Where there is some shelter, *Sargassum vulgare* grows well associated with the plentiful *Gracilaria ferox* (which occurs in richly branched bushy tufts), with *Laurencia obtusa,* and with species of *Bryopsis* and *Dictyota,* all larger than most of the other algae on the Cameroun coast. *Sargassum vulgare* predominates as shelter increases, accompanied by a good undergrowth of other weeds; but where the weeds die out, it becomes less common, the rock is covered with small algae, but without lithothamnia; areas of boulders have a covering of small felted algae (see Table 8.4). In some quiet places shaded by trees, a species of gorgonian prevails and *Sargassum* is absent.

The reactions of the biota to various degrees of exposure and shelter are less definite in Cameroun than in Ghana because the extremes of wave action are less pronounced (Figure 8.36). The vertical depths of zones are wholly inconstant, and the various levels of tides have little influence on the widths of zones where exposure is a marked factor.

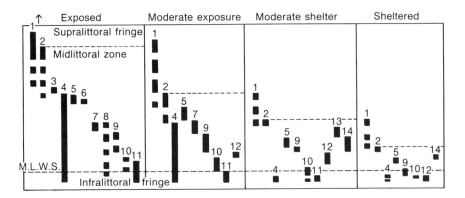

FIGURE **8.36**

The disposition of certain species on shores of varying exposures in Ghana: 1, *Littorina punctata;* 2, *Chthamalus dentatus;* 3, *Centroceras clavulatum;* 4, lithothamnia; 5, *Ulva fasciata;* 6, *Mytilus perna;* 7, *Chaetomorpha antennina;* 8, *Balanus tintinnabulum;* 9, red algal turf; 10, *Dictyopteris delicatula;* 11, *Echinometra lucunter;* 12, *Sargassum vulgare;* 13, *Enteromorpha* spp.; 14, *Pylaiella fulvescens.* Vertical depths are approximate. *M.L.W.S.,* mean low water of spring tides. [After Lawson (1954).]

In shelter, for instance, the supralittoral fringe is extremely narrow; but in exposed places, the extent upwards can only be indicated by a vertical arrow in a diagram of normal scale.

In complete shelter, the algal belts are telescoped into a very narrow strip along the water's edge. In such places, the biota may change strikingly in a few vertical inches. Generally speaking, completely sheltered situations are the least productive—their variety of species may be very small.

Compared with the corresponding area on the east coast of Africa, this western tropical area presents a very subdued picture, owing to the difference in richness between the Atlantic and the Indo-Pacific faunas and floras.

THE COAST FROM GHANA TO MAURITANIA

Lawson (1966) carried out further investigations of the west coast of Africa, from Ghana to Mauritania (Figure 8.37). For the greater part of this region, steep sandy beaches stretch along a relatively unindented coast. Such shores change their profiles with the movement of the sand—spits, bars, and lagoons make ever-changing patterns. At intervals, rocky headlands and outcrops appear; in places, cliffs rise directly from the sand or rest on platforms that are covered at high water.

Much of Lawson's work is concerned with sandy beaches, and is thus outside the range of this book, as are his experiments on the recolonisation of denuded areas.

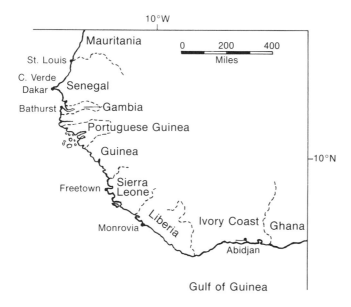

FIGURE **8.37**

A map of the African coast from Ghana to Mauritania, showing the places named in the text.

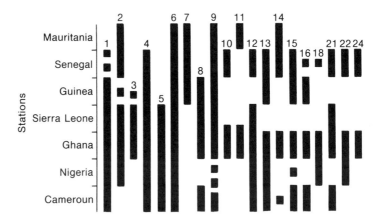

FIGURE **8.38**

The ranges of selected common species along the African coast from Cameroun to the south of Mauritania: 1, *Tectarius granosus;* 2, *Littorina punctata;* 3, *Bostrychia* spp.; 4, *Nerita senegalensis;* 5, *Chthamalus dentatus;* 6, *Siphonaria pectinata;* 7, *Chthamalus stellatus;* 8, *Ostrea* spp.; 9, *Centroceras clavulatum;* 10, *Ulva fasciata;* 11, *Mytilus perna;* 12, lithothamnia; 13, *Chaetomorpha antennina;* 14, *Laurencia* spp.; 15, *Balanus tintinnabulum;* 16, *Gigartina acicularis;* 18, *Bryocladia thrysigera;* 21, *Sargassum vulgare;* 22, *Dictyopteris delicatula;* 24, *Echinometra lucunter.* Note that *Siphonaria pectinata* is common from Cameroun to Mauritania; that *Tectarius granosus* is plentiful as far as Guinea, but is sparse in Senegal, disappearing before Mauritania; and so forth. Ghana seems to have the greatest abundance and variety of species, and Senegal the next. In Senegal, however, the number of individuals of a given species is typically not as great as it may be elsewhere.

The information relevant to this book has been set out in a graph (Figure 8.38) based on one made by Lawson in which he attempts to summarise the distribution of a selection of organisms common on rocky shores of West Africa. As he remarks concerning his figure, the graph is tentative; but from a careful study of it, a number of interesting facts emerge.

The area in question comprises the coasts of seven states—Cameroun, Nigeria, Ghana, Sierra Leone, Guinea, Senegal, and Mauritania—and some details of their biotas are given below.

The Supralittoral Fringe

Tectarius granosus and *Littorina punctata* are common in the supralittoral fringe along the whole length of the coast, but the former disappears between Senegal and Mauritania, and the latter is not present as far south as Cameroun. *L. cingulifera* spreads north and south, with its headquarters at Sierra Leone.

The Midlittoral Zone

The most ubiquitous and abundant animals of the midlittoral zone are *Siphonaria pectinata* and two species of *Chthamalus*—*C. dentatus* from Cameroun to Sierra Leone, and *C. stellatus* from there to Mauritania. *Nerita senegalensis* is common from Cameroun to Senegal. In the lower midlittoral, as usual, both animals and algae increase in quantity and in the number of species represented; this is particularly true of the algae. *Centroceras clavulatum* is widespread in association with *Chaetomorpha antennina* and lithothamnia.

The Infralittoral Fringe

Balanus tintinnabulum is usually an important animal in the infralittoral fringe, and may reach well up into the zone above; *Echinometra lucunter* is another characteristic creature.

The biota of Ghana (pp. 141–144) carries on fairly consistently west and north along the Atlantic coast of Africa. Careful consideration of the available records indicates that the most varied and plentiful biotas centre on the coasts of Ghana and Senegal, and that Guinea and Nigeria have fewer of the common intertidal species. Probably, this is because long stretches of these coasts have extensive mangrove swamps, and because the major part of the Nigerian coast consists of the great delta of the Niger River, with its shifting sands, and its tortuous waterways pouring fresh water into the gulf. Mauritania also has rather a different representative biota, because many of the "Ghana" type fall off in the region of Senegal, and are replaced by such significant species as *Mytilus perna* and the algae *Gracilaria folifera, Cystoseira* spp., *Ecklonia murati,* and *Taonia atomaria.*

CENTRAL EAST AFRICA

From 1952 to 1956, Kalk (1954, 1958, 1959a) and Isaac (1956) extended the ecological studies of the east coast of Africa to Inhaca Island in Delagoa Bay (lat. 26°0'S, long. 33°0'E) and to Mozambique (lat. 15°S, long. 40°45'E; see Figure 8.39). The nearest coast that had previously been surveyed is nearly 300 miles south at Isipingo, Natal, near Durban. Natal is a subtropical region in which the tropical component is fairly high, but Inhaca Island lies on the very fringe of the truly tropical area of northern Mozambique and therefore has far more tropical elements than Durban. Scolopendrina Cove on the south coast of the island is the warmest bay, and has the typical tropical associations (see Figure 8.40).

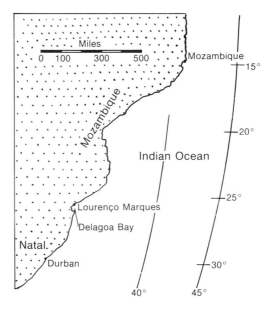

FIGURE **8.39**

A map of the African coast from Durban to Mozambique.

INHACA ISLAND, DELAGOA BAY

Inhaca Island lies about 21 miles by sea across Delagoa Bay from the town of Lourenço Marques. The east coast of the island is exposed to the Indian Ocean, and the shore slopes steeply down to the sea. Extensive sand flats surround the north, west, and south shores, which face the bay. There are numerous outcrops of calcareous sandstone on all shores. These may form wave-cut terraces with vertical cliffs, as at Isipingo. On the west coast, coral reefs lie at a depth of about 10 metres in the deep tidal channel. Extensive patches of coral débris cover the bottom near these reefs, and are almost always submerged. This region provides a wide range of marine habitats—rocky shores, sand and mud flats, scattered reef material on sand, true coral reefs (the most southern on African shores), and mangrove swamps. Only rocky shores are dealt with here.

The climate is not extreme. Rain falls all the year round, but even in the wetter summer (September–March), the rainfall is never very heavy. Moderate winds persist on the east coast; the other shores are sheltered. The average relative humidity for the area is about 77%. Table 8.5 gives comparative climatic data for Durban, Inhaca Island, and Mossuril, Mozambique.

Tides are semidiurnal; in Delagoa Bay, they have a range of about 9 feet for mean spring tides and 3 feet for mean neap tides. The salinity of the Indian Ocean and tidal channel is about 35.5‰, but may be reduced to 30.0‰ on sand flats, where fresh water is discharged from rivers.

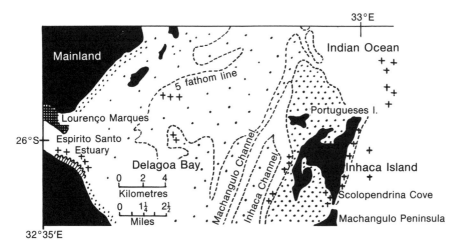

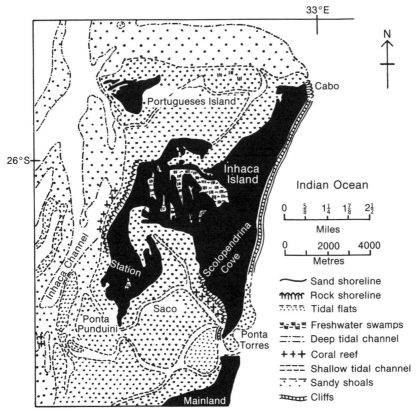

FIGURE **8.40**

Above, a general map of Delagoa Bay to show the position of Inhaca Island and the main physical features. *Below*, an enlarged and more detailed map of Inhaca Island. [Redrawn from various Admiralty charts.]

FIGURE **8.41**

The shore on the northeast coast of Inhaca Island. In the foreground is the algal mat of the infralittoral fringe; beyond this is the midlittoral platform, with cliffs in the far distance. [Photograph by M. Kalk.]

FIGURE **8.42**

A close view of the cliffs shown in Figure 8.41. Note the belt of the oyster *Crassostrea cucullata* at the foot of the cliff. [Photograph by M. Kalk.]

TABLE **8.5**
Climatic data for three localities on the east coast of Africa

Locality	Temperature (°C)			Rainfall (mm)			Records (yr)
	Max	Min	Mean	Summer	Winter	Total	
Durban	24.7	16.4	20.5	780	379	1159	29
Inhaca Island	26.9	18.7	22.8	823	274.5	1097.5	2
Mossuril	30.6	20.0	25.3	770	185	955	25

The Supralittoral Fringe

The organism highest in the supralittoral fringe is *Littorina obesa*, with *Tectarius natalensis* immediately below. The bulk of the barnacles occur in this zone, the lower limit of which is marked by the presence of *Chthamalus dentatus* in quantity.

The Midlittoral Zone

The midlittoral zone, as usual, can be divided into three parts. The upper can be recognized by the dense growth of *Chthamalus dentatus* mixed with *Littorina africana* and *Tetraclita serrata*. At its seaward edge, it is invaded by the top of a belt of the oyster *Crassostrea cuculalta*, in the less dense patches of which are an assortment of mussels, species of *Siphonaria*, gastropods (including *Nerita, Thais,* and *Cellana*), and perhaps a growth of corallines. Below this again, usually with a thick algal turf, sabellariid worms are dominant; and on gentle slopes, large sheets of zoanthids appear with various limpets and sponges, *Balanus amphitirite,* and *Cypraea* spp., which are outliers from the fringe below.

The Infralittoral Fringe

The infralittoral fringe is characterized by a dense and extensive covering of algae, in which *Sargassum elegans* and *Laurencia natalensis* may form pure communities. There are large areas of *Pyura stolonifera* with the usual small fauna and flora amongst them. Lowest of all are the corals with their associates.

The diagram of zonation patterns (Figure 8.43) will give an idea of the variations in the biota that take place because of differences in environment.

The Algal Population

A large majority of the algae of these regions occur also on the coasts of South Africa, particularly in Natal. Perhaps the most striking absentee is *Hypnea spicifera*, which is so dominant and typical of the lower zones in the Durban district.

Delagoa Bay, with its wide range of habitats, supports seaweeds of many types. Its numerous and extensive sand flats provide a suitable terrain for many species of sea grass. Prominent are *Thalassia hemprichii* and several species of *Cymodocea*.

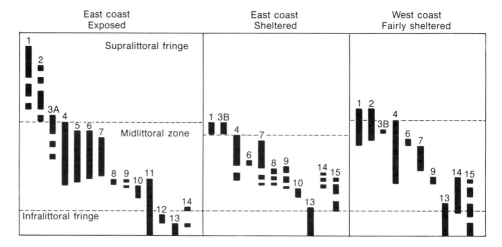

FIGURE **8.43**

A comparison of the vertical distributions of some typical species on different shores (two on the east and one on the west) of Inhaca Island: 1, *Littorina obesa;* 2, *Tectarius natalensis;* 3A, *Littorina africana;* 3B, *Littorina scabra;* 4, *Chthamalus dentatus;* 5, *Tetraclita serrata;* 6, *Siphonaria* sp.; 7, *Crassostrea cucullata;* 8, *Vermetus* sp.; 9, *Pomatoleios kraussi;* 10, zoanthids; 11, algal turf; 12, *Pyura stolonifera;* 13, corals; 14, *Cypraea* sp.; 15, *Balanus amphitrite; broken columns,* species present in patches; *unbroken columns,* species that form continuous belts. The usual telescoping of zones in sheltered conditions is clearly demonstrated.

Locally, certain algae (such as *Sargassum elegans* and *Laurencia natalensis*) can form more or less pure communities. Other representative species are *Padina commersonii* and *Udotea orientalis,* which occur in sheltered shady reaches.

On rocks, there is usually a covering of *Ralfsia expansa;* and in the lower midlittoral, in many places, there is a dense algal turf, of which some common constituents are *Caulerpa scapelliformis, Valonia macrophysa,* and *Gelidium reptans.* On vertical faces, *Pseudocodium devriesii,* and *Udotea orientalis* are common; in patches at lower levels, *Halimeda cuneata, Chamaedoris delphinii,* and *Colpomenia sinuosa* are prominent species, as are *Microdictyon kraussii* and *Dictyopteris longifolia* in low-level pools. Representative larger growths are *Galaxaura tenera, Plocamium glomeratum, Rhodymenia natalensis,* and *Kurtzingia natalensis.* At high levels and in shade, *Bostrychia binderi* is common. All the above species are found also on South African coasts.

Table 8.6 summarises some of the conspicuous floral and faunal differences between Natal (at Isipingo) and Mozambique (at Inhaca Island).

MOZAMBIQUE BAY

The study of the intertidal biota of this tropical area was carried out by Kalk and his colleagues (1959b) at various places around the shore and the islands in Mozambique Bay (see Figure 8.44A). This bay is well within the tropics, its northern boundary having a latitude of 15°S. As Table 8.5 indicates, the mean air temperature at Mossuril is 2.5°C

TABLE **8.6**

Some of the conspicuous differences between the biotas of Isipingo, near Durban, Natal, and Inhaca Island, Mozambique

Isipingo	Inhaca Island
Oxystele tabularis common.	Present but rare. *Planaxis* and *Cerithium* are at the end of their range from the tropics.
Patella granularis, P. variabilis, and *P. longicosta* all plentiful.	All absent.
Patella barbara abundant.	Small and rare.
Siphonaria annae, S. aspera, and *S. deflexa* common.	Replaced by *S. carbo* and *S. dayi.*
Siphonaria capensis common.	Present on exposed shores.
Siphonaria oculus common.	Extends further north with *S. dayi.*
Thais capensis, T. dubia, and *Cominella* spp. present.	All absent.
Thais intermedia, Drupa tuberculata, and *Urosalpinx heptagonalis* present.	All present (*Urosalpinx heptagonalis* is the most important carnivorous gastropod).
Mytilus perna forms belts.	Occurs only in clumps.
Brachyodontes variabilis is at the end of its range.	*Modiolus barbatus* replaces it in Mozambique Island.
Septifer bilocularis present in sheltered shaded places.	As in Natal, but not extending to northern Mozambique.
Pomatoleios kraussi abundant.	Sometimes replaced by blue *Vermetus* on exposed shores.
Cucumaria sykion common.	Very rare only on east coast and extends no further north.
Holothuria cinerascens and *H. leucospilota* common.	Many more tropical holothurians become plentiful.
Asterina exigua abundant.	Small and rare.
Tropical species of sea urchins and brittle stars common.	Very abundant; many more species present.
Sheets of zoanthids common.	Common, but the worm *Idanthrysus pennatus* competes with zoanthids.
Balanus amphitrite var. *dentatus* and *B. trigonus* are most successful species.	*B. amphitrite* var. *communis* present on semi-exposed unsilted shores.
Corals restricted, best colonies in rock pools.	Corals flourish and form reefs at suitable depths.
Hypnea spicifera dominant in infralittoral fringe.	Replaced by abundant *Sargassum elegans* and *Laurencia natalensis.*

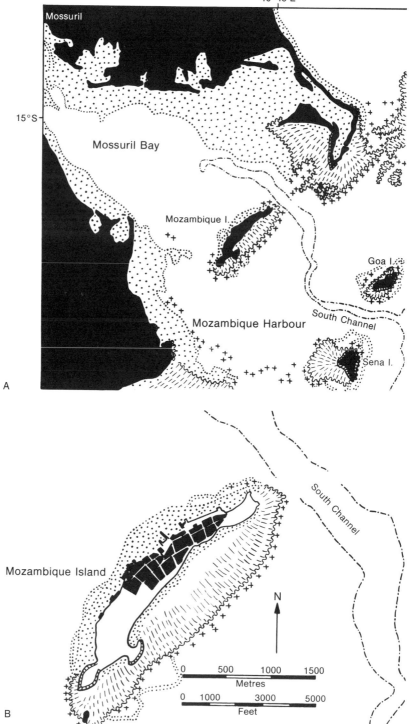

A

Mossuril

40°45'E

15°S

Mossuril Bay

Mozambique I.

Goa I.

South Channel

Mozambique Harbour

Sena I.

B

Mozambique Island

South Channel

N

| 0 | 500 | 1000 | 1500 |

Metres

| 0 | 1000 | 3000 | 5000 |

Feet

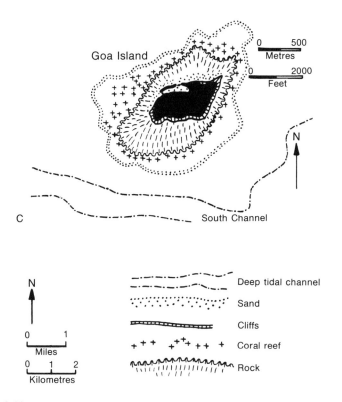

Goa Island

0   500
Metres

0   2000
Feet

N

C

South Channel

N

0   1
Miles

0   1   2
Kilometres

---·—·—·—·—·   Deep tidal channel

—·—·—·—·—·—·   Deep tidal channel

·:·:·:·:·:·:·:·:·:·:·:   Sand

▬▬▬▬▬   Cliffs

+ + + +⁺+ + + +   Coral reef

ʌʌʌʌʌʌʌʌʌʌ
ıılı¦ıl¦ıl¦ıl¦ı   Rock

FIGURE **8.44**

Mozambique harbour and Mossuril Bay, Mozambique: *A*, a general map of the area; *B* and *C*, enlarged and more detailed maps of Mozambique Island and Goa Island, respectively. The solid black areas in *B* indicate fortifications. [Redrawn from Admiralty charts.]

higher than at Inhaca Island, and 4.8°C higher than at Durban. There are no records of inshore sea temperatures, but the range of sea-surface temperatures in these latitudes is tropical, from 24°C in August to 28°C in February. Yearly rainfall is very heavy (see Table 8.5) and two-thirds of it occurs in the summer months.

As there are no permanent rivers flowing into the bay, the salinity of surface water can be taken to be roughly the same as that of the Indian Ocean, that is, 35‰; and the water is usually exceptionally clear due to this absence of river water. Tides are semidiurnal, and their range is approximately 13 feet (4.0 m); this is greater than the range in South Africa, which is about 6 feet.

Mozambique Bay contains numerous sandbanks and coral reefs, with swamps and mangroves in the network of creeks and inlets that form its upper reaches. The investigations were carried out mainly on Mozambique Island and Goa Island, and at various places on adjacent shores of the bay; but as these latter places often consist predominantly of swamps and mangrove thickets, they are not really relevant to this work on rocky shores, and will receive only passing reference.

Mozambique Island

The topography of Mozambique Island makes it very suitable for comparison with Inhaca Island. However, it possesses no fully exposed shore, owing to a well developed coral reef that lies some 3 miles (5 km) east of it, against which the full force of the oceanic waves is expended.

It is a small, densely populated, fortified island, almost in the centre of the bay, and only at low water are the rocky and sandy shores exposed—at other times, the masonry rises sheer from the sea. Figure 8.44B shows the arrangement of the various types of beach, and it will be noted that such wave action as there is—and it is never severe—affects mainly the northeast point, where the south channel flows in from the Indian Ocean.

The northeast semiexposed shore has a drop of 9.8 feet (3 m) in a distance of 1148 feet (350 m). It has a rather peculiar formation, in that, after a fairly steep sandy slope, there is a rampart or wall of eroded and turretted sandstone some 9.5 feet (3 m) high, from the seaward side of which a rocky platform slopes gradually to the sea, except for one sharp step at the bottom of the level of the lower midlittoral. The whole platform is a patchwork of sand, sand-covered rock, pools, and rocky ridges.

The Semiexposed Northeast Shore.   On the sandy slope behind the rocky rampart, numerous sand crabs (*Ocypode ceratophthalmus*) scamper about, and the small hermit crab *Coenobita rugosus* is common in *Nerita* shells.

On the tops of rocks, periwinkles are rare—only a few *Littorina obesa* and *Tectarius natalensis*—but several species of *Nerita* and *Siphonaria* are plentiful, the latter in all crevices. At the base of the rock, on both land and seaward faces, there is a band of rock oysters, and also a few barnacles (*Chthamalus dentatus*).

Along the sloping rocky platform are a wealth of animals. In depressions filled with deep sand is a thick growth of flowering plants (including species of *Cymodocea*); but further down, in less sandy places, the alga *Padina commersonii* takes over, and a thick turf of various algae surrounds the rims of pools and covers rocky ridges, being replaced yet farther down by dense masses of the mussel *Modiolus barbatus*. In the pools themselves are a wide selection of animals, increasing in variety as lower levels are reached (see Table 8.7). Colonies of *Zoanthus sansibaricus* may form a thick coating on rock surfaces here; but in the lower midlittoral, there are barnacles only at the northeast point, which is crowded with *Balanus amphitrite* where the current of the South Channel flows by into the bay.

The infralittoral fringe, at the outer edge of the platform, has a numerous and varied assortment of corals and their associates, including crabs, sipunculids, polychaets, alcyonarians, sponges, and anemones. Brightly coloured fish are abundant.

The Sheltered West Shore.   This has a much reduced population. Scattered on the rocks are *Crassostrea cucullata* below a belt of *Isognomon dentifera*. In the midlittoral zone, there are some isolated tubes of *Pomatoleios kraussi*, but there is no band of them. Holothurians, sea urchins, and brittle stars are plentiful, as on the east shore. From

TABLE **8.7**

Some animal and algal species common to the shores of Mozambique Island

Gastropods of the supralittoral fringe	Animals in pools
Nerita plicata	*Cerithium caeruleum*
Nerita polita	*Cerithium kochi*
Nerita textilis	*Cerithium morus*
Nerita undata	Crabs of many species
Siphonaria carbo	*Nerita albicella*
	Planaxis sulcata
Animals of the midlittoral zone	*Turbo marmoratus*
Actinopygia mauritania	
Echinometra mathaei	Species in algal turf
Holothuria atra	*Colpomenia* sp.
Holothuria difficilis	*Halimeda cuneata*
Holothuria leucospilota	*Turbinaria ornata*
Ophiactis modesta	*Ulva* sp.
Ophiocoma scolopendrina	*Valonia macrophysa*
Ophiocoma valenciae	*Valoniopsis pachynema*
Placophiothrix venosa	

the sand flats, the usual scavenger gastropods, such as species of *Nassarius* and *Natica,* roam about the surface, and hermit crabs clatter around in their vacated shells. The dominant worm is *Mesochaetopterus minutus,* whose sandy tubes form large hillocks.

The sand flats of the fringe have the usual covering of various sea grasses, among which live a number of small corals, occasional fungids with their commensal crabs, sponges, numerous large *Balanoglossus,* and a burrowing fauna, including anemones, worms, and gobies.

Goa Island

Goa Island, with its lighthouse, lies near the entrance to Mozambique Bay, about 5 miles (8 km) east of Mozambique Island. It is fully exposed, and waves roughly 9 feet high break on its shores. On its southeast coast is a vertical cliff.

Special features of this region are the presence of hundreds of large yellow crabs (*Grapsus maculatus*), numerous large chitons (*Acanthopleura brevispina*), and *Onchidium* sp., all of which may attain a length of 10 centimetres. The only gastropods at high levels are species of *Nerita,* periwinkles being rare even in crevices. A large pinkish *Tectarius,* which occurs as scattered individuals, is the most conspicuous animal of the midlittoral; there is no belt of oysters or mussels, and species of *Thais, Cellana,* and *Siphonaria* are infrequently found. Notable are the small anemones that occupy every crevice. Growths of closely knit green algae can form carpets on flat platforms. Generally speaking, animals are scarce. In the lower midlittoral, red and brown algae appear with corallines and sponges. In the lowest belt, the biota resembles that of the more exposed shore of Mozambique Island, but the infralittoral fringe is narrow, as the shore at this level soon falls away into deep water.

The Shores of Mozambique Bay

The zonation of the biota found on rocky formations on the sandy shore of Mozambique Bay, as at Inhambane, differs very little from that of places of a similar environment in Delagoa Bay.

The Supralittoral Fringe.   Here there are the same abundant *Littorina obesa* above and *Tectarius natalensis* below, along with individuals of tropical species of *Nerita*.

The Midlittoral Zone.   The upper part of this zone has a dense population of *Chthamalus dentatus* with an intermixture of *Tetraclita serrata*. However, the oysters are bigger and more conspicuous than at Inhaca Island. *Siphonaria carbo, S. dayi,* and *S. oculus* are common lower down.

The lower midlittoral is indicated by large colonies of *Vermetus* sp., followed by dense coatings of zoanthids (*Palythoa nelliae* and *Zoanthus sansibaricus*) with an intermixture of polychaets. Boulders may be encrusted with tubes of the sabellariid worm *Idanthrysus pennatus*. Hosts of commensals inhabit these tubes. *Patella barbara* can be found among the algae.

The Infralittoral Fringe.   This contains extensive growths of red and green algae, and patches of *Cymodocea ciliata* occur in suitable places. In some areas, *Pyura stolonifera* is abundant. There are occasional clumps of coral.

Durban, Delagoa Bay, and Mozambique Bay provide an interesting sequence of a biota that ranges from subtropical to fully tropical conditions. The differences between the inhabitants of Durban and Delagoa Bay are greater than between those of the latter and those of Mozambique Bay. The description of the biota of Mozambique Bay could apply almost equally well to islands of the Great Barrier Reef of Australia.

Detailed lists of animals and algae are to be found in the various papers by Isaac and Kalk. It is interesting to note that, at Inhaca, 85% of identified animals are tropical; 11% are warm-temperate, as found on the coast of Natal (the warmer shores of South Africa); and 4% are ubiquitous. Of the tropical species of Mozambique, 20% are also present on tropical Australian shores.

SELECTED REFERENCES

Buchanan, J. B., 1954. Marine molluscs of the Gold Coast, West Africa. *J. West Afr. Sci. Ass.* 1(1): 30–45.

———, 1959. The nature and position of the rock shore fauna of tropical west Africa in relation to the east Atlantic as a whole. *Ann. Rep. Challenger Soc.* (3) (11): 1–30.

Cleve, P. T., 1905. The plankton of South African seas, 1. Copepoda. *C. Good Hope Dep. Agr. Mar. Invest. S. Afr.* 3: 177–210.

Clowes, A. J., 1950. An introduction to the hydrology of South African waters. *S. Afr. Dep. Com. Ind. Fish. Mar. Biol. Surv. Div. Invest. Rep.* (12): 1–29.

Deacon, G. E. R., 1937. The hydrology of the southern ocean. *Discovery Rep.* 15: 1–124.

Dietrich, G., 1935a. Aufbau und Dynamik des südlichen Agulhasstromgebietes. *Veröff. Inst. Meereskunde Univ. Berlin* 27: 1–79.

———, 1935b. Zur Dynamik des atlantischen Zweiges Agulhasstromes. *Ann. Hydrogr. Maritimen Meteorol. Z. Seefahrt Meereskunde Deut. Seewarte Hamburg* 63: 383–387.

Gauld, D. T., and J. B. Buchanan, 1959. The principal features of the rock shore fauna in Ghana. *Oikos* 10(1): 121–132.

Gilchrist, J. D. F., 1902. Currents on the South African coast, as indicated by the course of drift bottles. *C. Good Hope Dep. Agr. Mar. Invest. S. Afr.* 2: 155–165.

Harris, T. F. W., 1961. Marine studies off the Natal coast. *CSIR Sym.* S2: 1–134.

Hart, T. J., and R. I. Currie, 1960. The Benguela Current. *Discovery Rep.* 21: 123–298.

Isaac, W. E., 1937. South African coastal waters in relation to ocean currents. *Geogr. Rev.* 27(4): 651–664.

———, 1956. Marine algae of Inhaca Island and of the Inhaca Peninsula, 1. *J. S. Afr. Bot.* 22(4): 161–193.

Kalk, M., 1954. Marine biological research at Inhaca Island, Mozambique. An interim report. *S. Afr. J. Sci.* 51(4): 107–115.

———, 1958. Ecological studies on the shores of Mozambique, 1. The fauna of intertidal rocks at Inhaca Island, Delagoa Bay. *Ann. Natal Mus.* 14(2): 189–242.

———, 1959a. The zoogeographical composition of the intertidal fauna at Inhaca Island, Mozambique. *S. Afr. J. Sci.* 55(7): 178–180.

———, 1959b. A general ecological survey of some shores in northern Mozambique. *Rev. Biol. Lisbon* 2(1): 1–23.

Lawson, G. W., 1955. Rocky shore zonation in the British Cameroons. *J. West Afr. Sci. Ass.* 1(2): 78–88.

———, 1956. Rocky shore zonation on the Gold Coast. *J. Ecol.* 44(1): 153–170.

———, 1957a. Some features of the intertidal ecology of Sierra Leone. *J. West. Afr. Sci. Ass.* 2(2): 166–174.

———, 1957b. Seasonal variation of intertidal zonation on the coasts of Ghana in relation to tidal factors. *J. Ecol.* 45: 831–860.

———, 1966. The littoral ecology of West Africa. *Annu. Rev. Oceanogr. Mar. Biol.* 4: 405–448.

MacNae, W. 1962. The fauna and flora of the eastern coasts of southern Africa in relation to ocean currents. *S. Afr. J. Sci.* 58(7): 208–212.

———, 1963. Mangrove swamps in South Africa. *J. Ecol.* 51(1): 1–25.

Stechow, E., 1925. Hydroiden der deutschen Tiefsee-Expedition. *Wiss. Ergeb. Deut. Tiefsee-Exped. (Valdivia) 1898–99* 17: 383–546.

Stephenson, T. A., 1936. The marine ecology of the South African coasts, with special reference to the habits of the limpets. *Proc. Linnean Soc. London* 148: 74–79.

———, 1939. The constitution of the intertidal fauna and flora of South Africa, 1. *J. Linnean Soc. London* 40: 487–536.

———, 1943. The causes of the vertical and horizontal distribution of organisms between tide-marks in South Africa. *Proc. Linnean Soc. London* 154: 219–232.

———, 1944. The constitution of the intertidal fauna and flora of South Africa, II. *Ann. Natal Mus.* 10: 261–358.

———, 1947. The constitution of the intertidal fauna and flora of South Africa, III. *Ann. Natal Mus.* 11: 207–324.

———, and N. A. H. Bokenham, 1938. The colonisation of denuded rock surfaces in the intertidal region of Cape Peninsula. *Ann. Natal Mus.* 9: 47–81.

———, ———, and F. L. M. Neugebauer, 1938. The vertical distribution of certain intertidal marine gastropods in False Bay, with notes on the development of two of them. *Ann. Natal Mus.* 9: 113–138.

———, and E. J. Eyre, 1938. The South African intertidal zone and its relation to ocean currents, V. A sub-tropical Indian Ocean shore. *Ann. Natal Mus.* 9: 21–46.

———, Anne Stephenson, and K. M. F. Bright, 1938. The South African intertidal zone and its relation to ocean currents, IV. The Port Elizabeth district. *Ann. Natal Mus.* 9: 1–20.

———, ———, and J. H. Day, 1940. The South African intertidal zone and its relation to ocean currents, VIII. Lamberts Bay and the West Coast. *Ann. Natal Mus.* 9: 345–380.

———, ———, and G. A. DuToit, 1937. The South African intertidal zone and its relation to ocean currents, I. A temperate Indian Ocean shore. *Trans. Roy. Soc. S. Afr.* 24: 341–382.

———, A. Zoond, and E. J. Eyre, 1934. The liberation and utilisation of oxygen by the population of rock-pools. *J. Exp. Biol.* 11: 162–172.

Sverdrup, H. V., M. W. Johnson, and R. H. Fleming, 1942. *The Oceans.* New York: Prentice-Hall.

Whillier, A., 1962. Ocean currents at East London. *Trans. S. Afr. Inst. Civil Eng.* 4(1): 1–7.

9

THE ATLANTIC COAST
OF NORTH AMERICA

FLORIDA

Our work along the Florida Keys and at Marineland was carried out during January and February of 1947 (see Stephenson and Stephenson, 1950, 1952).

TOPOGRAPHY

The Florida Peninsula and the chain of islands (keys) that continues southwestward from it lie wholly north of the Tropic of Cancer, which passes between Florida and Cuba. Beginning at the latitude of Miami (25°45′N), the keys form an elongated chain stretching southwestward along a regular curve, ending—except for isolated islets—at Key West, just north of latitude 24°30′N. This chain separates Florida Bay (part of the Gulf of Mexico) from Florida Strait, which, in its turn, separates southern Florida from Cuba on the south and from the Bahamas on the east (Figures 9.1 and 9.2).

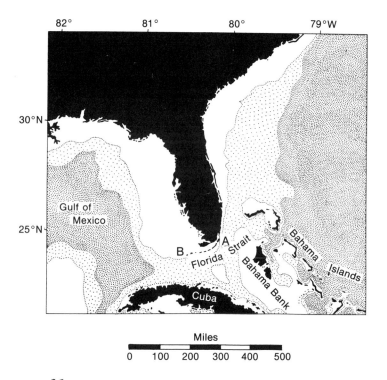

FIGURE **9.1**

The relation of the Florida Peninsula and the Florida Keys to other landmasses and to the adjacent submarine topography: *black*, land; *white*, shallow water (10–100 fathoms); *light stipple*, water 100–1000 fathoms deep; *heavy stipple*, water more than 1000 fathoms deep; *A–B*, the extent of the Florida Keys.

The submarine topography of this region is important biologically. A large submarine bank, the Floridian Plateau, projects southward from the North American mainland, and is covered by water less than 50 fathoms deep; the Florida Peninsula and the chain of keys are merely parts of it that arise above sea level. This means that the whole of the sea surrounding the keys for a few miles eastward and southward, and for a much longer distance westward, is very shallow; a great deal of it has less than three fathoms of water. The keys are not fully open to the Atlantic on their oceanic eastern side, because some distance to the south and east (more than 100 miles for the most part) lie the Bahama Islands and Cuba on their own large submarine banks. Therefore, even allowing for the separation afforded by the deep channel of the Florida Strait, the Floridian Plateau is not far removed from other large submarine banks that make the water round the keys relatively shallow for some distance in all directions.

At first sight, this great tongue-shaped bank seems to be an unlikely substratum for coral reefs; in fact, in the shallowest areas, the amount of coral—if it is present at all—is small and comprises few species. The most interesting region is the relatively narrow part of the bank that lies between the Florida Strait and the keys and follows their

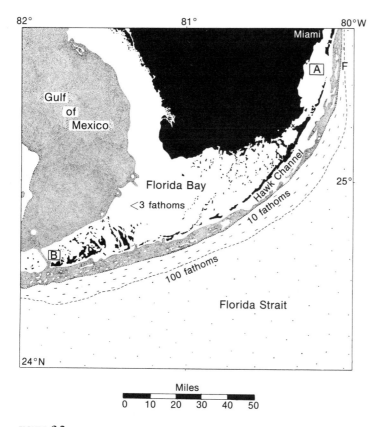

FIGURE **9.2**

The Florida Keys: *black*, land; *white*, water less than 3 fathoms deep; *heavy stipple*, water 3-10 fathoms deep; *dashes*, water 10-100 fathoms deep; *light stipple*, water more than 100 fathoms deep. The stations studied on the keys all lie between *A* and *B;* the Dry Tortugas and the Marquesas lie off the map beyond *B. F,* Fowey Rocks off Miami.

curve. This platform varies in width from more than 3 to nearly 7 nautical miles, and is covered by water less than 60 feet deep as far as the edge of the bank, which then slopes down to the deeper water of the Florida Strait. The detailed topography of this platform and bank is complicated; its surface is very irregular, with shoals for the most part under water less than 3 feet deep, and the slightly deeper Hawk Channel, navigable by small boats, between the keys and the outer part of the platform. In calm weather, the bottom is clearly visible; it varies from rock to mud, the latter often overgrown with *Thalassia*, algae, and populated by various animals. The hard bottom—which is often very rugged and irregular in its contours, with pinnacles and hollows and a white sediment—supports a good growth of coral, "reef patches," and plentiful gorgonians, including variously branched forms as well as sea fans. The landscape is that of a submarine forest, with all the lilac, purple, brown, and yellow "trees" (gorgonians)

FIGURE **9.3**

Under-water photograph taken at a depth of 10-20 feet off Elliot Key, Florida. Note the dominance of *Acropora palmata*. [Photograph by F. G. Walton Smith.]

waving gently in the swell. The corals are mainly species of *Siderastrea, Porites,* and *Manicina* that are relatively insensitive to the effects of variations in temperature, light, and sedimentation. Among the corals and gorgonians are the usual brightly coloured fish, sea urchins, sponges, and so forth.

Along the edge of the platform is a definite coral reef, in the form of an irregular broken ridge, with a steeper seaward than landward slope. The top of this ridge lies well below low-water level in some places; but in others, it comes near the surface, and may emerge slightly at low water of spring tides. Accumulations of coral rock, boulders, and débris sometimes form small islets. The reef consists of a healthy growth of coral with associated organisms of the usual type, and the corals show a distinct zonation with the ordinary amount of overlapping. The deeper and more solid parts of the reef are built of massive corals, whose individual colonies reach diameters of 5 feet or more, and among which *Montastrea (Orbicella) annularis* is particularly important as a reef builder. The elk-horn coral, *Acropora palmata,* with strong, flattened branches reaching a height of 10 feet or more, is another important form, and is particularly characteristic of the crest of the reef. Among these are smaller corals: the selection found along the landward side of the crest is, in part, different from that on the seaward side at low levels, and is supplemented by many gorgonians. The reef extends approximately from Fowey Rocks—just south of Miami (where it is petering out)—to a point off the Marquesas Keys in the southwest, with a limited recrudescence round the Dry Tortugas. Both the Marquesas and the Dry Tortugas lie beyond Key West. The reef has no great vertical extent, as the most flourishing coral growth occurs in less than 10 fathoms. Beyond this, the sea floor makes no sharp descent; it slopes gradually over a distance of 3–8 miles to the 100-fathom line. The heavy sedimentation on the bottom appears to be responsible for the shallow lower limit of the reef. The reef itself, lying along the edge of the bank, receives pure oceanic water driven against it by the prevailing easterly winds, and is not subject to an undue amount of sedimentation.

The Florida reef does not much resemble the sturdy structures to which the term "barrier reef" is applied in the Pacific—structures with a far stronger and more varied growth of coral, which freely reach and break the surface of the sea. Nor is the shallow channel behind the Florida reef much like the lagoon behind a typical barrier. At the same time, one can hardly deny entirely that the words "barrier" and "lagoon" are appropriate to this reef, as the reef does form a rather weak, mainly submerged barrier at the edge of a not very steep submarine slope, with a half-chocked lagoon behind it. It is, in fact, what Davis (1928) classifies as a "bank barrier" of the marginal belt of coral seas. Along the muddy bank facing the Gulf of Mexico, coral growth, though not absent, is reduced.

FIGURE **9.4**

Under-water photograph taken at a depth of 10–20 feet off Elliot Key, Florida, showing a mixture of *Acropora palmata,* gorgonians, and other forms. [Photograph by F. G. Walton Smith.]

THE FLORIDA KEYS

This account is based on investigations carried out over the keys from Key Largo, where the highway first reaches them, to Key West, at the end of the chain.

The rocky shores of the Florida Keys are composed of a series of dissected, clinkerlike, rocky platforms (many of them the remains of former coral reefs), which support a population of tropical marine plants and animals, typical of warm regions, but in some ways peculiar to this particular environment. The zonation conforms to the basic pattern that is so widespread over the world. Although details vary from key to key, there is a common structure of their rocky parts in the intertidal zone, which may be summarised as follows.

The Supralittoral Fringe

The most conspicuous intertidal feature is a low, well-marked platform of rock—the upper platform—which extends from the edge of the land vegetation towards the sea. This platform, which varies in width and in the abruptness of the slope of the seaward edge, constitutes, for the most part, the supralittoral fringe. It has four quite recognisable belts of discoloration running parallel to the shore: white, grey, black, and yellow. The first three of these belong to the supralittoral fringe, but the yellow zone belongs to the midlittoral (Figures 9.5 and 9.6).

The White Zone.   This is the actual meeting place of land and sea. It is hardly recognisable on some keys (Crawl Key and Vaca Key), as the belt is so narrow and bush-covered. No fully marine animals exist here, but notable maritime ones include hermit crabs, the isopod *Ligia baudiniana,* the rapid crabs *Sesarma cinereum* and *Cyclograpsus integer,* and *Tectarius muricatus,* the most regularly occurring gastropod of this zone. The vegetation is that typical of such areas—mangroves and other trees, including *Conocarpus* and *Thespesia;* fleshy composites, such as *Borrichia* and *Sesuvium;* and so forth. See Chapters 4 and 5 for further discussion of the white zone.

The Grey Zone.   This is less intertidal than the black, and its limits may be rather difficult to discern where it is narrow and invaded by the vegetation of the white zone. This is the highest zone in which any sign of macroscopic marine algae may be found, and a mosslike growth of *Bostrychia* appears in hollows and on the roots of trees and shrubs. The animals are fairly numerous, consisting of the usual snails—including species of *Littorina* (often *L. ziczac*), *Tectarius, Echininus,* and *Nerita*—and other small gastropods, such as *Melampus, Detracia,* and *Truncatella.* Present also are the crabs and isopods mentioned above, and spiders and earthworms. This type of community is not restricted to the immediate coastline—at Vaca Key, for instance, there are low-lying areas that extend inland far from the sea that have similar populations.

The Black Zone.   This zone is wetted completely at high water of the higher spring tides; at other times, it may be washed partly or not at all. Its surface is so fretted that the amount of immersion cannot really be calculated. In most places, there is no

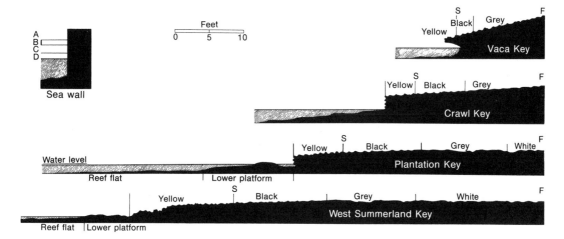

Sections across the rocky platforms of four keys and across a sea wall on one of them (West Summerland Key). Vertical and horizontal scales are the same. The water is shaded, and the water level for each section is that of the time of observation: except for that at Vaca Key, which is somewhat higher, all water levels lie near low water of spring tides. *S-F,* the extent of the supralittoral fringe. For further details, see Stephenson and Stephenson (1950, pp. 364 and 390).

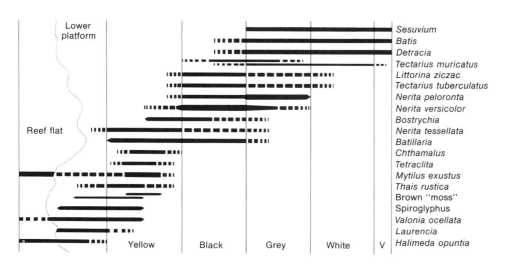

The intertidal distribution, in the Florida Keys, of twenty common plant and animal species (or groups of species) in relation to the various zones. This diagram is an approximation: it attempts to show average conditions for the Florida Keys as a whole. Both vertical and horizontal distribution are taken into account. *Tectarius muricatus* and the brown algal "moss" vary so much from one key to another that alternative distributions are shown. *V,* land vegetation.

vegetation of flowering plants. The discolouring agents are certainly, in part, the blue-green algae *Entophysalis deusta* and *Brachytrichia quoyi*, together with the chlorophycean *Tellamia intricata;* whether microscopic algae and lichens are involved or not—and if so, which species—we cannot tell. Many snails characteristic of the grey zone are common here, where some of them tend to reach their maximum development; *Littorina ziczac, Nerita versicolor, Tectarius tuberculatus,* and the elongate *Batillaria minima* occur here in vast numbers. The only macroscopic marine algae of regular occurrence form a mosslike growth, primarily in seaward hollows. This growth is dominated by *Bostrychia binderi* and *B. tenella.*

The Midlittoral Zone

The midlittoral zone is the truly intertidal region. It consists of two subzones—the yellow zone, which can be divided into two parts (a marginal one on the edge of the upper platform, and the actual edge of the platform and its seaward face) and a lower zone, which is a part of the rocky lower platform.

The Yellow Zone.   The upper part of the yellow zone has a population of small barnacles (*Chthamalus stellatus angustitergum*) and the *Bostrychia* moss already mentioned. The lower part of the zone is conspicuous for two dominant colonial organisms— sheets of the green alga *Valonia ocellata* and the sedentary gastropod *Dendropoma irregularis*. The number of algal species may be considerably increased locally.

The Lower Zone.   This is developed on some keys on the landward part of the lower rocky platform. It is characterised by a short—and possibly transient— reddish-brown mosslike growth of algae, including the familiar species of *Gelidium, Centroceras, Ceramium, Cladophoropsis, Herposiphonia,* and *Polysiphonia,* and with a short yellowish-green carpet of *Laurencia papillosa.*

As a whole, the midlittoral zone supports a varied animal population: numerous barnacles (species of *Chthamalus* and *Tetraclita,* in varying abundance, according to the exact locality); large numbers of gastropods, of which the most generally characteristic are probably neritids; and various bivalves, chitons, and crabs. Of the last three groups, *Mytilus exustus, Acanthopleura granulata,* and *Pachygrapsus transversus* are the commonest species. All the algae are small and encrusting.

The Infralittoral Fringe

Although the infralittoral fringe is recognisable, it is weakly and untypically developed, lacking any substantial growth of coral. It includes the edges of the lower rocky platform and parts of the reef flat itself, and it supports stragglers from both these places. Conspicuous features are the carpet of *Laurencia papillosa* previously mentioned, and the much sedimented mats of the calcareous green alga *Halimeda opuntia*. Also, in some places, there are patches of *Zoanthus sociatus* and anemones of the genera *Pymanthus* and *Condylactis*. Other species that deserve special mention are the seaurchin *Echinometra lucunter* and the mussels *Arca barbata* and *A. umbonata.*

The Reef Flat

The reef flat is a different world altogether, supporting a far greater number of species. For a detailed description, see Stephenson and Stephenson (1950).

The reef flat has two zones: (1) such parts of it as emerge from the sea at the lowest tides and belong to the infralittoral fringe; and (2) those parts that are permanently submerged and belong to the infralittoral zone proper.

For a description of this region, see Chapters 4 and 5, and the Barrier Reef section of Chapter 11, where the reef flats of tropical and subtropical shores are described and compared.

Distinctive Features of the Florida Keys

It is important to emphasize that the region of the Florida Keys is tropical, not subtropical. Although the keys do not actually fall within the tropics, they lie but a short distance north of the Tropic of Cancer. The Florida Current (the main southwestern part of the Gulf Stream system) flows through the Florida Strait, coming close inshore and following the curve of the plateau on which the keys are situated. This northward displacement of tropical sea-surface conditions undoubtedly affects the keys. Hence, the Florida Keys are virtually tropical because of their geographical position, though lying, of course, in the "marginal belt" of the tropical area. Indeed the constituent species of the intertidal biota of the Florida Keys clearly indicated that the area is tropical: The situation of the corals has already been described; and if we turn to other shore-dwelling organisms, a long list could be made of the algae and invertebrates characteristic of the coasts of the Florida Keys that are typical inhabitants of tropical areas. The reef flats of the keys are closely comparable to the flat tops of many Pacific reefs, just as the inhabitants of the rocky intertidal platforms of the keys resemble the inhabitants of raised rocky areas of those reefs. In other words, the northeastward keys formed a coral reef in the past, situated at a lower level; at present, they are surrounded by a tropical marine area in which the conditions of a reef flat are widely extended, and the rich sublittoral growth of coral is confined, by sediment and shallows, to an offshore position. In consequence, the area is more closely comparable to the coral reefs of other regions, such as the Great Barrier Reef of Australia, than to shores in subtropical transitional regions, such as the coast of Natal in South Africa. This conclusion is borne out by temperature ranges (Figure 9.7).

We were very much struck, during our work on the Florida Keys, with the low organic production of the marine plants and animals between tidemarks. Although a great many species are present, and some are represented by very large numbers of individuals, they are nearly all small, and the total weight of plant and animal material that they produce per unit area must be regarded as low. It is probable that, if a direct comparison were made between the weight of organic matter produced at the lower intertidal levels on the keys and that developed by the communities occupying similar positions in other parts of the world (a typical *Laminaria* bed, for example, or a rich growth of reef corals and associated forms, or a *Pyura* bed in South Africa or Australia), it would be found that the keys produce at a very meagre rate. This may be, in part, because of the unfavourable effects of sediment, taken together with a low average supply of plant

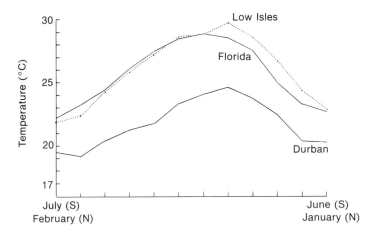

FIGURE **9.7**

A comparison of the annual variation in sea-surface temperature of the Florida region with those of Durban, South Africa, and Low Isles, on the Great Barrier Reef of Australia. *S*, Southern Hemisphere; *N*, Northern Hemisphere. Compare Figure 5.2 (p. 60).

nutrients in the surrounding warm, shallow sea, whose water is not refreshed with nutrient-rich water by upwelling from the depths of the adjacent ocean, and, in part, because the higher parts of the intertidal belt become very hot and dry when exposed to the air on a large number of days in the year.

MARINELAND

Marineland is situated on a long, straight, sandy beach that is fully exposed to the Atlantic and runs approximately north-northwest to south-southeast (Figure 9.8). It lies about 18 miles south of St. Augustine and 35 miles north of Daytona Beach. The summer maximum temperature is about 27–28°C, and the winter minimum is near 13–14°C.

Our studies were carried out during March of 1947 on outcrops of rock lying on the sandy beach south of the Marine Studios; but these outcrops were then being removed for commercial purposes, and now, 25 years later, the whole area may be radically changed. The rock, which is known as "coquina," is chiefly composed of coarse shell gravel with pockets of sandstone, and is found also on certain other Florida beaches. It weathers quite differently from the spiky, jagged coral rock of the Florida Keys, for it forms wide, flat surfaces dissected by deep, smooth-sided pot-holes that are from a few inches to more than 5 feet in depth, many with more or less vertical sides. There are also many overhanging surfaces. In some places, the rocky outcrops are lost in the sand; in others, they are almost continuous to low-water level. Naturally, the arrangement of organisms is dependent on the detailed configuration of the sand and rock.

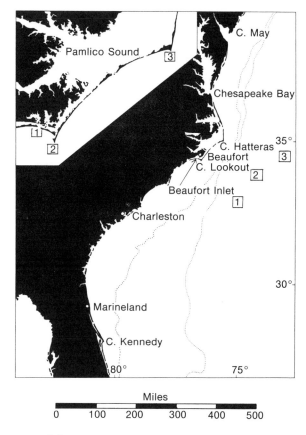

FIGURE **9.8**

Part of the Atlantic coast of the United States: *black,* land; *dotted lines,* approximate 100-fathom and 1000-fathom contours; *inset,* the region between Cape Hatteras and Beaufort Inlet.

The coast is very exposed, so that—except for the slight mitigation effected by a belt of shallow inshore water—it receives the full force of the waves. Consequently, the whole of the rocky area may be wave-washed. It is vigorously scoured by sand caught up by the sea, and sand may be shifted en masse with the result that the amount of visible rock is extremely variable, the whole appearance of the shore being changed from time to time.

However, in spite of this variability, the slope of the beach tends to follow a predictable contour (Figure 9.10). Above the beach is a flat, sandy plateau, with areas of rock embedded in sand, leading—by way of a short, steep slope—to a long rocky platform. The gentle angle of this rocky platform grows more acute until the rock begins to break up into boulders; these increase in size to form rather rectangular turretlike forms, with smooth curves and deep pot-holes, creating a much broken upper platform. Typically,

FIGURE **9.9**

The northern end of the tract of coquina rock at Marineland, northern Florida. The view is to the north, with the Marine Studios in the distance.

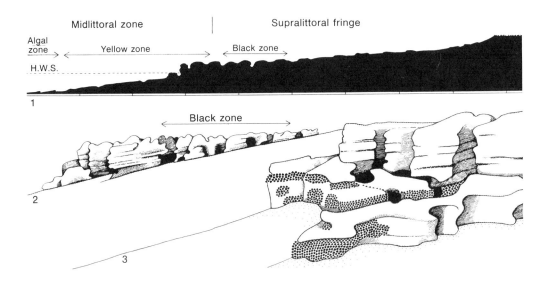

FIGURE **9.10**

Marineland, Florida; *1*, a measured section across the shore (scale is 15-foot divisions, vertical and horizontal scales the same); *2*, the turretlike rocks forming the upper platform; *3*, on a larger scale, a portion of the seaward edge of the upper platform, showing the distribution of *Tetraclita*. H.W.S., high water of spring tides.

the edge of this platform terminates in a short, precipitous drop to a wide slope of smooth rock, which may be either continuous to the sea or broken up into isolated blocks.

The zonation at Marineland has some striking differences from that of the Florida Keys. This is not surprising, if it is borne in mind that the normal wave action at Marineland is very strong, the position being oceanic and the rocks being subject to constant scouring and burying by shifting sand, and that the Florida Keys have to contend with the perpetual deposition of soft, calcareous sediment.

The Supralittoral Fringe

The supralittoral fringe is normal, in that it possesses: (1) myxophycean blackening, which is particularly conspicuous on vertical faces, such as the sides of pot-holes; (2) a growth, at the highest level, of low maritime plants and bushes; and (3) a population of *Ligia* (in this case, *L. exotica*). But the only other common animals, none of which are found at high levels, are the small limpet *Siphonaria pectinata* and the barnacle *Chthamalus fragilis*. There seems to be an almost complete dearth of snails. Intensive search revealed no *Tectarius*, and only 14 individuals of *Littorina ziczac*, which is probably close to the end of its range at Marineland. It lives at a lower level than is probably characteristic of the species, which is frequently true of other shore animals at the ends of their ranges. There also seems to be a seasonal growth of *Enteromorpha*, which occurs locally in the lower parts of the supralittoral fringe and overlaps into the zone below.

The Midlittoral Zone

The midlittoral zone at Marineland is more typical in appearance than the supralittoral fringe. Its uppermost edge is overlapped by the black zone of the supralittoral fringe—a common occurrence elsewhere.

The Upper or Yellow Zone. This zone occupies the marginal turrets of the upper platform and the greater part of the smooth, sloping rock below them. It has considerable areas of bare, sand-scoured rock, with a population of but few species; these, however, are abundant where conditions are favourable. Two species of open-rock barnacles are abundant: *Chthamalus fragilis* and *Tetraclita squamosa stalactifera*—the latter a much larger species that concentrates on vertical and overhanging surfaces lower down than *C. fragilis*. There are countless thousands of *Siphonaria pectinata, Crassostrea virginica,* and *Mytilus exustus*. The zone is often bare of macroscopic algae, but there can be patches of short *Ulva lactuca* and *Enteromorpha* and, in damp and shaded places, the small, dark red, mosslike forms of such species as *Porphyra leucosticta* and *Gelidium pusillum. Ocypode albicans,* the sand-burrowing crab, may be seen in the summer only.

The Lower or Algal Zone. The rocks here are extensively covered with sheets of green and dark reddish-brown algae that are much caked with sand and gravel. The algae are smaller than those on the reef flats of the Florida Keys. Typically, the higher levels have more chlorophyceans (mainly *Ulva* and *Enteromorpha*) than rhodophyceans. There are some clearings inhabited by barnacles, mussels, and oysters, but algae are dominant; at the lowest levels, these animals disappear. Strong, cold winds, to which *Ulva* and *Enteromorpha* are particularly sensitive, have a very destructive effect on most algae, but *Bryocladia cuspidata* seems little affected by them. This species, with branches like furry tails, is the commonest rhodophycean. It occurs chiefly below the main concentration of *Ulva*, and continues down to the lowest low water and beyond, where it coats the low-level rocks and is typically clotted with sand. At the time of our visit, algal growth was at its minimum. Algae seem better represented in the summer (*Padina vickersiae*, for example, appears in significant amounts in the warm months).

The Infralittoral Fringe

The infralittoral fringe may be considered to be the lowest strip of the algal zone described above—that is, those parts of it uncovered only at the lowest tides. At the time of our visit, it was wholly dominated by *Gracilaria folifera*, both on the rocks and growing up through the sand. Other algae were *Bryocladia cuspidata*, the coralline *Jania rubens*, the green threads of *Chaetomorpha linum*, and the small *Gelidiopsis intricata*. Lithothamnia reach their best local development here; but though the growth is better than that on the Florida Keys, it cannot compare with growths on tropical shores.

The Fauna of Overhangs

This shore has a great many deep overhangs. Although we had little time for detailed study of their cryptofauna, a brief note on that subject is here fitting. One of the most curious features of overhangs and shaded surfaces is what looks like a layer of immovable gravel. On investigation, this proves to be a sheet of thousands of small anemones with gravel attached to their verrucae: these are *Anthopleura varioarmata*, similar to the British *A. thallia*. There are also smaller numbers of the large *Bunodosoma cavernata*, which corresponds, ecologically, with the British *Tealia*. Also noted were sponges, polyzoans swarming with caprellids, and hydroids. Species found both here and on the open rock include *Crassostrea virginica*, *Mytilus exustus*, and *Thais floridana*. There are various species of barnacles forming heavy encrustations, and the very distinctive *Tetraclita squamosa* f. *floridana*. The crab *Plagusia depressa* lurks in crevices under the overhangs.

Distinctive Features of Marineland

There is nothing really tropical about the fauna and flora of Marineland. Except for *Littorina ziczac*, which is at the northern limit of its range, the most stenothermic warm-water species of the Florida Keys are conspicuously absent. It is remarkable that there is so sharp a change in the distance of less than 300 miles that separates Marineland from the reefs around Fowey Rocks off Miami. It would appear that a winter barrier against the northward migration of warm-water species exists in the neighbourhood of Cape Kennedy, which is less than 60 miles south of Marineland (see Chapter 3, pp. 40–43). At any rate, the species that are common both at Marineland and in tropical regions are some of the more eurythermic warm-water forms, not the most stenothermic ones. The Marineland biota must be classified as warm-temperate, not subtropical, for the latter classification would imply that there is a reduced but conspicuous representation of stenothermic tropical forms, as on the Natal coast of South Africa. There is, of course, a seasonal variation in the composition of the Marineland population between the warmer and colder months of the year.

Our work deals chiefly with a selection of the commonest macroscopic plants and animals on any shore that constitute a large proportion (by weight) of the whole population of the shore. We find that, in any normally populated region, our collections refer, on an average, to well over 100 such species. This number may represent only a fraction of the total number of species actually present, but we believe it to be an indicative fraction, with a reasonably constant relation to the total. If the number of species we list for a particular region falls markedly below 100, we feel justified in assuming that, at that place, the number of species is subnormal. At Marineland, we listed only 37 forms—an unusually low figure, though many of those are represented by a colossal number of individuals. It is really not surprising that only a relatively small number of forms can flourish on that shore, subjected as it is to strong wave action, shifting and scouring sands, considerable heat in summer, and occasional lethal frosts and cold winds in winter.

THE CAROLINAS

THE CHARLESTON REGION

The city of Charleston, South Carolina, lies on a peninsula bounded by two rivers (Figures 9.8 and 9.11). The only open seacoast is on the outer shores of Sullivans and Morris islands; the estuary behind them is a maze of reedbeds, sand, and mud flats, intersected by narrow inlets and waterways. The oceanic beaches are sandy, with no natural intertidal rock; but there are a number of artificial rocky structures that are so like natural reefs that they are able to support an intertidal population. North Jetty and South Jetty are the two largest, and provide, on their offshore parts, a fairly normal open-sea marine environment. We also examined some rocky breakwaters on the southwestern side of Sullivans Island, and the sea wall on the southern end of the Charleston Peninsula. We examined, too, the oyster localities at Bears Bluff, southwest of Charleston on Wadmalaw Island. Our work in the area was carried out during April of 1947 (see Stephenson and Stephenson, 1952).

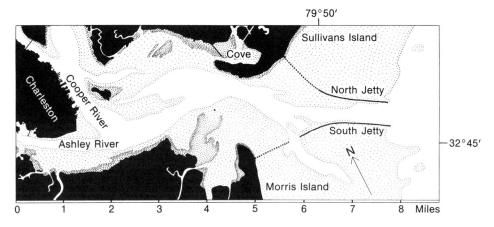

FIGURE **9.11**

The Charleston Estuary: *black*, land; *heavy stipple*, sand beaches and mud-banks; *light stipple*, shallow water; *white*, deeper water. The two large offshore jetties are shown, their emerged parts as *continuous irregular lines*, and their submerged parts as *dotted lines*. Some features shown in this map are variable. The scale at the bottom is in nautical miles.

FIGURE **9.12**

Part of the shoreward end of North Jetty off Sullivans Island, near Charleston, South Carolina, showing the similarity of this type of breakwater to a broken reef of natural rock. [Photograph by G. R. Lunz.]

The sea near Charleston is very shallow for some distance offshore: the 100-fathom line is more than 50 nautical miles out. Approximately 3 miles from the mouth of the estuary (nearly as far as the seaward ends of the jetties), the sea, at mean low water, is mainly less than 20 feet deep. The extreme (predicted) tidal range in 1947 was 7.9 feet, and the average range of spring tides was 6.0 feet (6.1 feet for Sullivans Island and the North Jetty). Average sea temperatures for inshore waters of the coast outside the harbour are 10°C and 27°C for January and August, respectively. The normal inshore salinity is rather low (32–33‰), and can be still lower in times of flood.

Much mud and silt is brought down by the rivers, and there is a considerable quantity of oil in the water: these foul the lower intertidal regions in varying degrees. All rocky structures are fouled by oil to some extent; but no part of the jetties seems to be depopulated by the action of mud, silt, or variable salinity, though these conditions have a restraining influence.

The intertidal zonation can be observed most satisfactorily on North Jetty and South Jetty, where it is quite clear, even though complicated by its broken nature, with caves, overhangs, and lumps of rock with an enormous variety of open surfaces at all angles (Figure 9.13).

The Supralittoral Fringe

There is the usual blackening caused by myxophyceans and lichens in the supralittoral fringe, although it is not readily visible as a distinct band, and there are some stragglers from the barnacle zone below. *Ligia* is common and, under foul conditions, may be killed in large quantities by oil. *Littorina irrorata*, which is characteristic of salt marshes and reedbeds, and seems to be a straggler rather than "at home" on the open rock, was the only member of its genus that we found.

The Midlittoral Zone

The midlittoral zone is strongly developed. It may be subdivided into a *Chthamalus* zone at the top, an oyster zone in the middle, and a muddy zone, which fades off into the infralittoral fringe below.

The *Chthamalus* Zone. The main growth is a dense one of *Chthamalus fragilis*, limited rather sharply above, but tending to overlap below, where it meets or overruns the upper limit of another barnacle, *Balanus improvisus*. In most places, the growth is clean, but it may be invaded by small, pale-coloured plants of *Porphyra* and a grassy coating of *Enteromorpha* at its lower edge. There are occasional large specimens of *Tetraclita squamosa stalactifera*.

The Oyster Zone. For the sake of brevity, we call this the oyster zone, but mussels, barnacles, and *Porphyra* are equally characteristic. We are restricting our comments on oysters (which occur in numerous places along the whole estuary) to those on open rock, which belong to the middle levels and occur lower down in crevices and caves. The usual species is *Crassostrea virginica*. The commonest mussel is *Mytilus*

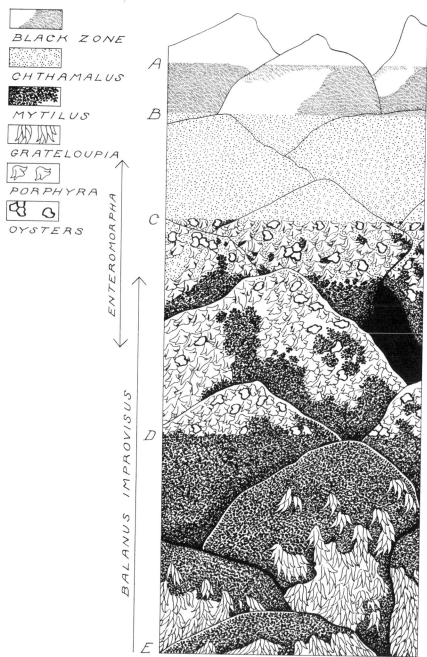

FIGURE **9.13**

The zonation on the south side of the offshore part of the South Jetty at Charleston, as seen in April, 1947: *A–B*, black zone; *B–C*, *Chthamalus* zone; *C–D*, oyster zone; *D–E*, muddy zone and infralittoral fringe. The arrow labeled *"Enteromorpha"* refers to the maximal representation of this genus at this time and place. *Enteromorpha* spp. are seasonal algae.

exustus, which occurs in huge numbers on the jetties; in the oyster zone, they are clean, but they continue down to and below low water, where they may be overgrown by algae or barnacles and covered by mud and silt. The barnacles are the large, whitish *Balanus eburneus,* and *B. improvisus,* occuring in a mixture; the numbers of these animals fluctuate, and we were informed that they were particularly high at the time of our visit. We found no limpets on the jetties.

The growth of algae is extensive, but *Porphyra* was the largest form we found. Below the cleaner *Chthamalus,* the algae form a highly coloured patchwork of brighter and darker greens, reddish browns, and dark mahogany reds. The leading forms are *Porphyra* and *Enteromorpha. Ulva lactuca* is also of general occurrence, but it does not extend as high as *Enteromorpha,* and is at its best on low-lying crests of rock.

The Muddy Zone.   Because this zone fades off quickly, it may be considered as the upper part of the infralittoral fringe.

The Infralittoral Fringe

The infralittoral fringe begins as a distinct belt below the oyster zone. In this belt—the top of which can be considered the muddy zone of the midlittoral—*Mytilus exustus* and *Balanus improvisus* continue in quantity; *Porphyra, Enteromorpha,* and oysters have practically vanished. Then, in sand and mud, an assortment of algae appear, the commonest of which is *Grateloupia felicina,* with its short olive-green stipes. Other forms are *Gelidium, Rhodymenia,* and *Chaetomorpha.*

The notable animals of the infralittoral fringe include the boring whelk *Urosalpinx cinereus* (which corresponds ecologically to the British *Thais lapillus*) and various species of anemones, ascidians, hydroids, and polyzoans.

Distinctive Features of the Charleston Region

There is no doubt that the intertidal population of Charleston, like that of Marineland, is warm-temperate. We believe that the number of species at Charleston is subnormal for a rocky marine area (see p. 177): our records show only 34 common macroscopic species. The combination of conditions is different, however: Marineland has a sandy, clean, open, oceanic shore, whereas the Charleston area is estuarine, and its shores are dirty at their best, filthy at their worst.

The supralittoral fringe is abnormal in the possession of few *Littorina,* these being very locally distributed. The oyster zone of the midlittoral is distinguished by barnacles continuing down into it in fair quantity, but this has already been noted as possible a temporary feature. The infralittoral fringe is not well marked.

The most striking contrast to Marineland is the complete absence, at Charleston, of any species of common limpet on open rock, whereas Marineland has millions of siphonarids. Marineland has an abundance of *Tetraclita* and *Anthopleura,* of which the former are scarce and the latter absent at Charleston. The common Marineland whelk appears to be *Thais floridiana,* and that at Charleston, *Urosalpinx cinereus. Littorina ziczac,* present at Marineland, is replaced by *L. irrorata* at Charleston. But Charleston is more than 200 miles farther north, and it is not surprising that certain warm-water species have disappeared. Although the pollution and variable salinity at Charleston must contribute to these differences, sea temperature may be regarded as the primary determinant.

Despite such differences, a careful study of the fundamentals of the zonation of these two regions shows that they are more similar than they first appear to be, and that, ecologically and geographically, both belong to the same faunistic province, but to a different one than the Florida Keys.

THE BEAUFORT REGION

The coastline around Beaufort, North Carolina, is extremely intricate (Figures 9.8 and 9.14). Offshore from the small towns of Moorhead City and Beaufort, which lie opposite each other on two tongues of the mainland connected by a bridge, rise two narrow sandy banks, named Bogue Bank and Shackleford Bank, respectively. A gap between these, Beaufort Inlet, gives access to the open ocean. Cape Hatteras lies some 70 miles to the northeast. Near Beaufort is the small Pivers Island, on which the Marine Station of Duke University is situated.

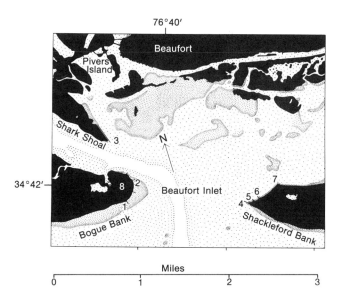

FIGURE **9.14**

The Beaufort area: *black,* land and marshes; *heavy stipple,* sand beaches and mud-banks; *light stipple,* water 0–20 feet deep at mean low water (occasional deeper holes in this area are not shown); *white,* deeper water; *1–7,* breakwaters; *8,* Fort Macon. The arrow is on the meridian 76°40'. Some of the features shown are variable.

FIGURE **9.15**

An intertidal reef composed of peat on the shore of an island near Beaufort, North Carolina. [Photograph by H. L. Blomquist.]

We could find no natural intertidal rock, although there are intertidal reefs of peat that weather into shapes similar to those of rocky reefs. However, as at Charleston, there were several rocky breakwaters supporting typical marine intertidal populations; and these were the object of our study, which was carried out during April and May of 1947.

The sea off Beaufort Inlet around Cape Lookout Jetty is shallow—it is less than 60 feet deep within a radius of 5 nautical miles—and there are shoals offshore that create breakers; the 100 fathom line lies more than 40 nautical miles from Beaufort Inlet. The sea around the breakwaters is much less turbid than it is at Charleston, and at Cape Lookout, it is much cleaner than the enclosed water. Although the intertidal habitats on the breakwaters are fairly clean, the lower parts are much affected by varying amounts of sand, or mud, or both, according to prevailing conditions.

The salinity in the Beaufort region is extremely variable. There is much less change in the water near Beaufort Inlet, where most of our work was done, than in the more enclosed waters affected by rivers, streams, and heavy rain, as around Pivers Island. Sea temperatures vary greatly, according to the particular locality: they are naturally more extreme in inland waters. The annual range at Cape Lookout is 10-27°C. During the coldest months, a gradient of 5-12°C exists between the warmer Cape Lookout water and the colder enclosed water around Beaufort, where, apparently, many algae are killed by heat and light in the summer and by cold in extreme winters.

The extreme (predicted) tidal range for Hampton Roads (the reference station for Beaufort) was 4.3 feet for 1947, and the average spring range was 3.0 feet. It may be noted that the joint vertical depth of the most fully intertidal zones (oyster zones plus muddy zones) approximates 2.75 feet, which is about the same as the average (predicted) distance between the higher water of spring tides and the mean low water of neap tides at Hampton Roads.

Typical marine intertidal populations were studied on the breakwaters, which largely consisted of rough and irregular blocks of granite and mica schist of various sizes, like those at Charleston. Seven of them were examined, all within a radius of 3 miles of Pivers Island—two at Fort Macon, one on Shark Shoal, and four on Shackleford Bank. They have so much in common that one general description suffices.

Plate 11 shows the average arrangement of zones of common animals and plants on the breakwaters near Beaufort.

The Supralittoral Fringe

There is the usual distinct zone of blackening in the supralittoral fringe, although it may be irregular in some places because of the broken substratum. On some breakwaters, at the higher levels, there are two species of rapid animals, the isopod *Ligia exotica* and the small crab *Sesarma cinereum.* Burrowing in the sand (in great numbers, in some places, particularly around Shark Shoal breakwater) are the gregarious *Uca pugilator* and the ghost crab *Ocypode albicans,* although the only estuarine features where these animals occur are a few scanty patches of reeds. Another estuary dweller, *Littorina irrorata*, appears where reeds grow near the rocks. *L. saxatilis* var. *tenebrosa* occurs erratically; it is common on Shackleford Bank, but chiefly in the midlittoral zone.

The Midlittoral Zone

As at Charleston, the midlittoral zone is well developed and can be similarly subdivided into three subzones.

The *Chthamalus* Zone.   This is similar to the corresponding subzone at Charleston, but we did not find any *Tetraclita; Balanus improvisus* was present, but not as prevalent as at Charleston. On the other hand, the larger *B. amphitrite niveus* was equally characteristic of the *Chthamalus* zone and the oyster zone at Beaufort, but we did not see it at Charleston. Intertidal barnacles include both transient and persistent species, however, and prolonged study is required to sort out the peculiarities of distribution that this causes. *Crassostrea virginica* was present as an invader from the subzone below, as were small yellow-brown plants of *Porphyra leucosticta.*

The Oyster Zone.   This is entirely covered at high waters of the greater tides, and uncovered by all low waters in ordinary weather. Most of the population of *Crassostrea virginica* and fully grown *Porphyra leucosticta* (which is probably seasonal) occurs here, and *Mytilus exustus* and *Balanus amphitrite* are very prominent. *M. edulis* lies sparingly among *M. exustus,* becoming common in some places. *B. eburneus* overlaps from below. The large ribbed mussel *Modiolus demissus* is present, but is at its best in more estuarine situations. We were unable to find any limpets on open rock surfaces on the breakwaters. Animals characteristic of the undersides of rocks, such as *Diodora* and *Crepidula*, are not uncommon.

The short, mosslike *Gelidium pusillum* grows locally on both mussel and oyster shells. The most conspicuous algal growth at the time of our visit was the bright green drapery of *Ulva* and *Enteromorpha*. The luxuriance of this green growth undoubtedly varies with the season, as does, no doubt, the assortment of species present.

From a geographical point of view, one of the most interesting features is *Fucus vesiculosus,* which extends from the top of the oyster zone into the muddy zone. Its most flourishing colony was on Shark Shoal breakwater. Apparently, the species has at times been more widespread here than when we saw it; Beaufort is its known southern limit on the Atlantic coast.

The Muddy Zone. This can be entirely exposed at the more extreme low waters, and may, at first sight, appear to be bare. Nevertheless, including the stragglers from the zones below and above, and the species that live under stones, the biota is considerable. In places, mussels can cover large areas of rock, and other attached bivalves are common, particularly *Ostrea equestris* and *Anomia simplex*. Barnacles are on the wane in this subzone; the most common species is *Balanus eburneus.* Among the other animals present are sponges, whelks, and worms.

The Infralittoral Fringe

The infralittoral fringe has an extremely rich and varied growth of algae, producing the effect of a silky Persian carpet. The growth consists of small to medium-sized forms—there are no really large forms present—and is of the kind typical of warm-temperate regions: it is rich in chlorophyceans and rhodophyceans, but lacks the bulky phaeophyceans of colder areas.

Animals are plentiful, both as species and as individuals, but are less conspicuous than the algae, although areas of considerable size are inhabited entirely by the anemone *Aiptasia pallida*. Notable, too, are the large quantities of the sea urchin *Arbacia punctulata,* the polyzoans living both above and below stones, the colonies of the gorgonian *Leptogorgia virbulata,* and the colonies of the dull-coloured branching coral *Oculina arbuscula.* For a detailed list of the biota of this area, see Stephenson and Stephenson (1952).

The Cape Lookout Jetty

The Cape Lookout jetty merits a brief description, as it lies some distance from Beaufort and provides a different environment. The water here is clearer and more affected by the Gulf Stream than the water at Beaufort, the tidal range is greater, the seas are rougher, salinities are more stable, and the winter sea-temperature minimum is higher by approximately 7°C.

There are two principal zones here, the upper zone, and the continuous algal zone. The upper zone is characterized by *Balanus amphitrite niveus* and *Mytilus exustus,* with some *Chthamalus* and a sprinkling of small *Crassostrea virginica.* It is relatively free

FIGURE **9.16**
A jetty extending out to sea from Cape Lookout, North Carolina.
[Photograph by L. G. Williams.]

from algae except for some *Porphyra* and small *Enteromorpha*. In its lower parts, mussels may form a distinct band. Below the mussels, in the continuous algal zone, the rock is coated with an almost unbroken velvety covering of small green and reddish-brown algae of many kinds (17 species were recorded). The algae in the lowest parts may be coated with a blackish silt, and are sparse under overhangs.

At the time of our visit, some large algae were found at the lowest levels, including *Sargassum filipendula* and *Gracilaria folifera*. Distinctive animals are the large whelk *Thais floridana* and the barnacle *Balanus tintinnabulum antillensis*. Special features are the sparseness of *Crassostrea virginica*—there is no visible oyster zone—and the shortness of the algae. Probably, if the rock were higher, there would be a typical *Chthamalus* zone. Were there more oysters, their maximal concentration would probably be on the rock near the lower extent of the barnacles and the upper extent of the algae. When closely examined, the continuous algal zone is seen to present a certain amount of subsidiary zonation within its width. At the time of our visit, this zone had a massive concentration of *Ulva* and *Enteromorpha* in its upper part. Its lower part, so far as can be judged, corresponds to the infralittoral fringe and muddy zone of Beaufort. The muddy zone is obliterated at Cape Lookout by an overgrowth of algae, which has the usual silt among basal parts, although it is not apparent at the surface.

Distinctive Features of the Beaufort Region

The intertidal population of the Beaufort region is clearly warm-temperate. The population appears to be normal, rich, and varied; our collection yielded about 105 common forms, as against 37 for Marineland and 34 for Charleston. This is because the environment here is much less rigorous than at Marineland and much cleaner than at Charleston. The zonation is normal. The infralittoral fringe here is better marked than at Charleston, Marineland, or the Florida Keys, and at its best has the typical population of a warm-temperate region, being abnormal only in the sparseness of lithothamnia.

The apparent absence of common limpets on the open rock, a remarkable feature shared by Charleston, is of particular interest. Seemingly, the common *Siphonaria* of the south have disappeared, and the common northern Atlantic limpet, *Acmaea testudinalis testudinalis,* does not reach this far south.* Animals such as *Crepidula* and *Diodora,* which live in concealment in crevices and stones, are common.

We have shown that there is a close resemblance between Charleston and Marineland in the distribution of certain important zone-forming species, but the resemblance between Charleston and Beaufort is even more obvious: Beaufort shows the full and typical development, in a region where an extensive and varied population can flourish, of the type of zonation that exists at Charleston, but is modified there by more restricting environmental conditions.

NOVA SCOTIA AND
PRINCE EDWARD ISLAND

INTRODUCTION

Having studied tropical and warm-temperate regions on the Atlantic coast of North America, we chose Nova Scotia and Prince Edward Island as suitable areas for the study of a cold temperate one. The work was carried out in July, August, and September of 1948 (see Stephenson and Stephenson, 1954a, 1954b).

The intertidal ecology of the places selected along the shores of these areas will be considered in four separate groups: (1) the Atlantic coast of Nova Scotia, including the shores of Peggy Cove, Halifax Harbour, and the inner reaches of St Margaret Bay;

---

*Perhaps the limited amount of solid substrate in the intertidal areas of the Carolinas may have some effect, as might the lack of a prevailing ocean current from the north that would act to spread larval forms southward. See Figure 3.3 (p. 28).

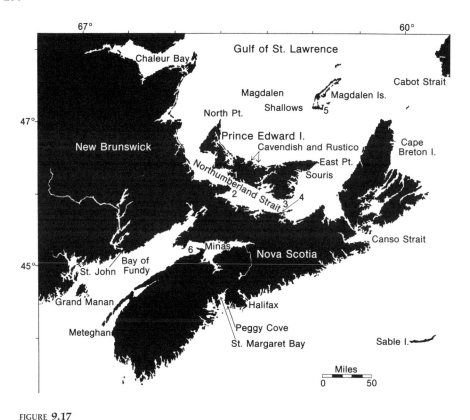

FIGURE **9.17**

Map of the Nova Scotia area: *1,* Fernwood and Seacow Head; *2,* Cape Tormentine; *3,* Wood Island; *4,* Caribou Island; *5,* Entry Island; *6,* Hall Harbour.

(2) the shores of the Bay of Fundy, including those of Hall Harbour and Meteghan, Nova Scotia; (3) the coasts of the Gulf of St Lawrence, including those of North Point, Cavendish, Rustico, and East Point, Prince Edward Island; and (4) the shores of Northumberland Strait, including Fernwood, Prince Edward Island, and Caribou Island, Nova Scotia (see Figure 9.17).

It is impossible to understand the biota of this region without having a clear picture of its physiography. (See Chapter 7 for further discussion of the climate of this region.)

The Bay of Fundy

The main channel of the Bay of Fundy is more than a hundred miles long, and has an average width of over 30 miles. The margins and the inner end of the bay are less than 75 feet deep. The tidal range, which is less than 20 feet at the mouth, increases to a maximum of 53.5 feet—the greatest known range in the world—in parts of the Minas Basin.

Owing chiefly to the exceptional tidal action, which causes an unusual amount of mixing and upwelling, extreme daily variation in the temperature of the surface water is prevented; thus, the annual range is moderate, the minimum being 1.8°C in February, and the maximum, 11.4°C in September (see Figure 9.18). There are variations, of course, because of unexpected conditions in particular years. Means for the warmest months (August and September) vary between 10°C and 14.7°C, and means for the coldest month, which is usually February, are 1.7–3.0°C. In most years, ice forms along the shores, but in many parts of the bay, the temperature almost never falls below zero.

Salinity at the surface, although below that of the open ocean, is high enough (usually about 30–33‰) to support a well developed marine population, except near sources of fresh water.

The Atlantic Coast of Nova Scotia

The long, indented Atlantic coast of Nova Scotia faces generally southeast. Its waters overlie a wide submarine platform, the Scotian Shelf, which is mainly less than 100 fathoms deep; an inshore strip and various offshore banks rise to less than 50 fathoms.

Typically, there are three layers of water over the shelf, all of them subject to variations in thickness, temperature, and so forth, in response to changes in the wind, weather, and currents. The surface layer has a salinity below 32.0‰. In winter, its temperature lies below 5.0°C, but it may approach 20.0°C in the summer. The intermediate cold layer, with a salinity of 32.0–33.5‰, has a temperature below 5.0°C in winter and summer alike. The bottom layer has a salinity higher than 33.5‰, and its temperature is usually above 5.0°C.

Consequently, intertidal populations are subjected to very cold water (sometimes below 0°C) in winter, and to relatively warm water (often near 20.0°C, or locally even higher) in summer. There may also be sudden changes of temperature, even in warm weather, if for some reason there is an upwelling of water from the intermediate colder layer.

Fogs can occur at any time of year, but sunny weather is common in summer. Wave action is often very powerful, but the great irregularity of the coast provides many sheltered habitats.

The range of tides, about 6.0–7.5 feet, is not unusual. In winter, there is little or no ice on the southern part of the open coast, but solid sheets of ice may form in some of the inlets.

The Gulf of St Lawrence

The Gulf of St Lawrence is a great enclosed sea with profound complications in its tides and depths. We are not concerned here with the working of the whole tidal system of the Gulf, but with the effects produced in Magdalen Shallows, particularly along the northern coast of Prince Edward Island and in Northumberland Strait (see Figures 7.2 and 9.17).

Figure 7.2 (p. 84) shows that there is a submarine shelf less than 50 fathoms (91.4 m) deep that is so wide in its southern part that it contains Magdalen Island, Prince Edward Island, Cape Breton Island, the Magdalen Shallows, and Northumberland Strait. This means that there is a surface layer of water of some stability in this area.

Figure 9.18 shows the annual variation in sea-surface temperatures in the area of Nova Scotia. The mean summer peak for Entry Island of the Magdalen group in the Gulf is higher than that for Halifax, Nova Scotia, and is higher yet than that for Grand Manan in the Bay of Fundy. On the other hand, winter temperatures fall markedly below zero at Entry Island, but not at Halifax or Grand Manan. While we have already noted that winter temperatures in the Bay of Fundy and on the Atlantic coast of Nova Scotia may fall below zero, such low temperatures are not nearly so widespread or of such long duration in these areas as they are in the Gulf of St Lawrence. Correlated with this is the fact that ice usually covers the Magdalen Shallows in winter, which has a much more serious effect on intertidal life than do the conditions in the Bay of Fundy and on the Atlantic coast of Nova Scotia.

Another feature of the shallows that must always be borne in mind is that sharp changes in temperature and salinity occur from day to day or over short irregular periods when the layers of water are under the influence of strong winds and other climatic factors.

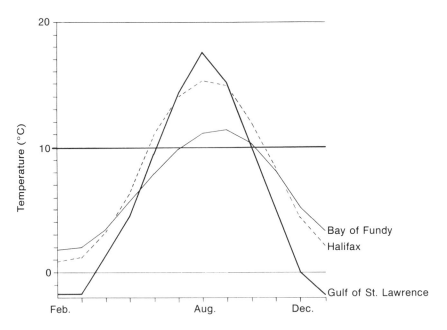

FIGURE **9.18**

The annual variation in sea-surface temperatures at three places in the Nova Scotia area: the Bay of Fundy (Grand Manan); Halifax, Nova Scotia; and the Gulf of St Lawrence (Entry Island in the Magdalen Islands). Values are monthly means over periods of five or more years. [After Hachey (1939, table 1).]

FIGURE **9.19**

An aerial view (from 250–300 feet) of ice off the north coast of Prince Edward Island, Gulf of St Lawrence, on March 7, 1951. The roughness of the ice is well illustrated. [Photograph supplied by Capt. W. J. Balcom, Ice Information Officer, Canadian Federal Department of Transport.]

FIGURE **9.20**

Piled-up ice on Gull Rock Reef, Caribou Island, in April. [The year and the photographer are unknown.]

As in all such complicated regions, the tidal variations in the Gulf of St Lawrence are markedly different from place to place, and must be studied individually as they occur. For instance, on the north coast of Prince Edward Island there are several consecutive days that occur twice in each month when there is only one distinct high and one distinct low water, instead of two of each. Also, the tides may sometimes remain at the same level for hours at a time, as we ourselves confirmed when working on this coast. The range of the spring tides is roughly 3–4 feet.

THE ATLANTIC COAST OF NOVA SCOTIA

Peggy Cove

The stretch of rocky coast adjacent to the harbour at Peggy Cove shows the standard marine population of the open Atlantic coast of Nova Scotia in its most fully developed form. The coast consists of immense masses of grey and buff-tinged granite that slope into the sea at all angles, providing a great variation in the extent of exposure and shelter.

The Supralittoral Fringe.   The uppermost limit of the fringe is white, and represents the transition between the maritime vegetation—grass, herbs, bushes, and lichens—and the shore proper (see Plate 12). Below this is a zone where the granite has a buff tint, which extends from about the level to which waves wash in moderate weather (well above actual sea level at high water) down to the upper limit of the occurrence of either barnacles or *Fucus* in quantity. Within this zone, the usual blackening produced by microphytes (including species of *Calothrix, Plectonema, Codiolum,* and *Entophysalis*) is strongly marked in some localities, vaguely in others. Typically, it has an irregular upper limit and a sharp lower one, with a gap of white granite between it and the uppermost *Fucus*. Certain slopes upon the lower part of the buff zone are inhabited by innumerable snails, *Littorina saxatilis,* which extend into the midlittoral at least as far as the upper growth of *Fucus.*

The Midlittoral Zone.   This zone extends from the upper limit of the occurrence of barnacles in quantity to the upper limit of the laminarian zone. Typically, there are three subdivisions: a narrow belt of barnacles, a wider one of *Fucus,* and a narrow one dominated by *Chondrus,* or *Alaria,* or both.

All of the barnacles appear to be *Balanus balanoides.* In some places, there is a neat belt of them above the *Fucus,* but it varies in extent in different localities. The barnacles extend freely into the *Fucus* zone, more plentifully into its upper parts, but may be absent altogether in areas where the weed growth is dense.

The upper limit of fucoid growth may be spectacularly and sharply defined. The four species present are *Fucus spiralis, F. vesiculosus, F. edentatus,* and *Ascophyllum nodosum.* In many places, *F. spiralis* forms a narrow, fairly distinct, orange-brown band at the top of the zone (with the highest plants of *F. vesiculosus* and *A. nodosum* occurring within it). It does not shun surf-beaten places, but there forms a part of a wider belt of abundant but rather impoverished plants. *F. vesiculosus* and *F. edentatus* together form a wide band below *F. spiralis.* Where there is moderate exposure, *F. vesiculosus* predominates above and *F. edentatus* below, and eventually, as exposure increases, the former disappears altogether; with increasing shelter, on the other hand, it is *F. edentatus* that fades out, but both species really attain their largest and best growth in relatively sheltered places. Full-sized *A. nodosum* seems to occur in quantity only in moderate shelter.

Under the fucoids, and in clearings amongst them, there is a rich growth of smaller algae, including *Chondrus crispus, Halosaccion ramentaceum, Pylaiella littoralis, Cladophora rupestris, Leathesia difformis* and a large amount of *Hildenbrandia prototypus.* The last species, at its best, covers the rock with a blood-red encrustation. *Polysiphonia lanosa* occurs on *Ascophyllum* here as it does in Britain.

The animals include *Littorina obtusata* and *L. littorea,* the former occurring from above the *F. spiralis* (and often attached to these plants) to the bottom of the *Fucus* zone. *L. obtusata* is reduced or absent in exposed places but, wherever wave action is less strong, it appears in force throughout the lower zone, and may even extend up to the lower part of the supralittoral fringe. Also characteristic of the fauna are hydroids (*Sertularia pumila* is common), the limpet *Acmaea testudinalis testudinalis,* beds of *Mytilus edulis,* the dog-whelk *Thais lapillus,* the sea urchin *Strongylocentrotus dröbachiensis,* and the anemone *Tealia crassicornis.*

The lowest strip of the midlittoral zone lies below the greatest growth of *Fucus,* and is best developed in areas not exposed to maximal wave action. In such places, it is dominated by a dense carpet of golden-green and brownish *Chondrus crispus,* and we have therefore termed it the *Chondrus* subzone. Within it are areas turfed with short plants of *Corallina officinalis* that shelter vast numbers of the small white bivalve *Hiatella (Saxicava) arctica.* The bladelike *Alaria esculenta* replaces *Chondrus* on very exposed rocks, where it is often found growing on beds of mussels. In such places, the general colour of the area is a deep crimson-brown. In some places, solid sheets of *Mytilus edulis* replace the turf. Other common animals are *Modiolus modiolus* and *Asterias vulgaris.* Other common algae are *Rhodymenia palmata, Chordaria flagelliformis, Halosaccion ramentaceum,* and *Leathesia difformis.*

The Infralittoral Fringe. There is a strongly marked and fully developed laminarian zone extending from well above low water of spring tides in calm weather to well below the lowest tide levels. Two species of tough, leathery algae form a waving forest of whiplike blades that are lashed about in the swells. They are *Laminaria agardhii,* which has the broad ribbonlike form of the European *L. saccharina* and reaches a length of 9 feet or more, and the palmate *L. digitata,* which attains a length of more than 5 feet. Among the plants of this forest are an undergrowth of plants and various animals from the *Chondrus* subzone.

There is a coating of lithothamnia, well developed in some places, which can extend in pools right up into the supralittoral fringe.

As previously stated, the granite slopes near Peggy Cove southwest of Halifax support an intertidal population that is typical of the North Atlantic coast. That is to say, the zonation of the shore includes a normally developed supralittoral fringe; a midlittoral well supplied with barnacles, with subzones of *Fucus* and *Chondrus,* together with a considerable population of various accessory species; and an infralittoral fringe consisting of a strongly developed laminarian zone.

Halifax Harbour

Halifax Harbour is an area with a biota intermediate between a fully developed open-coast population and one in a deep inlet. Our work in the area was carried out in August of 1948; since that time, the shore on which we worked has been obliterated by naval operations. It had reduced wave action, but was not a protected deep-set inlet. The surface temperature of the sea varies from less than $1.0°C$ in February to more than $15.0°C$ in August (see Figure 9.18).

The top of the supralittoral fringe was bare but for many *Littorina saxatilis* and a few *L. littorea,* the latter becoming abundant lower down. The midlittoral had large quantities of mussels and barnacles, including, in many places, unusually large individuals of *Balanus balanoides.* There was a good deal of the normal *Fucus vesiculosus* and *Ascophyllum nodosum,* and *Thais lapillus, Littorina obtusata,* and small *Acmaea testudinalis* were common. A special feature here was the great abundance of *Ralfsia verrucosa* on stones and mussels at mid–low-water levels (at that time, between low water of neap tides and low water of spring tides). Young plants of *Laminaria agardhii* were common, but not enough to make a laminarian zone. There was a good selection of other algae, some of which were large, under and among stones, and there were extensive mussel beds with the usual assortment of starfish, polyzoans, polynoids, amphipods, sponges, and so forth. It was clearly a population midway between the fully developed one of the open coasts and the reduced one of the deep inlets. (For detailed lists of the biota, see Stephenson and Stephenson, 1954b.)

St Margaret Bay

Wherever one follows the coastline into an inlet from the open sea (and this coast is deeply indented by hundreds of inlets), the effect of increasing distance from open water can be observed. An excellent example of this effect may be seen in the inlets that give off from St Margaret Bay (Mason Cove, for example), some of which extend inland 13-14 miles (see Plate 13). Many of these inlets are flanked by woodland, and the intertidal zone of most of them is narrow. The slopes of their shores are variously steep and gentle, and are studded with scattered boulders, some of which are very large.

The Supralittoral Fringe.   The white zone here corresponds to that of Peggy Cove, but it is much narrower and more distinctly marked, and lies very close to the level of high water of spring tides, rather than well above it (see Figure 9.21).

The buff zone, too, is narrower and lower down, and both its boundaries are very clearly distinguished. The band of blackening, when fully developed, is so straight in some places that it appears to have been drawn with a ruler. As at Peggy Cove, a considerable number of species are responsible for the blackening. There is no conspicuous belt of littorinids above the midlittoral.

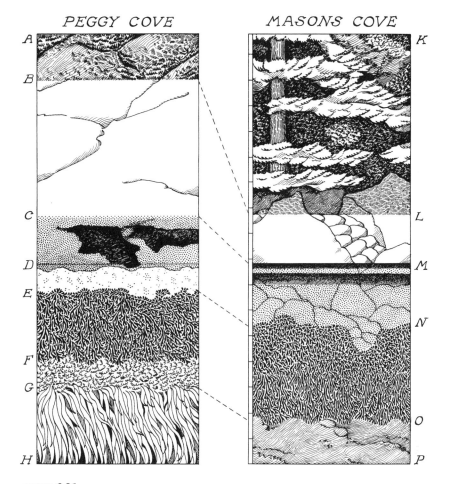

PEGGY COVE   MASONS COVE

FIGURE **9.21**

A comparison of the zonation at the entry of Masons Cove, an inlet of St Margaret Bay, Nova Scotia, with that at Peggy Cove, on the open Atlantic coast of Nova Scotia. The diagonal broken lines connect comparable levels in the two columns. *Peggy Cove: A–B,* land lichens, herbs, grass, and bushes; *B–C,* nearly bare rock (white zone); *C–D,* buff-coloured rock with patches of *Codium* and myxophyceans; *D–E,* barnacles; *E–F,* fucoid zone; *F–G, Chondrus* zone; *G–H,* laminarian zone. *Masons Cove: K–L,* forest of conifers, with other trees and bushes and land lichens on rocks; *L–M,* nearly bare rock (white zone); *M–N,* buff-coloured rock with neat bands of myxophycean blackening; *N–O,* fucoid zone; *O–P,* mud, gravel, and stones. A scale in feet from actual measurements is shown against Masons Cove; it can be applied to Peggy Cove, to which the shore at Masons Cove was correlated as closely as possible.

The Midlittoral Zone. Most noticeable, by comparison with the midlittoral zone at Peggy Cove, is the absence here of barnacles and the presence of a very large population of littorinids from the buff zone to below the level of low water of spring tides. *Littorina saxatilis* is most plentiful among the fucoids, and *L. littorea* is common below tidemarks as well. There is a dense growth of fucoids thoroughly overgrown by epiphytes. *Ascophyllum nodosum* (in two very different forms) and *Fucus vesiculosus* (in three different forms) are the most abundant species. Long trailing plants of *Entero-morpha prolifera* are also commonly seen. Mussels are abundant in the lower parts, as are amphipods and small fishes under the growth of fucoids.

The Infralittoral Fringe. The population that characterises this zone on normal rocky shores is completely lacking here. This lack is not wholly due to the distance from the sea, but is also due to the nature of the bottom, which may be of muddy sand, gravel, stones, or yielding mud. The plant growth consists primarily of areas of *Zostera marina*, often smothered in the alga *Pylaiella littoralis*. The common surface animals are *Littorina littorea*, starfish, whelks, two plentiful bivalves, and two species of *Natica*. In all, this is a rather dismal area.

One notable feature of the inlets of St Margaret Bay is the presence of unattached "rafts" of living tangled weed. These are composed mainly of the three forms of *Fucus vesiculosus* that grow attached in the midlittoral zone, and are often felted together by the byssus of mussels (*Mytilus edulis*) that grow among them. The mussels also occur on occasional rocks, and may also be found in heavy, loose masses, whose weight keeps them anchored to the substratum.

THE BAY OF FUNDY

The Bay of Fundy is notable because, in one of its inner branches from the Minas Basin, the tidal range is 53.5 feet, greater than any other known tidal range. Our work here was mainly at Hall Harbour, where the range exceeds 30 feet. Neither here nor at Meteghan, however, was there any particularly marked zonation that could be correlated with the great tidal range.

Meteghan

The shore of Meteghan, Nova Scotia, offers an example of conditions intermediate between those of the shores of Peggy Cove and Hall Harbour. The Meteghan shore is not truly oceanic, but is exposed to a large body of water (the Bay of Fundy, which is more than 70 miles across at that point, and gives onto the open Atlantic a short distance southwestward) and is, therefore, subject to considerable wave action. It has a steeply cliffed coast of shalelike rock, and the strata are commonly tilted up on edge.

The zone of particular interest is the midlittoral—the supralittoral and the infralittoral fringes have the usual characteristic littorinids and laminarians, respectively. The midlittoral zone at Meteghan has the same essential features that it has at Peggy Cove, but with some important modifications. Barnacles are more abundant here, and are crowded into a distinct belt in the upper part. There are the usual plants of *Fucus vesiculosus, F. edentatus,* and *Ascophyllum nodosum,* which show the customary reactions to degrees of exposure, and in sheltered places can suppress barnacles altogether; they extend so freely downwards into the *Chondrus* zone that this zone is often indistinct, and in sheltered, quiet places, the *Chondrus* becomes a mere undergrowth of the *Fucus*. *Gelidium stellata* is more abundant here than we ever saw it elsewhere.

Hall Harbour

The shore at Hall Harbour was studied during August of 1948, but was only seen at low water of the lesser spring tides; greater spring tides would fall decidedly lower: the tidal range at this locality exceeds 30 feet.

The area is a small harbour on the Minas Channel; a little to its northeast is a headland of hard grey basalt. From its foot, a massive rocky reef slopes unevenly to low-water level. This reef has irregular platforms studded with large and small boulders, and drops off precipitously, in some places, near sea level. The higher rocks are angular in shape and contrast with the rounded contours of the lower ones, which are worn by the friction of pebbles and ice.

This coast is not oceanic nor is it exposed to as much open water as Meteghan is, but it does have 60 miles of open water to westward, and is subject to considerable wave action and to strong currents.

FIGURE **9.22**

A view of the headland at Halls Harbour, Bay of Fundy, on August 20, 1948, at low water of a lesser spring tide. The angular shapes of the higher rocks contrast with the rounded shapes of the lower ones, which have been worn by pebbles and ice. The smooth rocks are thickly powdered with barnacles; on their lower parts is a dense, dark growth of *Porphyra* and other algae.

FIGURE **9.23**

The headland shown in Figure 9.22 on February 15, 1949. Note the ice foot. [Photograph by E. Armstrong.]

The Supralittoral Fringe. This is a well-marked and extensive belt, inhabited by immense numbers of *Littorina saxatilis*. These snails, many of which are very large, extend plentifully through the midlittoral and are common to within a few feet of low water. The black zone is difficult to discern, as the basalt itself contains black patches, but there is a wide, ill-defined blackening, produced in part by impoverished plants of *Hildenbrandia prototypus*.

The Midlittoral Zone. This zone has a vertical depth of 30 feet. There is an enormous development of *Balanus balanoides*, sometimes reaching down to low-water level. *Acmaea testudinalis* is common, and *Littorina littorea*, *L. obtusata* and *Thais lapillus* are all abundant and normal in their distribution. *Mytilus edulis* is uncommon.

Fucoids are abundant: *Fucus vesiculosus* and *Ascophyllum nodosum* extend over a wide range, and the latter, which may be over 2 feet long, is commonly overgrown by *Polysiphonia lanosa*. Another notable alga is *Porphyra umbilicalis*, which exists in two forms: a handsome dark purple plant that occurs mainly from low water upwards, and a shorter brownish one that is plentiful from the upper midlittoral to high levels in the supralittoral fringe. The overlapping of these two forms in the middle of the shore indicates the position of the upper and lower parts of the midlittoral zone. Numerous other species together make a very substantial cover of seaweeds, but *Chondrus crispus*, though present in pools, occurs nowhere in quantities sufficient to constitute a subzone.

The Infralittoral Fringe. Much of this fringe could be seen only through water, as the tide was not low enough to expose it, but there were no signs of a laminarian zone, nor even isolated laminarian plants in the pools higher up. There seemed to be no more than a ghost of the normal fringe population of animals and algae. This may be due partly to the action of ice and partly to the nature of the substratum, which consists more of pebbles and boulders than of continuous rock.

THE GULF OF ST LAWRENCE

When we visited various localities on the Gulf of St Lawrence in July and August of 1948, we found a most barren and unusual coastline. Its exceptional features are connected not only with ice, but also with the nature of the rock. The coast faces northward or westward, and is exposed both to strong wave action from the huge Gulf (due north, open water stretches for 200 miles) and to a harsh climate.

The predominantly red sandstone of the shore (it is greenish in places) is soft enough to be scraped off with a knife; but there are areas where the stone is harder, breaking down into boulders and pebbles. There are low, receding, red cliffs, from the foot of which extend rocky platforms alternating with sandy patches. The platforms are broken up in some places and continuous in others, but all of them have numerous crevices, hollows, pot-holes, and so forth, and much of the surface between the irregularities has an extremely smooth, clean appearance. The general aspect of the platforms is one of extraordinary bareness (see Plate 14). The first definite landmark is near the level of low water of spring tides, where a dense and unusual growth of algae begins.

Cavendish and Rustico

The environment at Cavendish and Rustico is so abnormal that it is not easily described in terms of the usual zones, although it may be easily related to those zones when it has been adequately studied.

Supralittoral and Midlittoral Regions. Above the infralittoral fringe, there is nothing resembling the zonation typical of other shores, although certain zones are discernible (see Figure 9.24). These include: (1) a green zone, wetted by wave action and by high water when it reaches the rock below, that is discoloured by microphytes, mainly *Calothrix scopulorum;* (2) a zone—of varying width—of *Littorina saxatilis* that marks the local maximal density of this species; and (3) a lower *L. littorea* zone—not everywhere present—that is typically slippery with microphytes and encrusting algae (including *Rivularia atra* and *Ralfsia clavata*), and in which a short, orange-brown growth of *Pylaiella littoralis* is common.

There is no true balanoid zone, but *Balanus balanoides* occurs up to the *Littorina saxatilis* zone, always in crevices and small irregularities that give protection from the scraping of ice. Because they are not overcrowded, these barnacles are mostly large and well-formed.

Fucus vesiculosus is the only fucoid present. It appears locally above low water as small plants in secluded places, but below the level of low water of spring tides, it can form dense and healthy growths. Though fertile and healthy, these plants are subnormal in size, and may represent a variety of the species.

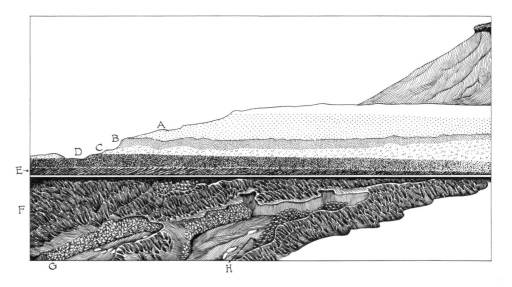

FIGURE **9.24**

Zonation on a rocky spur near Cavendish, Prince Edward Island, in July, 1948. Above the water are the following zones: *A,* a blackish green zone of *Calothrix scopulorum* (*light stipple*); *B,* a zone of *Littorina saxatilis* (*heavy stipple*); *C,* the upper part of the *Littorina littorea* zone (*irregular stipple*); *D,* a belt of dark microphytes (the lower part of the *L. littorea* zone); and *E,* a narrow band of collapsed plants belonging to the *Chordaria* growth below. Below the water: *F,* an 18-inch growth of whiplike annual brown algae (*Chordaria* and others); *G, Chondrus;* and *H,* short *Fucus* on protected surfaces or in hollows.

Microphytes and encrusting forms are, of course, particularly noticeable on the otherwise bare rock, and intensive study would undoubtedly reveal a considerable number with a wide variety of distributions. Our samples contained *Ralfsia clavata, Rivularia atra, Hildenbrandia prototypus,* and *Calothrix scopulorum.* The last species is well developed in many places, and may form a zone around low-water level.

Other algae are rare, especially on open rock. In a few places, brilliant green patches of small *Enteromorpha clathrata* and *Blidingia minima* may be found. The most generally common species is a short orange-brown form of *Pylaiella littoralis,* which grows directly on the rock surface from low water to the *Littorina saxatilis* zone. It is at its best at lower levels, and occurs in dense sheets under ledges and in crevices.

Other than the barnacles and littorinids already mentioned, animals are rather scarce. There are a limited number of *Mytilus edulis* in crevices, and a few amphipods. No limpets or *Thais* were recorded here.

The Infralittoral Fringe. In marked contrast to the bareness of the upper shore, the shore below low-water level supports a dense and healthy growth of algae. The uppermost plants collapse on the rock if they are exposed to the air, but the submerged ones form a soft, dark, waving forest of coarse, brown, beardlike forms. There is no general dominance of fucoids and we found no laminarians, though a few have been recorded by other investigators. In this regard, it must always be borne in mind that, in any particular year, there may be an excessive amount of ice or unusually severe storms, either of which can completely remove large algae. The dominant alga is *Chordaria flagelliformis*, which attains a length of up to 18 inches. Common of subsidiary plants are *Delamerea attentuata* and *Dictyosiphon foeniculaceus;* also present are species of *Cladophora, Ulva,* and *Polysiphonia.* Lithothamnia and *Corallina officinalis* occur only in small patches or tufts on harder surfaces, such as shells and stones.

Littorina littorea and the small crab *Cancer irroratus,* together with *Mytilus edulis,* live amongst weed or under crevices and ledges. *Balanus balanoides* extends a short distance below low-water level, and multitudes of amphipods live amongst some of the various algae.

North Point and East Point

It is possible, at North Point and East Point, to wade out a considerable distance, as the beach shelves shallowly in many places. The biota of these localities is quite similar to that of Cavendish; one notable difference is the presence here of the whiplike alga *Chorda filum.* A particularly noticeable phenomenon at North Point is that, from a little above the water level of low spring tides downwards, the stones are covered with a dense fur of short hydroids that form a definite zone extending along the shore as far as the eye can see. Many of the hydroids—all of those collected were *Obelia dichotoma*— shine green in the sunlight, perhaps owing to an overgrowth of diatoms. There is also a tremendous flora of algae that form soft, delicate, impalpable beards and tufts of various colours; some of these grow on stones, but large numbers are epiphytic on the larger algae.

NORTHUMBERLAND STRAIT

Three localities on Northumberland Strait—two on the south coast of Prince Edward Island and one on the north coast of Nova Scotia—were visited during August 1948. The shores of Northumberland Strait showed a considerable contrast to those of the Gulf of St Lawrence, both in the development of barnacles, limpets, and algae, and in the number and kind of species present.

Souris

The intertidal rocks at Souris, Prince Edward Island, though of a red sandstone similar to that at Cavendish, do not have the same planed appearance, and have a far richer covering of algae. As at Cavendish, a green zone and a *Littorina saxatilis* zone are present, forming a supralittoral fringe of sorts; but the *Balanus balanoides* at Souris, unlike the same species at Cavendish, forms a normal, though irregular, belt. Just above the level of low water of low spring tides, there is a good turf of *Chondrus,* among which grow plants of *Fucus vesiculosus.* The latter species continues to the top of the barnacle zone, where there is also some *F. spiralis*. At the lowest level of the shore, some *F. serratus* appears. At the ends of the reefs, where the sand begins, the rocks are conspicuously green with a healthy and extensive growth of *Enteromorpha linza* and *Ulva lactuca*.

From the shelving seafloor, it is possible to get a good view of the infralittoral fringe, which offers a complete contrast to the typical infralittoral fringe of the north shore of Prince Edward Island. Here, it is *Chondrus*, not *Chordaria*, that is dominant, making a continuous carpet of gold and green. Among it are well developed plants of *Fucus serratus*, low-level forms of *F. vesiculosus*, and some *Chordaria*. *Corallina* is also present, as on the north coast.

Fernwood

The rock formation at Fernwood, Prince Edward Island, is similar to that at Souris. There is a green zone and a *Littorina saxatilis* zone, the latter overlapping the topmost barnacles. A general growth of microphytes is present, but makes no slippery zone as at Cavendish. Barnacles occur in vast quantities and make a well-marked zone with a fairly definite upper limit. The barnacle zone, which extends downwards to a foot or so below low water,* is populated mainly by *Balanus balanoides*, with a sprinkling of *B. improvisus*: in its lower part, fucoids are plentiful, with fairly well grown *Fucus vesiculosus* of the rather slender, small, intertidal type; there is a considerable growth of *Chondrus*; and *Corallina officinalis* was more plentiful here than elsewhere on Prince Edward Island.

Below low water, the commonest algae is *Chondrus*. Groves of *Fucus serratus* and *F. vesiculosus*, most of the plants extensively epiphytised, are also common. There are no laminarians, and lithothamnia occurs sparsely only on hard objects, such as stones and shells.

The fauna is somewhat more interesting. *Crassostrea virginica* is common, as are *Crepidula fornicata* and *C. plana*. A species of *Thais* is also present. In addition to *Cancer irroratus*, the crabs include the small, dark *Neopanope texana sayi* and the hermit crab *Eupagurus acadianus*, both of which occur at lower levels and in secluded places. Here, too, we found the only anemone we saw on Prince Edward Island, a small *Metridium senile* var. *pallidium*. Also seen were extensive growths of the short form of *Obelia dichotoma* that we found also at North Point.

Caribou Island

With the exception of Caribou Island, the other places we visited on Northumberland Strait showed but small variations in the general features described for Souris and Fernwood. Caribou Island showed a number of noteworthy peculiarities.

The top zones are not fully represented, both because the cliffs are not high enough,

---

*The low-water level referred to here is near mean low water of the lower spring tides, possibly about a foot above the level of the lowest individual spring tides.

and because they are much affected by runnels of fresh water bringing down earth from the tops of the cliffs. The midlittoral zone is wide and has three belts: (1) a *Littorina littorea* zone, in which *Rivularia atra* is prominent and barnacles and *Fucus vesiculosus* are sparsely represented; (2) a belt with numerous barnacles and a moderate quantity of *Fucus;* and (3) a strip in which *Fucus vesiculosus* and *F. serratus* are present in great quantity, and in which barnacles, plentiful toward the top of the strip, diminish in number toward low water. *Ascophyllum nodosum* appears in this last subzone, as does *Littorina obtusata*, which can be found both on the rock and on the *Fucus* and *Ascophyllum.*

The submarine forest seems to be composed of the two *Fucus* species mentioned above, *F. serratus* probably being dominant. There is no *Chordaria* forest or *Chondrus* turf, but there is the usual extensive undergrowth, and oysters and *Crepidula* are common.

It would appear that, to a certain extent, the Northumberland Strait is a distinct biological entity: the supralittoral fringe alone is normal—the midlittoral is singularly bare, being denuded of most of its barnacles and fucoids as well as other organisms, and the laminarian zone has been replaced by a shorter growth of annual algae that is dominated by *Chordaria flagelliformis*. There are other distinctive features as well, but the usual widespread zones are recognizable everywhere, even though environmental factors have caused them to undergo some striking modifications. For a detailed discussion of the geographical features of this region, see Stephenson and Stephenson (1954b).

SELECTED REFERENCES

[In addition to the references given below, see those in the bibliography to Chapter 7, "Shores Affected by Ice" (p. 95).]

Davis, W. M., 1928. *The Coral Reef Problem* (AGS Special Publication 9). New York: American Geographical Society.

Hachey, H. B., 1939. Surface water temperatures of the Canadian Atlantic seaboard. *J. Fish. Res. Board Can.* 4(5): 339–348.

Stephenson, T. A., and Anne Stephenson, 1950. Life between tide-marks in North America, I. The Florida Keys. *J. Ecol.* 38: 354–402.

———, and ———, 1952. Life between tide-marks in North America, II. Northern Florida and the Carolinas. *J. Ecol.* 40: 1–49.

———, and ———, 1954a. Life between tide-marks in North America, IIIA. Nova Scotia and Prince Edward Island: description of the region. *J. Ecol.* 42: 14–45.

———, and ———, 1954b. Life between tide-marks in North America, IIIB. Nova Scotia and Prince Edward Island: the geographical features of the region. *J. Ecol.* 42: 46–70.

10

THE PACIFIC COAST
OF NORTH AMERICA

In our survey of the Pacific coast of North America, field studies were made at various places on Vancouver Island, British Columbia, and along two stretches of the California coast—the Pacific Grove region (with the Monterey Peninsula and Point Lobos) and the La Jolla region, which is very near the Mexican boundary. Discussion of the coast of Chile, however, is based not on field studies of our own, but on studies of other investigators, notably Guiler (1959a, 1959b).

VANCOUVER ISLAND

Surveys were carried out between June 26 and August 21, 1947, at five localities on Vancouver Island. Our attention was particularly concentrated on Brandon Island, in Departure Bay, near Nanaimo, a locality that seemed, and proved to be, one of unusual ecological interest. We studied the shore at Horswell Bluff and the reef at False Narrows, which provided us with useful comparisons with Brandon Island—all three stations lie in relatively sheltered places. We also examined two localities in the vicinity of Victoria, both of which are near to open seas, although in no way oceanic.

INTRODUCTION

Vancouver Island lies between latitudes 48° and 51°N. Both the island and the adjacent mainland are mountainous, and are separated from each other by the complicated Strait of Georgia (Figure 10.1). The southern end of this strait is connected to the Pacific Ocean by Juan de Fuca Strait. The latter continues offshore across the continental shelf as a submarine canyon, which assists the influx of cold bottom water. This entrance is sufficiently wide and unobstructed to permit a considerable amount of wave action on the inner coast of southern Vancouver Island, and the tidal currents there are strong.

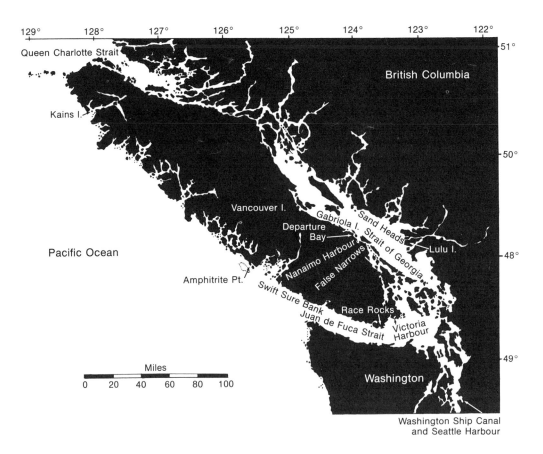

FIGURE 10.1
A map of the Vancouver Island area.

On the other hand, the northern section of the Strait of Georgia is practically an inland sea, remote from the ocean, and with water deeper than that in most of the approaches to it, which are also partly blocked by numerous islands. What wave action exists is primarily of local origin; tidal currents are strong, and the salinity of the water is affected by rivers (the largest of which is the Fraser River), especially in the summer. The water in this region appears to be unusually rich in nutrient salts, partly due to the upwelling of cold bottom water at the mouth of Juan de Fuca Strait.

The sea surrounding Vancouver Island is of relatively low salinity (the monthly means all fall below 33‰) and its temperature range is cold-temperate (monthly means at the sea surface fall below 10°C, but do not approach zero, in the winter, and usually rise to between 10° and 20°C in the summer). The variations in salinity and sea temperatures are of interest because the west coast of Vancouver Island is oceanic, but much of the east coast borders the partially enclosed Strait of Georgia. Figure 10.2 shows the variations in salinity and sea temperatures at three stations around Vancouver Island, based on monthly means averaged over the years 1948-57.

The annual salinity curve for Kain's Island, which is oceanic, is fairly regular. It lies between 28‰ and 32‰, and shows a decided rise from July to September (the same is true of Amphitrite Point). At Race Rocks near Victoria in Juan de Fuca Strait,

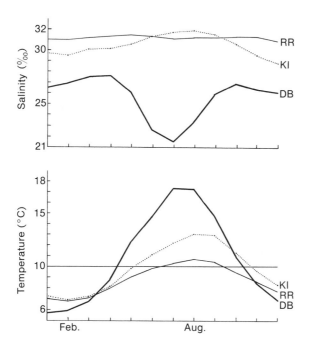

FIGURE **10.2**

The variations in salinity and temperature at three stations round Vancouver Island, based on monthly means averaged over the years 1948-1947. *RR,* Race Rocks, in the strait of Juan de Fuca; *KI,* Kains Island, off the northwest coast; *DB,* Departure Bay, near Nanaimo (see Figure 10.1).

the curve is not far from a straight line near the level of 31‰; evidently, evaporation and dilution by fresh water approximately balance each other. The situation at Departure Bay is entirely different. The whole curve lies at a lower level, below 28‰; and it shows an extremely marked fall during the summer, reaching a minimum of 21.5‰ in July. Is it certain that similar curves would be found for the many localities on the northern, much more broken, sheltered coasts of the Strait of Georgia, as they are far from the ocean, and are periodically subject to major influxes of fresh water.

The monthly mean sea temperatures for Kain's Island vary between 6.9°C in February and 13.0°C in August. Conditions are similar at Amphitrite Point, Swiftsure Bank, and Race Rocks, although the maximum at Swiftsure Bank is only 12.0°C, and that at Race Rocks is slightly lower. Winter means rarely fall below 4.4°C, and never reach zero; consequently, winter ice is exceptional and of very little ecological importance. Departure Bay presents quite a different picture, however: winter temperatures fall a little lower here than they do at places on the oceanic side, but rise far higher in the summer—to 17.3°C in July, for example (although individual monthly means may be higher). Because Departure Bay is not particularly shallow, this temperature range must be due less to the depth of the water than to the sheltered position of the bay and the prevailing temperature of the water in the Strait of Georgia.

BRANDON ISLAND

Brandon Island lies at lat. 49°12′26″N and long. 123°57′18″W in Departure Bay, just opposite the Pacific Biological Station (Figure 10.3). It is the main island of a small group. The island is more than 700 yards (640 m) long, by less than 100 yards (91 m) wide, and its long axis runs roughly east to west. The water is fairly deep all around the island. The south coast consists of smooth gradual slopes of conglomerate with occasional boulders of volcanic rock, and there is a good deal of sand below low water. In a fine summer, this shore gets many hours of sunshine. Most of the north coast is steep, the slope varying from gradual to vertical or overhanging. It is much more shaded than the south coast, the sunshine being limited to early morning and evening.

The south and north coasts are alike in being subject only to slight wave action, and the sea is usually negotiable in a rowing boat. On the whole, the island is most exposed at its eastern end. Differences in wave action are amongst the most effective influences controlling zonation; when such differences are absent, variations must be attributed to other causes.

The island is crowned with a good selection of maritime dry-land vegetation, including bushes and trees, various garden escapes, and weeds of cultivation. All round the island, just above high water, is a belt of terrestrial lichens and mosses that reaches into the vegetation above.

We found that the distribution of plants and animals on this island so clearly demonstrated the relationship between particular features of the environment and the reactions of organisms that it repaid detailed study. We had the use of an accurate tide gauge, which we found of great advantage in studying zonation and tide levels.

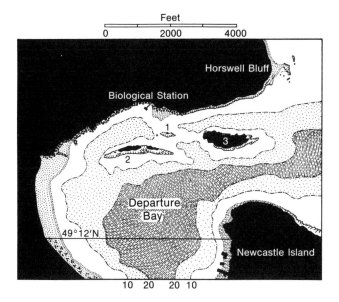

FIGURE **10.3**

Departure Bay, near Nanaimo: *black*, land; *white*, water less than 10 fathoms deep; *stipple*, water 10–20 fathoms deep; *shading*, water more than 20 fathoms deep; *1*, Inskip Rock; *2*, Brandon Island; *3*, Jesse Island.

FIGURE **10.4**

The south coast of Brandon Island, looking westward, in July, 1947. The wooded hills in the background belong to the coast of Departure Bay. On the rocky slope in the foreground are two dark zones of algae separated by a wide gap. The lower of these is the "beard" zone, populated chiefly by colonial diatoms; the upper is the *Fucus* zone. The gap between these is occupied by the bare zone of the infralittoral fringe below, and the lower barnacle zone of the midlittoral above.

FIGURE **10.5**

Part of the north coast of Brandon Island in July, 1947. Visible as a continuous grey band is the pronounced zone of *Verrucaria striatula* and its associates, followed below by the very sharp white line representing the upper limit of the majority of the barnacles. Below the barnacle line are dark patches of *Fucus* overgrowing the barnacles; the upper limit of the *Fucus* is only a few inches below that of the barnacles. Above the *Verrucaria* zone, the pale lower limit of the larger terrestrial lichens can also be seen.

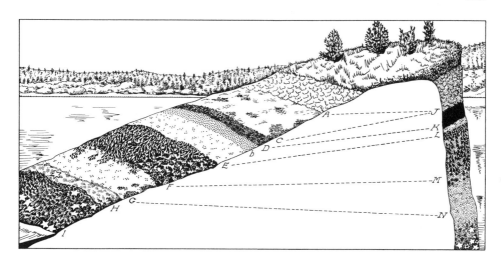

FIGURE **10.6**

A diagrammatic section through Brandon Island, with the coast of Departure Bay in the background; *A-J,* lower limit of terrestrial lichens; *C-J,* upper limit of densest blackening; *D-K,* upper limit of main barnacle population; *b,* lower limit of *Chthamalus* subzone; *E-L,* upper limit of main *Fucus* Population; *G-N,* lower limit of most barnacles; *H,* upper limit of "beard" zone; *I,* low water of a low spring tide. The drawing is not to scale. Compare with Figure 10.7.

The Supralittoral Fringe

The South Coast. The supralittoral fringe on the south coast is a broad, bare, well-marked strip lying between the lower limit of the terrestrial lichens and the upper limit of the main barnacle population (Figure 10.7, *A–D*). It has a colossal population of small *Littorina scutulata* in its lower part, but few of them reach to the top of the fringe (Figure 10.7, *B*). They extend freely down through the midlittoral zone as far as the infralittoral fringe, becoming scarcer lower down. Another common animal is the rapid isopod *Ligia pallasi.* There is a distinct but not strongly marked black zone immediately above the upper limit of main barnacles (Figure 10.7, *C–D*), and over-lapping them; higher up, the black zone occurs only in irregular patches. The blackening is chiefly due to impoverished *Hildenbrandia prototypus,* and probably also to *Verrucaria* and myxophyceans.

The North Coast. The gap between the terrestrial lichens and the main popula-tion of barnacles (Figure 10.7, *J–K*) is much narrower on the north coast, and is often filled with a very intense blackening (due, mainly, to *Verrucaria*), which extends, some-what, into the areas above and below. *Littorina scutulata* is common, and continues right up to the terrestrial lichens, where individuals are much larger, possibly because of the rich black "fodder."

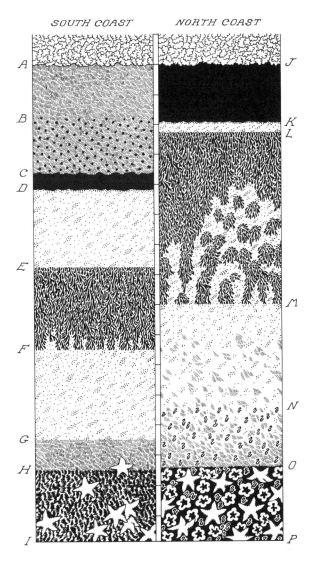

SOUTH COAST NORTH COAST

FIGURE **10.7**

A comparison of the zonation on the south and north coasts of Brandon Island. The scale is shown in feet between the two columns. The figure treats the main populations of selected species only. *A* and *J*, lower limit of terrestrial lichens; *B*, upper limit of most littorinids; *C* and *J*, upper limit of densest blackening; *D* and *K*, upper limit of main barnacle population; *E* and *L*, upper limit of main *Fucus* population; *F* and *M*, lower limit of main *Fucus* population; *G*, lower limit of main barnacle population; *H*, upper limit of "beard" zone; *N*, approximate upper limit of most *Serpula*; *O*, upper limit of *Metridium* zone; *I* and *P*, approximate level of low water of a low spring tide. See text for further explanation.

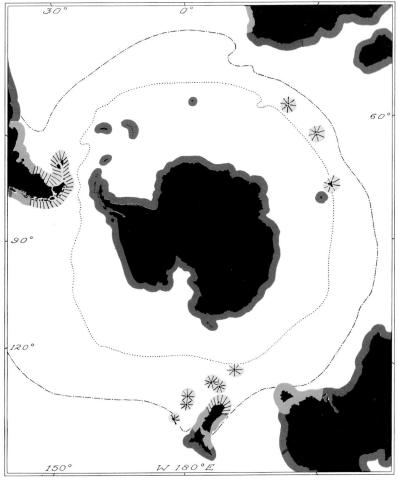

ANTARCTIC AND SUBANTARCTIC (S.S.)
COLD-TEMPERATE (ANTIBOREAL) MIXED
WARM-TEMPERATE
TROPICAL AND SUBTROPICAL
ANTARCTIC CONVERGENCE
SUBTROPICAL CONVERGENCE

PLATE 1. The general distribution of southern littoral faunas. The cold-temperate faunas occur primarily between the Antarctic and Subtropical convergences; exceptions, due to cold currents, are the west coast of South Africa and part of the south coast of Australia. Although antarctic littoral faunas appear genuinely circumpolar, warm- and cold-temperate ones appear to be independent faunas inhabiting similar conditions, with a limited number of southern or circumpolar forms representing them.

PLATE 2. The zonation between tidemarks on Gibbet Island, Bermuda, a typical open coast. Note the black band of myxophyceans marking the supralittoral fringe. In the upper part of the midlittoral, there is a belt of barnacles with small patches of algae, and below, the rock is veneered by a growth of the vermetid *Spiroglyphus* and a sparse brown pile of small algae. The bountiful growth of the algae of the infralittoral fringe lies below water level. [Reprinted, by permission of the publisher, from T. A. Stephenson and Anne Stephenson, "The Bermuda Islands," *Endeavor,* vol. 13, no. 50, fig. 5 (copyright © 1954).]

PLATE 3. A funnel-shaped boiler lying off the west coast of Bermuda. [Reprinted, by permission of the publisher, from T. A. Stephenson and Anne Stephenson, "The Bermuda Islands," *Endeavor,* vol. 13, no. 50, fig. 11 (copyright © 1954).]

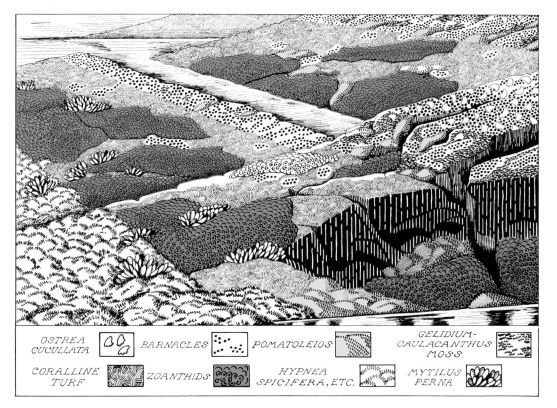

| OSTREA CUCULLATA | | BARNACLES | | POMATOLEIOS | | GELIDIUM‑ CAULACANTHUS MOSS | |
| CORALLINE TURF | | ZOANTHIDS | | HYPNEA SPICIFERA, ETC. | | MYTILUS PERNA | |

PLATE **4.** Variations in the characteristic zonation of the coast of Natal, South Africa, produced by exposure to different types of wave action. This drawing is based on conditions observed at Umhlali (the mussels have been added); compare with Figure 8.5 (page 104). The slope on the left is exposed to strong wave action: the Hypnea zone is well developed, zoanthids (*blue*) are confined to a belt above it, and *Pomatoleios* (*yellow*) occurs mainly higher still. In the right foreground, the rock slopes steeply to a creek with modified wave action: there is no *Hypnea* zone, and the zoanthids and *Pomatoleios* reach to low-water level. On the sheltered slope in the upper part of the figure, the *Hypnea* zone is replaced by coralline turf, zoanthids, and *Pomatoleios*, which organisms normally occur above the *Hypnea* zone in more exposed places. The *Littorina* zone and infralittoral fringe are not included.

PLATE **5.** A type of zonation common on boulders and smooth slopes in positions exposed to fairly strong wave action on the west coast of South Africa. The whole of the rock is carpeted by lithothamnia (the coralliform species above with others below it) from level *A* downwards. In this example, barnacles are reduced to a group in a cleft on the right. The bare zone is clearly shown between the *Porphyra* above and the lithothamnia below.

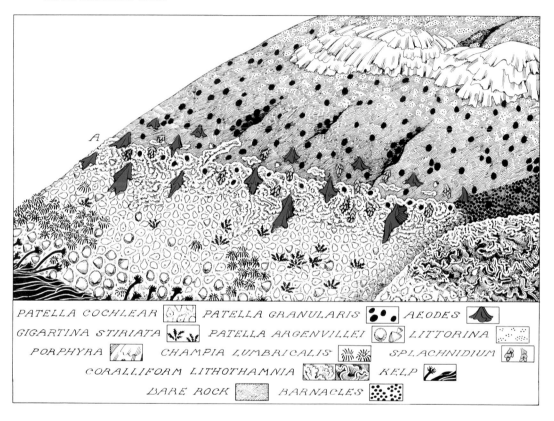

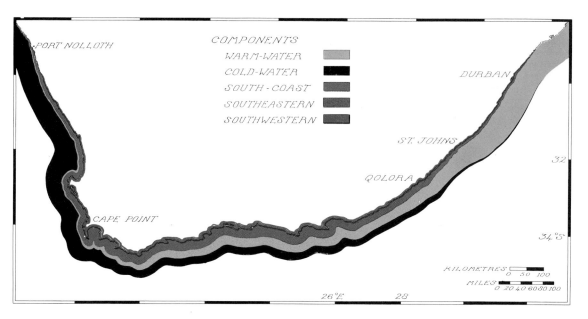

PLATE 6. The arrangement of the components of the intertidal fauna and flora of South Africa. The ubiquitous component, because it extends around the whole coast, is omitted, as are a number of minor local components. Of the remaining five principal components, three are more important than the others. The map illustrates not only the complexity of the overlapping, but also the way in which the components combine, so that, although there is some east-west change everywhere, it is more rapid along some parts of the coast than along others, with the result that each of the three different component patterns persists, in a fairly typical form, along a considerable stretch of coast. Thus, the south coast shows one pattern of colour bands, the coast of Durban shows quite a different one, and the west coast shows yet another; whatever the complexity of the components (it is probably greater than is here shown), there are still three main faunas.

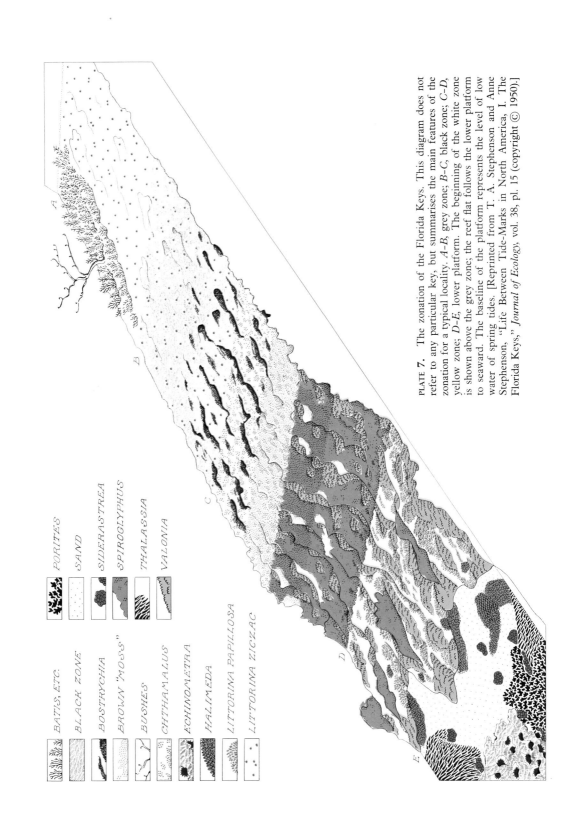

PLATE 7. The zonation of the Florida Keys. This diagram does not refer to any particular key, but summarises the main features of the zonation for a typical locality. *A–B*, grey zone; *B–C*, black zone; *C–D*, yellow zone; *D–E*, lower platform. The beginning of the white zone is shown above the grey zone; the reef flat follows the lower platform to seaward. The baseline of the platform represents the level of low water of spring tides. [Reprinted from T. A. Stephenson and Anne Stephenson, "Life Between Tide-Marks in North America, I. The Florida Keys," *Journal of Ecology*, vol. 38, pl. 15 (copyright © 1950).]

PORITES

SAND

SIDERASTREA

SPIROGLYPHUS

THALASSIA

VALONIA

BATIS, ETC.

BLACK ZONE

BOSTRYCHIA

BROWN "MOSS"

BUSHES

CHTHAMALUS

ECHINOMETRA

HALIMEDA

LITTORINA PAPILLOSA

LITTORINA ZICZAC

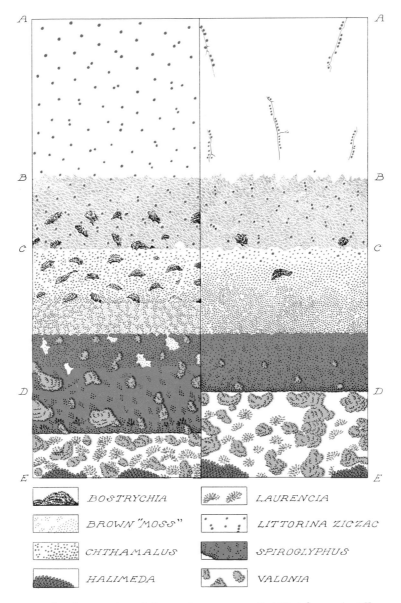

BOSTRYCHIA		*LAURENCIA*	
BROWN "MOSS"		*LITTORINA ZICZAC*	
CHTHAMALUS		*SPIROGLYPHUS*	
HALIMEDA		*VALONIA*	

PLATE **8.** A comparison of the zonation on a sea wall at West Summerland Key (*right*) and that on a natural rocky platform (*left*). The depths of the zones on the wall are drawn to scale (*A-E* represents 4 feet). The zones in the column are reduced to comparable depths. *A-B,* grey zone (zone *A* of wall); *B-C,* black zone (zone *B* of wall); *C-D,* yellow zone (zone *C* of wall); *D-E,* lower platform (zone *D* of wall). [Reprinted from T. A. Stephenson and Anne Stephenson, "Life Between Tide-Marks in North America, I. The Florida Keys," *Journal of Ecology,* vol. 38, pl. 14 (copyright © 1950).]

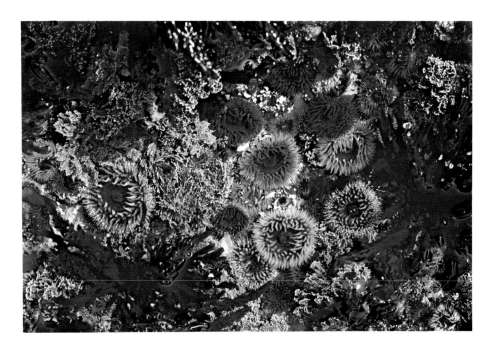

PLATE **9.** A group of *Bunodactis reynaudi* on the coast of the Cape Peninsula, South Africa. [Photograph by Charles Field, Wynberg, Cape Town.]

PLATE **10.** The upper platform on the oceanic side of Plantation Key, Florida, as seen at low water of a spring tide. The black and yellow zones are clearly visible; the grey and white zones appear narrow because they are foreshortened. The reef flat is under shallow water on the right. The lower platform is reduced to spurs running seaward from the upper platform. [Reprinted, by permission of the publisher, from T. A. Stephenson, "Coral Reefs," *Endeavor,* vol. 5, no. 19, pl. 3 (copyright © 1946).]

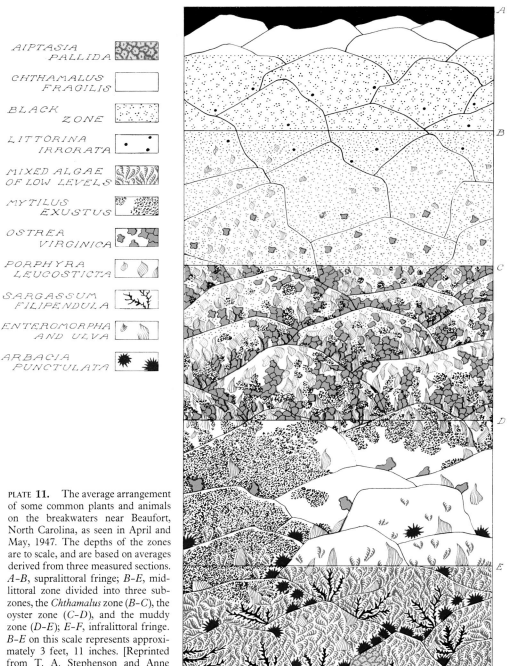

AIPTASIA PALLIDA	
CHTHAMALUS FRAGILIS	
BLACK ZONE	
LITTORINA IRRORATA	
MIXED ALGAE OF LOW LEVELS	
MYTILUS EXUSTUS	
OSTREA VIRGINICA	
PORPHYRA LEUCOSTICTA	
SARGASSUM FILIPENDULA	
ENTEROMORPHA AND ULVA	
ARBACIA PUNCTULATA	

PLATE **11.** The average arrangement of some common plants and animals on the breakwaters near Beaufort, North Carolina, as seen in April and May, 1947. The depths of the zones are to scale, and are based on averages derived from three measured sections. *A-B*, supralittoral fringe; *B-E*, midlittoral zone divided into three subzones, the *Chthamalus* zone (*B-C*), the oyster zone (*C-D*), and the muddy zone (*D-E*); *E-F*, infralittoral fringe. *B-E* on this scale represents approximately 3 feet, 11 inches. [Reprinted from T. A. Stephenson and Anne Stephenson, "Life Between Tide-Marks in North America, II. Northern Florida and the Carolinas," *Journal of Ecology*, vol. 40, pl. 6, (copyright © 1952).]

PLATE **12.** Granite rocks of the coast outside the harbour at Peggy Cove, Nova Scotia, as seen on September 6, 1948. Immediately above sea level is the laminarian zone (partly exposed at low water of a spring tide), and above this, a reddish brown belt of *Chondrus crispus* (with some *Alaria*), followed by a wider, blackish band of *Fucus*. Above the *Fucus* in the foreground is a narrow white strip of barnacles; above the barnacles is the buff zone, with its maplike grey patches of *Codiolum* and myxophyceans. [Reprinted, by permission of the publisher, from T. A. Stephenson and Anne Stephenson, "Life Between Tide-Marks in North America," *Endeavor,* vol. 9, no. 33, fig. 1 (copyright © 1950).]

PLATE **13.**   The shore at the entrance to Mason's Cove, Nova Scotia, in 1948. Below the trees, a pale band is visible; it is followed by a well-marked black zone, striped in places, produced by myxophyceans and intertidal lichens. Below this, there is a belt in which the granite rock is discoloured to a buff tint, and a broad zone of fucoids that is subdivided into upper (yellower) and lower (browner) parts. [Reprinted, by permission of the publisher, from T. A. Stephenson and Anne Stephenson, "Life Between Tide-Marks in North America," *Endeavor,* vol. 9, no. 33, fig. 2 (copyright © 1950).]

PLATE **14.** The coast near Cavendish, Prince Edward Island, on July 2, 1948. The tide is low, showing the bareness of the red sandstone. The dark seaweed on the sandy beach in the foreground is washed up, not growing.

PLATE **15.** The coast in a small bay on Northumberland Strait, near Souris, Prince Edward Island, on August 5, 1948. The tide is low, showing the sandstone substratum with a well-developed covering of growing algae. Note the contrast with the shore shown in Plate 14.

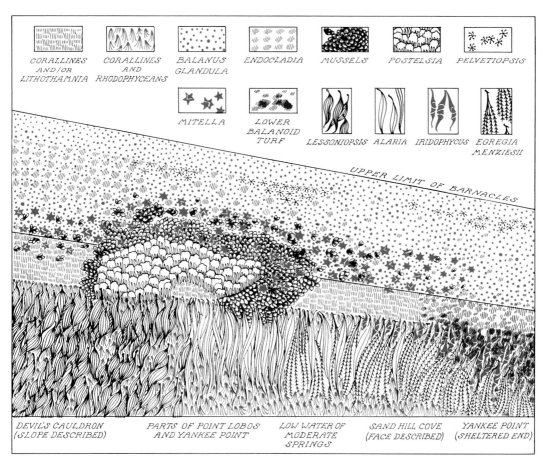

Key symbols (top labels):

CORALLINES AND/OR LITHOTHAMNIA — CORALLINES AND RHODOPHYCEANS — BALANUS GLANDULA — ENDOCLADIA — MUSSELS — POSTELSIA — PELVETIOPSIS

MITELLA — LOWER BALANOID TURF — LESSONIOPSIS — ALARIA — IRIDOPHYCUS — EGREGIA MENZIESII

UPPER LIMIT OF BARNACLES

DEVIL'S CAULDRON (SLOPE DESCRIBED) — PARTS OF POINT LOBOS AND YANKEE POINT — LOW WATER OF MODERATE SPRINGS — SAND HILL COVE (FACE DESCRIBED) — YANKEE POINT (SHELTERED END)

PLATE **16.** A comparison of the zonation of steep slopes at different localities in the Pacific Grove region. The various slopes are subject to different types and degrees of wave action. The line across the centre of the figure indicates the boundary between the upper and lower balanoid zones. Compare Plate 17.

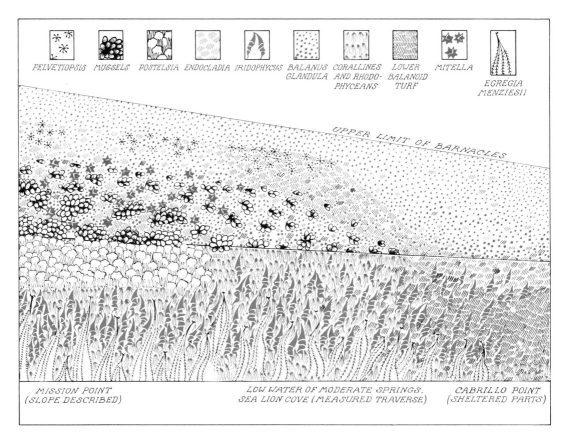

PELVETIOPSIS MUSSELS POSTELSIA ENDOCLADIA IRIDOPHYCUS BALANUS
GLANDULA
CORALLINES
AND RHODO-
PHYCEANS
LOWER
BALANOID
TURF
MITELLA
EGREGIA
MENZIESII

UPPER LIMIT OF BARNACLES

MISSION POINT
(SLOPE DESCRIBED)

LOW WATER OF MODERATE SPRINGS.
SEA LION COVE (MEASURED TRAVERSE)

CABRILLO POINT
(SHELTERED PARTS)

PLATE **17.** A comparison of the zonation of moderate slopes at different localities in the Pacific Grove region. The various slopes are subject to different types and degrees of wave action. In this drawing, as in Plate 16, the exposure to strong wave action is on the left and sheltered conditions are on the right.

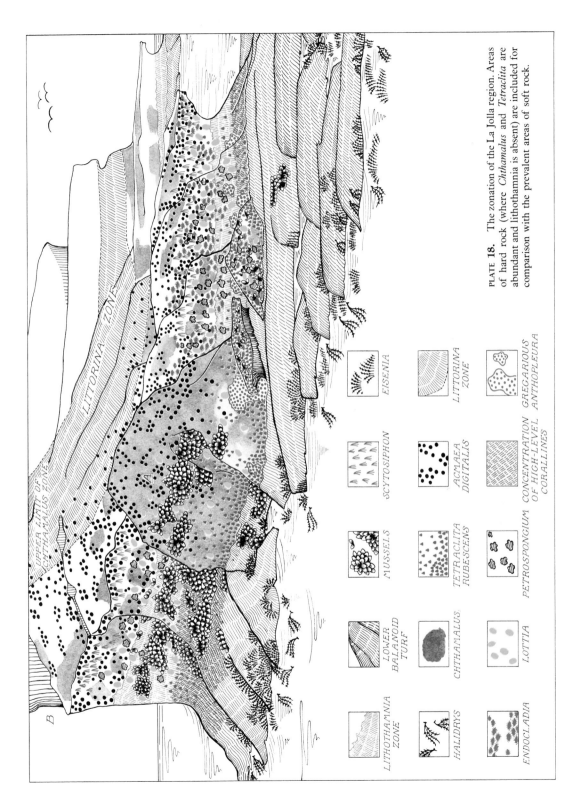

LITTORINA ZONE

LITTORINA ZONE

UPPER LIMIT OF
CHTHAMALUS ZONE

B

LITHOTHAMNIA
ZONE

HALIDRYS

ENDOCLADIA

LOWER
BALANOID
TURF

CHTHAMALUS

LOTTIA

MUSSELS

TETRACLITA
RUBESCENS

PETROSPONGIUM

SCYTOSIPHON

ACMAEA
DIGITALIS

CONCENTRATION
OF HIGH-LEVEL
CORALLINES

EISENIA

LITTORINA
ZONE

GREGARIOUS
ANTHOPLEURA

PLATE **18.** The zonation of the La Jolla region. Areas of hard rock (where *Chthamalus* and *Tetraclita* are abundant and lithothamnia is absent) are included for comparison with the prevalent areas of soft rock.

PLATE **19.** View of a rocky coast at La Jolla, seen at low water. The lower rocks are covered by green *Phyllospadix* or a reddish brown turf of short algae. Above the turf can be seen the pinkish lithothamnia zone with the bare rock above (this is an area in which most of the rock is soft sandstone).

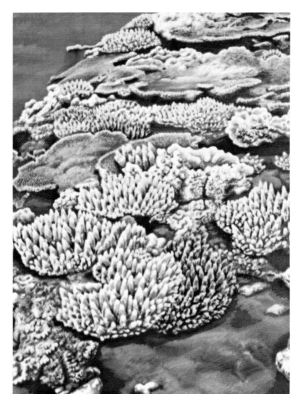

PLATE **20.** A growth of living coral on the top of a small coral islet on the landward side of Yonge Reef, in the outer portion of the Great Barrier Reef, Australia. All the corals belong to *Acropora*, the main genus of reef-building species.

The Midlittoral Zone

The South Coast. On the southcoast of Brandon Island, the midlittoral zone comprises three distinct subzones: the upper barnacle zone, the lower barnacle zone, and the *Fucus* zone.

Barnacles form a dense growth from the top to the bottom of the midlittoral (Figure 10.7, *D–G*), and show no diminution among the *Fucus*. Their upper limit is clearly marked, and varies only a few inches in level all along the shore; their lower limit is less sharp, but still fairly constant. Four species are involved: *Chthamalus dalli, Balanus glandula, B. crenatus* and *B. cariosus. C. dalli* is abundant throughout the midlittoral, but it is primarily a high-level form (Figure 10.6, *b*). On the smooth, even slopes of the upper barnacle subzone there are two divisions: an upper strip, where *C. dalli* is by far the most plentiful species, and another strip immediately below, where adult *B. glandula* is the most plentiful. Both species can appear in the supralittoral fringe. *B. cariosus* is a spiny barnacle, and is much the largest of the four species. It occurs in the lower parts of the *Fucus* subzone and below. Although plentiful, it is far less abundant than the other species of barnacles. The main midlittoral barnacle population is dominated by *B. glandula* and *B. crenatus,* the latter being particularly important in the lower part of the *Fucus* subzone and below. In the vast population of barnacles in the lower zone, the great majority of the individuals are young and densely crowded. Because *B. glandula* and *B. crenatus* look very much alike at this stage, it was impossible to determine in the field which of the two is dominant.

Fucus forms a conspicuous zone with a sharp upper edge. The growth is densest at the high levels, with outlying plants extending into the zone above. At its lower limit, the growth is much more irregular (Figure 10.6, *E–F*).

Apart from barnacles and *Fucus, Mytilus edulis* is by far the most prevalent midlittoral organism. These mussels are most numerous and largest among the *Fucus*, where they may occur attached to the *Fucus* or in dense sheets attached to the rocks. They are reduced in numbers in the higher regions, though they are sometimes relatively plentiful in higher pools. Small specimens are locally abundant lower on the shore, and form beds in some places.

Limpets are common, but small, and are much less conspicuous here than on the coasts of Britain and South Africa and on more exposed Pacific shores. *Acmaea pelta* is the commonest limpet, and is found among *Fucus* and the lower-zone barnacles. *A. persona* is common on boulders in the upper barnacle zone.

The periwinkle *Littorina sitchana* is fairly common. It is generally somewhat larger than the *L. scutulata* already mentioned, and extends throughout and beyond the *Fucus* zone, the biggest specimens occurring highest up. One peculiarity of this shore is the absence of whelks from the midlittoral. Small crabs, of course, are common.

In 1947, there were few oysters; but Dr D. B. Quayle, in a letter to us in 1960, described a solid mass of the oyster *Crassostrea gigas*, introduced from Japan, that had come to occupy a belt between the four-foot and the eight-foot tidal levels (Figure 10.8). The general appearance of Brandon Island in 1960, therefore, must have been very different from what it was in 1947.

FIGURE **10.8**

The oyster *Crassostrea gigas,* introduced by the Japanese, occupying a belt between the 4-foot and 8-foot tidal levels on the south shore of Brandon Island. The photograph was taken in 1960. [Photograph by D. B. Quayle.]

The rock in the lower midlittoral is much coloured by encrusting species of algae, principally *Hildenbrandia,* with species of *Ralfsia, Peyssonellia,* and *Rivularia. Gigartina papillata* is common among the *Fucus.* The rather few pools contain an assortment of algae, including growing plants of *Sargassum, Ulva, Gigartina,* and various corallines. Patches of lithothamnia occur in some of the pools.

The North Coast. The subdivision of the midlittoral zone of the north coast is fundamentally similar to that of the south coast, but details and tidal levels are remarkably different.

On the shaded northern cliffs, the barnacles extend en masse much higher than they do on the southern shore. There is no *Chthamalus* strip above the *Balanus glandula;* instead, the two species intermingle and the latter is dominant and larger. Both species continue sparsely or intermittently down to the lower levels of the zone, and show no distinctly visible lower limit (Figures 10.6 and 10.7, *K–N*). *B. cariosus* is common at low levels, where many other barnacles exist among the short algal turf.

Here, the strongest growth of *Fucus* is much higher than it is on the south coast. Its upper limit is clearly marked; the density of the growth, however, is extremely variable, being scanty on the steepest slopes and virtually nonexistant under the over-hangs (Figures 10.6 and 10.7, *L–M*). The lower limits of the *Fucus* are vague and very irregular. Myriads of small, densely packed mussels occur fairly high up among *Fucus* and barnacles, but mussels are generally scarce in the lower zones.

The lowest part of the midlittoral is the least populated, having a mere scattering of barnacles, mussels, and mosslike growths of various small algae. Notable among the algae are *Endocladia muricata* (the ecological counterpart of the similar *Caulacanthus ustulatus* of South Africa) and *Enteromorpha clathrata*, which forms a distinct band in some places. Purple starfish, *Pisaster ochraceus*, are occasional invaders of this region; they appear more numerous and seem to climb higher here than on the south coast.

The Infralittoral Fringe

South Coast.   The infralittoral fringe of the south coast is unusual, as it is peculiarly deficient in large plants. It consists of two subzones—one bare but for occasional barnacles and the ubiquitous encrustations of *Hildenbrandia* and similar algae (Figures 10.6 and 10.7, *G–H*), and a lower belt exposed only at the low spring tides (Figure 10.6, *H–I*). In the lower belt, the rock is coated densely with a soft, slimy, silt-ridden covering of brown beardlike and mosslike algae, many of them identifiable only by microscopic examination (see Stephenson and Stephenson, 1961a). The upper limit of this belt is sharp and fairly constant, and makes only occasional incursions into the belt above. The bottom is sandy in many places below low water, with beds of *Zostera* in some places, and patches of lithothamnia in pools.

By far the most conspicuous and plentiful animals are the starfish, first and foremost *Pisaster ochraceus* (many individuals are a foot or more in diameter). The other common starfish are *P. brevispinus*, *Demasterias imbricata*, *Evasterias troschellii*, and *Pycnopodia helianthoides*. Crabs, too, are common.

The North Coast.   As on the south coast, the infralittoral fringe here has an upper and lower subdivision, but the population of the latter is very unlike that of the south coast. Although there is a bare belt corresponding approximately to that of the south coast (Figure 10.7, *M–O*), the mosslike algae are weakly developed and the tube worm *Serpula vermicularis* is common (Figure 10.7, *N–O*).

The lower belt of the infralittoral fringe of this coast is remarkable (Figure 10.5, *O–P*). On the one hand, the population of small algae is meagre, but there are stumps of *Sargassum* and a weak fringe of *Laminaria* at the extreme lower edge of the belt, and a limited amount of *Nerocystis*, *Laminaria*, and *Zostera* can be seen under deep water. A certain amount of lithothamnia appears in caves, clefts, and shady places, and *Codium fragile* and *Ceramium gardneri*, among other algae, are also present. On the other hand, animals are prolific and striking. The most notable are the large anemones *Metridium senile fimbricatum*, which extended in mass formation as far as one can see into deep water. There is an abundance of *Pisaster ochraceus*, and *Serpula vermicularis* with its crimson and variegated fans is also conspicuous. There is a variety of other animals, including large whelks, echinoderms, the limpet *Diodora aspera*, ascidians, and chitons. (For detailed list, see Stephenson and Stephenson, 1961a, p. 19). This varied animal population far exceeds that of the infralittoral fringe of the south coast.

Figure 2.4 (p. 13) is a diagrammatic comparison between the zonation, correlated to eight mean tide levels, on the north and south coasts of Brandon Island.

OTHER LOCALITIES ON VANCOUVER ISLAND

The kind of zonation described in the preceeding account of the north and south coasts of Brandon Island is widespread in the sheltered waters of this region. The same general plan, with marked differences in detail due to changes in various physical factors, prevails all along the coast of Vancouver Island. The degree to which rock surfaces are broken has an effect on zonation, and the degree of exposure to wave action and tide rips seems to have even more. The angle of the slope of the rocks and the amount of exposure to sunshine also produce considerable changes in the general plan of zonation.

In our general study of the coast of Vancouver Island, we paid special attention to Horswell Bluff, the reef at False Narrows, and an area around Victoria. Figure 10.9 indicates the zonation of these particular localities for which only a general description and comparison need be given.

Horswell Bluff is an irregular headland that provides a much more broken and complicated substratum than Brandon Island. The shore is rather steep, but flattens out to a certain extent on its lower reaches, and is subject to much more wave action than Brandon Island.

The reef at False Narrows runs almost due east and west between the mainland and Gabriola Island (Figure 10.1). It is composed of overlapping ledges of shale that slope to the water on the north side and end abruptly or overhang on the south side. Normally, the reef does not suffer strong wave action, but extremely strong tide rips sweep past it. The reef, which is totally submerged at high water of spring tides, is not high enough to have a supralittoral fringe.

In the neighbourhood of Victoria, we worked on two headlands to the east of the entry to the harbour. The shore at Victoria is directly open to Juan de Fuca Strait, and the rocks are about as unprotected as any in the region, but the situation is certainly not oceanic (Figure 10.10). The tidal range here is less than at Departure Bay, and the low waters occur mostly during daylight from March to August, and during darkness for the rest of the year. The differences in tidal behaviour between Departure Bay and Victoria no doubt affect the zonation of those localities, but our data was not sufficient to enable us to make a satisfactory comment about this.

The Supralittoral Fringe

The supralittoral fringe is lacking on the reef at False Narrows, but is typically well developed at the other localities. *Ligia pallasi* and *Littorina scutulata* are ubiquitous, and at Victoria *Littorina sitchana* is present also. The black zone ranges from continuous to patchy, and includes *Verrucaria striatula*, *V. maura*, and impoverished *Hildenbrandia prototypus*.

The Midlittoral Zone

The midlittoral zone is wide and well marked at all places, but is not, as at Brandon Island, clearly divided into three subzones (upper barnacle zone, broad *Fucus* zone, and lower barnacle zone). On some coasts, mussels are dominant and replace the conspicuous

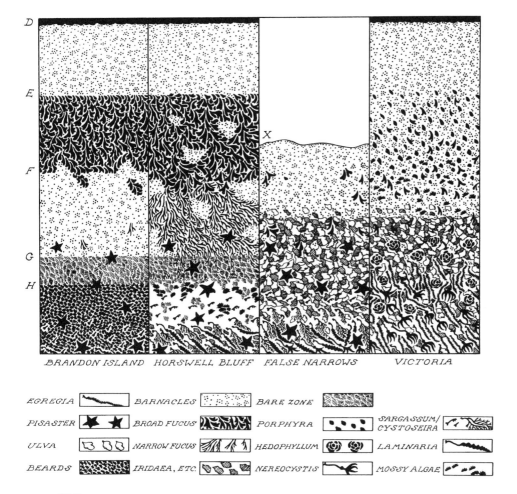

D

E

F

G

H

BRANDON ISLAND   HORSWELL BLUFF   FALSE NARROWS   VICTORIA

EGREGIA		BARNACLES		BARE ZONE	
PISASTER		BROAD FUCUS		PORPHYRA	SARGASSUM/CYSTOSEIRA
ULVA		NARROW FUCUS		HEDOPHYLLUM	LAMINARIA
BEARDS		IRIDAEA, ETC.		NEREOCYSTIS	MOSSY ALGAE

FIGURE **10.9**

A comparison of the zonation at several localities described in the text. The supralittoral fringe is omitted. *X* indicates the top of the reef in False Narrows. Only the first column (the south coast of Brandon Island) is to scale: *D-E,* the upper barnacle zone; *E-F,* the *Fucus* zone; *F-G,* the lower barnacle zone; *G-H,* the bare zone of the infralittoral fringe; *H,* the upper limit of the "beard" zone of the infralittoral fringe. The difference in tidal range between Victoria and Departure Bay is ignored.

Fucus belt; on others, neither prevails, and the midlittoral, which is covered uniformly with barnacles, is not divisible into subzones.

At Horswell Bluff, the two higher subzones are distinguishable, but there is no distinct lower barnacle zone, as it is overgrown with *Fucus* and other algae. The barnacles in the upper zone consist of a fairly even mixture of *Chthamalus dalli* and *Balanus glandula,* with some *Fucus* growing among them. The barnacles continue into the *Fucus* zone under the algal growth, where they are joined by *B. crenatus* and *B. cariosus.* Apart from *Fucus,* there is a considerable population of algae: *Gigartina papillata* and other rhodophyceans make a carpet of short or mossy plants, encrusting *Hildenbrandia* extends above and below the zone, and *Ulva, Sargassum, Leathesia, Prionitis,* and corallines, among others, also occur in varying quantities. *Acmaea pelta, A. persona,* and *A. scutum* are common, as are various crabs—*Hemigrapsus nudus* is inordinately abundant, and *H. oregonensis* is plentiful at low levels.

On the False Narrows reef and at Victoria, there is not enough *Fucus* to form a distinct zone; there, the maximal growth of algae is in the lower part of the midlittoral, where *Fucus* is merely one constituent of a mixed growth of weed. Consequently, these midlittoral zones have but two subdivisions: an upper one with few algae and a lower one with many, the latter sometimes invaded by species from the infralittoral fringe.

The highest part of the reef at False Narrows is at about the same level as the lower parts of the *Fucus* subzone of adjacent shores, and its main feature is a dense population of barnacles and mussels. At Victoria, the only noticeable algae on open rock are the black, mosslike *Endocladia muricata* and a dusky *Porphyra* that is quite distinct from the low-level epiphytic one. There is a good growth of weed in pools. The barnacles are closely massed, and large specimens of *Balanus cariosus* are particularly notable (Figure 10.11). Notable too is the small peaked limpet *Acmaea digitalis.* In the densely weeded zone of both Victoria and False Narrows, *Ulva expansa*—sometimes 3 feet (91 cm) across—is common, as are *Gigartina binghamiae* (reminiscent of the South African *G. radula*), *Iridaea heterocarpa, I. splendens, Leathesia difformis, Sargassum,* and various corallines. Notable at Victoria are the unusual laminarian *Hydrophyllum sessile* and *Rhodomela larix,* a black form that is an ecolocial counterpart of the South African *Plocamium cornutum.*

FIGURE **10.10**

Part of the rocky area near Victoria in August, 1947. A mixed algal growth is visible on some of the lower rocks, and a pale band—the upper barnacles—can be seen at a higher level. On the extreme left are some floating stipes of *Nereocystis.*

FIGURE **10.11**

A close-up view of an area of barnacle-covered rocks near Victoria in August, 1947. The large barnacles are the spiny *Balanus cariosus.* Growing on and between them are innumerable smaller barnacles of other species, chiefly *Balanus glandula* and *Chthamalus dalli.* A few plants of *Fucus* are visible.

Animals common to both Victoria and False Narrows are the usual species of *Acmaea* (but *A. digitalis* is abundant only at Victoria, mainly high on the shore), *Mytilus edulis,* and various whelks (including *Thais lamellosa* and the smaller *T. emarginata* and *T. lima,* ecological counterparts of the British *T. lapillus* and the South African *T. dubia*).

Animals common at Victoria are the gregarious polychaet *Dodecaceria fewkesi,* the very conspicuous fan worm *Eudistylia vancouveri* (the counterpart of the African *Sabellastarte longa*), and the small anemone *Anthopleura elegantissima,* which is moss-green in colour with a pinkish sheen and occurs in enormous numbers. Giant individuals of *Mytilus californianus* occur here and in the laminarian zone. *M. edulis* is fairly common at most levels, but forms no continuous beds.

The Infralittoral Fringe

At Brandon Island, there are two subzones in the infralittoral fringe—an upper bare strip and a lower belt—with markedly different populations on the south and north coasts. At Horswell Bluff, these subzones are represented by a zone of reduced barnacles that contains *Fucus,* and a zone of mixed but impoverished algal growth. At the lowest level of the lower belt there are some poorly developed laminarians, but the dominant kelps are moderate-sized *Sargassum muticum* or *Cystoseira geminata* or both. There are patches of lithothamnia and mosslike growths of algae of various sizes. Limpets, chitons, bivalves, *Metridium, Serpula,* and *Thais* occur in moderate quantities. *Pisaster* sprawls about plentifully, and frequently climbs into the upper zones.

The False Narrows reef, on its northern, sloping side, is covered entirely by a rich growth of algae, those of the midlittoral merging with those of the infralittoral with no apparent break. Consequently, the lower limit of barnacles in quantity is probably forced higher up the slope. There is a healthy growth of laminarians, but no impressive amount of it is exposed to the air. The undergrowing algae are numerous and varied, and include *Iridophycus, Gigartina,* and *Odonthalia.* The animals in the undergrowth, including sponges, polyzoans, and hydroids, are similarly abundant.

The steep slopes and overhangs of the southern side of the False Narrows reef correspond to the *Metridium*–starfish belt of Brandon Island, where a similar configuration of the coast strongly affects the population. Notable are the large numbers of the giant barnacle *Balanus nubilis* (some individuals have a basal diameter of 5 inches, or 12.7 cm), the very conical *B. rostratus,* and the flattened anomuran crab *Petrolisthes eriomerus,* which lives under stones. Numerous holes in the rock are occupied by borers or by subsequently arriving nonboring species. The fauna of these cavities belongs to several groups, bivalves being the most important; particularly plentiful is the large white *Penitella penita.*

At Victoria, the biota is very similar to that of the north shore of the False Narrows reef; but the laminarians are larger and of greater variety, and include quantities of the giant *Nereocystis luetkeana,* long fronds of *Egregia menziesii,* and the cabbagelike *Hedophyllum sessile.* There is a great quantity of *Ulva* and the epiphytic *Porphyra,* plenty of corallines, and sheets of lithothamnia in some places. Animals of the fringe include some from higher zones (such as species of *Acmaea* and *Thais*) in addition to sea urchins, scarlet starfish, and sea cucumbers.

CONSTITUTION OF THE POPULATION

Although the biota of this region includes many species that are common to all localities, there are many interesting differences between the biotas of the various localities.

The laminarian zone, so characteristic of cold northern waters, is variously developed: it is very poor at Brandon Island, where the *Metridium*–starfish fauna extends well below water; at Horswell Bluff, the growth is a little better; it is strong in False Narrows, though it emerges little from the sea; and it is best developed at Victoria, where it emerges somewhat more.

Pisaster ochraceus was very scarce at the time of our visit to Victoria. This may be connected with the fact that the mussels then did not form extensive beds. *Metridium* was absent at Victoria, perhaps owing to the lack there of its particular habitat.

Certain characteristics of the biota at Victoria are probably related to the increased wave action there. The large spiny barnacle *Balanus cariosus* is most abundant and widespread (however, the size and distribution of this species also appears to be controlled by variations in salinity). Limpets are often larger, more numerous, and more conspicuous here than in sheltered places, and the presence of *Acmaea* is a characteristic of Victoria. Another characteristic of the area is the great abundance of the anemone *Anthopleura elegantissima*. Increased wave action is also partly responsible for the prevalence of the laminarian *Hedophyllum sessile*, the presence of *Mytilus californianus,* and the reduction of *Fucus*.

The marine fauna and flora of the area around Nanaimo and Victoria are truly remarkable, notably in the large size attained by so many of their members. Starfish over a foot (30 cm) across and weighing more than 2 pounds (roughly 1 kg) are abundant; the largest barnacles are 5 inches (12.7 cm) in diameter and correspondingly heavy; and anemones and chitons a foot or more in length are common. Among the seaweeds, *Nereocystis* exceeds 40 feet (12 m) in length; *Sargassum, Cystoseira, Egregia,* and *Alaria* exceed 12 feet (3.7 m); and there are smaller algae, the colonies of which attain a diameter of 2-3 feet (60-90 cm), including species of *Ulva, Iridaea,* and *Gigartina*.

Tables 10.1 and 10.2 give some idea of the sizes of representative plants and animals, but because they are based chiefly upon our own notes, they do not necessarily indicate the maximum size possible for each species. For some species, sizes are given in square brackets: these figures have been derived from the work of other authors.

This abundance of large forms is all the more striking because it is correlated, in at least some parts of the area (in Departure Bay, for example), with a low range of salinity (see Figure 10.2). This area supports a fully marine population that is characterised not only by particularly large species, but by a rich population of echinoderms

TABLE **10.1**

Some plants common to the shores around Nanaimo and Victoria and their typical sizes

Plant	Size*
Alaria valida	Up to 12 ft (3.7 m) or more†
Cystoseira geminata	Exceeds 13 ft (4 m)
Egregia menziesii	Exceeds 7 ft (2.1 m) [25 ft (7.6 m)]‡
Gigartina binghamiae	2 ft 3 in (68 cm)§
Iridaea heterocarpa	1 ft 6 in (45.7 cm)§
Iridaea splendens	2 ft 1 in × 1 ft 8 in (63.5 cm × 50.8 cm)§
Laminaria cuneifolia	Exceeds 6 ft 6 in (2 m)
Nereocystis luetkeana	Exceeds 40 ft (12.1 m), with 12 ft (3.7 m) blades [125 ft (38 m)]‡
Sargassum muticum	Exceeds 13 ft (4 m)
Ulva expansa	Up to 3 ft (91.4 cm)
Zostera marina	Leaves exceed 5 ft (1.5 m)

*Greatest dimension, unless otherwise indicated.
†Some species of *Alaria* exceed 70 ft (21 m).
‡Figures in square brackets derived from work of other authors.
§Figures from measurement of single typical individual.

TABLE 10.2

Some animals common to the shores around Nanaimo
and Victoria and their typical sizes

Animal	Size*
Balanus nubilis	Basal diameter up to 5 in (12.7 cm)
Crassostrea gigas	6 in (15.2 cm) [12 in (30.5 cm)]†
Cryptochiton stelleri	12 in (30.5 cm)‡
Dermasterias imbricata	8 in (20.3 cm)‡
Evasterias troschelii	19 in (48.3 cm)‡ [2 ft (60 cm) or more]†
Metridium senile	12 in (30.5 cm)‡
Pisaster brevispinus	12 in (30.5 cm)‡
Pisaster ochraceus	13 in (33 cm) or more§
Pycnopodia helianthoides	1-2 ft (30-60 cm) or more

*Greatest dimension, unless otherwise indicated.
†Figures in square brackets derived from work of other authors.
‡Figures from measurement of single typical individual.
§Large *P. ochraceus* may weigh more than 2 lb (1 kg).

as well. This is most unexpected, and it widens our conception of the extent to which a marine population can adapt itself to subnormal salinities. It is presumed that the unusually favourable supply of nutrients in the region is important in this connection.

In many basic respects, the intertidal zonation of this area is much like that of similar areas in other parts of the world. Its main characteristics—including a strong representation of barnacles, mussels, *Fucus,* and laminarians, arranged in the typical manner—are typical of northern cold-temperate regions, and the usual effects of exposure to different amounts of wave action are seen. Among the *un*usual features, in addition to the ones mentioned in the preceding paragraph, are the marked abundance and variety of starfish, the presence of the gregarious *Anthopleura elegantissima* in vast numbers, the unusual nature of some of the barnacles (the prickly texture of *Balanus cariosus,* for example, and the excessive size of *B. nubilis*), and the interesting selection of laminarians that inhabit the area. Such features as the relation between zonation and tidal levels, the contrast between the north and south coasts of Brandon Island, and the occurrence of the curious *Metridium*-starfish community, are effects of a special development of the local population in relation to the particular topography of the area they inhabit. The sheltered parts of the region lack certain features of the open Pacific coast, such as an abundance of stalked barnacles (*Pollicipes polymerus*) and ribbed mussels (*Mytilus californianus*), and a zone of the laminarian *Postelsia palmaeformis.* These species do enter the Strait of Juan de Fuca, and some of them extend to the semiexposed coasts round its inner end, but they all disappear wherever there is increased shelter.

Finally, our description of the region agrees well with previous accounts of similar or adjacent areas published by other authors, but extends these and considers the district in relation to the other parts of the world that we have studied. We agree strongly with the opinion that many local variations in populations of barnacles, for example—and the same applies to other organisms—can be fully understood only if one knows the exact history of a particular shore for a short period of years preceding the time of observation.

THE PACIFIC GROVE REGION

Our study of the region round Pacific Grove, California, was carried out in September and October of 1947. Most of our time was spent at Cabrillo Point, on which the Hopkins Marine Station of Stanford University is situated, but we also studied Yankee Point, Point Lobos, Point Cypress, and Mission Point in Carmel Bay (Figure 10.12). These localities cover the whole range of habitats to be found on the rocky shores of the region.

INTRODUCTION

The shores near Pacific Grove and those round Nanaimo belong to different worlds. Instead of quiet borders of inlets and straits, many below wooded slopes and far from the ocean, we now have a fully exposed coastal region of bays and headlands. On the most prominent of these headlands, the wave action is probably as powerful as anywhere in temperate latitudes. Mountainous waves surge in from the Pacific, often terrifying in aspect when seen at close quarters, and especially impressive when they come tearing in from a fog, as they so often do on this coast. When they explode on the rocks and cliffs in cascades of foam, sending solid spouts of spray into the air, they create a turmoil of broken water, sometimes white throughout the whole of a small bay. Even when the surface seems smooth, the swell is often heavy enough to create large breakers between tidemarks.

The topography is far from simple. The coastline is not straight; the major headlands are dissected into a succession of minor prominences separated by small bays and inlets (Figures 10.12 and 10.14). The rocky shore varies from steep, vertical, or overhanging cliffs to gentler slopes; some parts of the shore are protected from the full force of the waves by outlying reefs, others are not. The interaction between rocks and waves produces a whole series of intertidal habitats that differ from one another not only in that one will be subject to maximal wave action while another close by is sheltered, but also in that one will be open to a direct approach by the waves, while another is

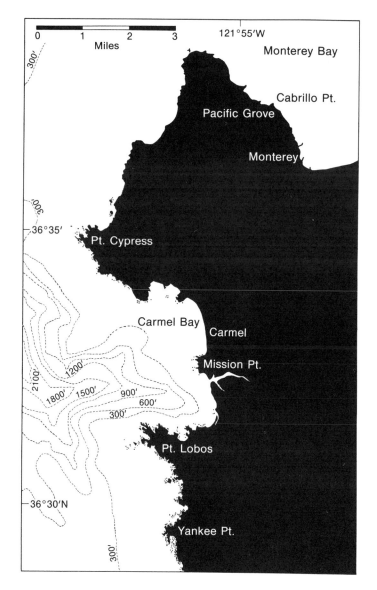

FIGURE **10.12**

A map of the Monterey Peninsula, California, and the coast to the south of it. Submarine contours are shown as broken lines.

swept obliquely by them, or dashed by waves rebounding from the foot of a neighbouring cliff. Consequently, different types of wave action—explosive shock, downward pull, upward surge, sideways sweep, and so on—are to be found in small areas close to one another. In even the most sheltered places, wave action is stronger than at Nanaimo; moreover, wave action in the tidal zone is frequent, and genuine calm unusual, occurring (according to Bolin, 1949) only in autumn. On very rough days, in many areas, work between tidemarks is impossible.

FIGURE **10.13**

Rough seas on a California coast. [Photograph by D. P. Abbott.]

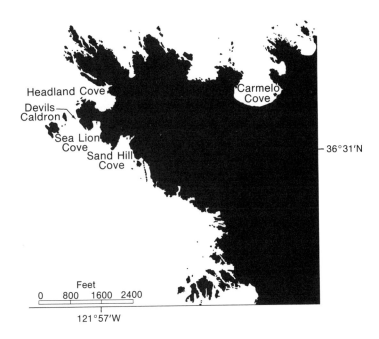

FIGURE **10.14**

A map of Point Lobos, California.

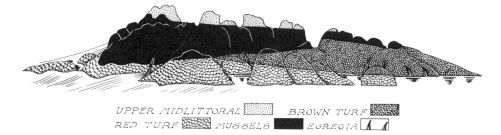

UPPER MIDLITTORAL BROWN TURF
RED TURF MUSSELS EGREGIA

FIGURE **10.15**

Rocky reefs at Cabrillo Point, Pacific Grove, showing the relation between the distribution of certain plants and animals and the amount of wave action. The wave action is strong on the left of the figure and reduced on the right. The mussels become confined to a narrow band, at a higher level, as the shelter increases, and finally disappear. The red turf is more strongly developed on the left, the brown on the right.

Nothing in this region is more impressive than the sensitive reaction of the plants and animals to the action of the waves (Figure 10.15). The population to be found on a wave-swept promontory of Point Lobos is very unlike that of a sheltered inlet at Cabrillo Point (Figures 10.16 and 10.17), but every conceivable variation and intermediate combination of species may be discovered, in one place or another, between the exposed and sheltered populations. Some of these variations are illustrated in Plates 16 and 17 and Figures 10.18 and 10.19.

FIGURE **10.16**

The shore at Point Lobos, California.

FIGURE **10.17**
Rocks at Cabrillo Point, Pacific Grove, showing the lower midlittoral turf on a calm day.

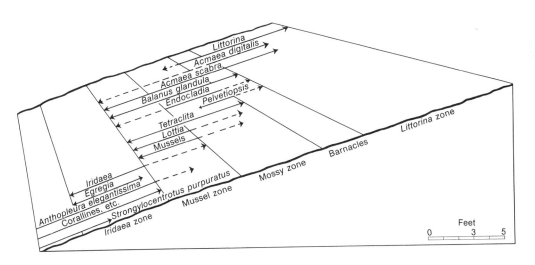

FIGURE **10.18**

The extents of the main populations of several common plants and animals on a smooth slope in Sea Lion Cove, Point Lobos. The thick diagonal line in the foreground represents a measured level section. Vertical and horizontal scales are the same. This diagram supplements Plate 17.

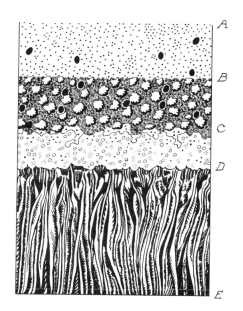

FIGURE 10.19

Part of a vertical rocky face in Sand Hill Cove, Point Lobos, subject to submaximal wave action. No *Postelsia* was present here, and few mussels. *A–B,* zone of *Lottia* (*black ovals*) and *Balanus glandula* (*dots*); *B–C,* zone of *Lottia, Pollicipes,* and a maximal encrustation of *Hildenbrandia* (*dark shading*); *C–D,* zone of continuous lithothamnia with *Tetraclita* (*small circles*) and some *Balanus glandula; D–E,* zone of *Alaria, Egregia,* and their associates.

Our first impression of the shore round Pacific Grove was that we were back in South Africa (see Comparison of Pacific Grove with South Africa, p. 236). Although very few South African species occur in California, there are so many in the one country that recall similar ones in the other, and so many common features in their arrangement, that there is a genuine resemblance. As in South Africa, the shore population is very rich in species, and this richness is no doubt encouraged by two circumstances. The first is that the lowest tides in the Pacific Grove region (which differ little in range and time from those at the Golden Gate, San Francisco) occur before the hottest hours of the day in summer, the time receding steadily from April to September. The second is the frequency of fog along this coast, especially in the morning, which also protects emerged plants and animals from the sun. The powerful wave action of which we have spoken also has a major effect.

Summary of the Zonation

The intertidal zonation of this region shows all the usual widespread features, though modified by local conditions. The supralittoral fringe has a typical population of periwinkles, limpets, and high-level isopods. Although a continuous black zone is rarely seen, irregular patches of blackening are fairly common.

The midlittoral zone is markedly divided, in the more sheltered places, into upper and lower subzones. The upper of these had the usual population of barnacles and limpets, with relatively few, and mostly small, algae. The lower subzone is heavily carpeted by a dark brownish turf composed of a mixture of rather small species, primarily rhodophyceans. In many places, this turf has a sharply defined upper limit. Accompanying the turf are certain large algae. Where wave action is stronger, this turf is replaced by a pink growth of lithothamnia and corallines, or by a growth of other red algae, overgrown, in places, by *Postelsia*.

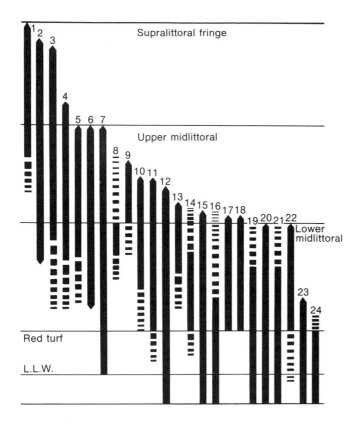

FIGURE **10.20**

The distribution of some common plants and animals, in relation to the zones, at Mussel Point, Pacific Grove: 1, *Littorina planaxis;* 2, *Littorina scutulata;* 3, *Acmaea digitalis;* 4, *Acmaea scabra;* 5, *Balanus glandula;* 6, *Chthamalus;* 7, *Thais emarginata;* 8, *Endocladia;* 9, *Lottia;* 10, *Tetraclita;* 11, *Mytilus;* 12, *Anthopleura elegantissima;* 13, *Acmaea limatula;* 14, *Iridaea;* 15, *Haliotis cracherodii;* 16, *Pisaster ochraceus;* 17, *Acmaea pelta;* 18, *Acmaea scutum;* 19, lithothamnia; 20, *Hildenbrandia;* 21, corallines; 22, brown turf; 23, *Egregia;* 24, red turf. The information applies primarily to areas of moderate shelter, but *Mytilus* and *Pisaster* (more abundant in rougher areas) are included for comparison. The diagram is an approximation, illustrating the relations of one common combination of species, and is not to scale.

The infralittoral fringe is essentially a laminarian zone. In many places, it is unusual in appearance because of the steepness of the rocks, on which the laminarians, many of them very long and trailing, hang like a drapery. The holdfasts of the laminarians may continue into the lower midlittoral, particularly those of *Egregia,* whose long streamers may be seen among the lower midlittoral turf even in sheltered areas.

Some of the offshore islets are inhabited by sea lions, and by such birds as pelicans, cormorants, and gulls. Even at the Hopkins Marine Station, we found the constant barking of the sea lions to be one of the pleasant accompaniments of human existence.

A diagram showing the distribution of some common species in the principal zones at Cabrillo Point is provided in Figure 10.20.

DISTINCTIVE FEATURES OF THE REGION

Limpets

Limpets are found on most of the world's seashores. On some coasts, as on many of those of the North Atlantic, the limpet population of the open rock may comprise very few species, although those species may be represented by vast numbers of individuals. On certain other shores, as on those of California, both species and individuals are numerous. Although limpets as a group are cosmopolitan, the limpet population of one part of the world may consist largely of genera not found in that of another. For example, the limpet population of South Africa is comparable with that of California in its variety and richness, but differs from it completely in composition: The leading genera of the South African shores are *Patella, Helcion, Cellana,* and *Siphonaria; Acmaea* is present, but much less important. In California, on the other hand, *Acmaea* is the dominant genus, and includes at least 17 species (most of them intertidal) whose ranges include the Pacific Grove region; and, with the exception of the species *Siphonaria thersites,* the genus *Siphonaria* (so abundant round the coast of South Africa) appears on the Pacific coast of North America only south of Point Conception. Further, there are no species of *Patella* in California; and *Lottia,* present in California, is entirely absent from South Africa.

Acmaea is in full force at Pacific Grove, and shows a very decided zonation on the shore, much as *Patella* and the others do in South Africa. *Acmaea digitalis* extends higher up the rocks than any other limpet, being characteristic of the supralittoral fringe and upper midlittoral; it is followed by *A. scabra,* which is especially abundant in the second of these belts. Next come *A. pelta, A. limatula, A. scutum* and *A. mitra,* all of them fairly common in the lower midlittoral (Figure 10.20). A small hybrid, *A. pelta* × *A. digitallis,* which superficially resembles a *Siphonaria,* is common throughout the midlittoral. Numerous *Lottia gigantea* occupy a fairly restricted zone in the lower part of the upper midlittoral. Curiously enough, several of these species extend lower down on mussels than they do on the rock itself; and the small, blackish *Acmaea asmi* is habitually found on the shells of the snails *Tegula funebralis* and *T. brunnea.* Detailed studies of the Pacific Grove forms have been made by several authors, but the distribution just described is derived from our own observations.

Turfs of the Lower Levels

The most conspicuous landmark on sheltered rocks near Pacific Grove is the upper limit, sharply marked in many places, of the turf of short seaweeds covering the lower parts of the shore (See Plate 17, *Cabrillo Point*). In the lower midlittoral, this turf is typically very dense, resembling a coarse-piled velvet of a dark, even blackish-brown colour relieved by areas of vivid green. The green may be due either to mosslike cushions of *Cladophora trichotoma,* or to various species of *Ulva* (including *U. lactuca, U. taeniata,* and *U. lobata*). The brown is not due to the presence of phaeophyceans: the turf is composed primarily of rhodophyceans that are blackish or brownish in colour rather than red, in which respect it resembled turfs of similar levels on some of the

South African coasts. This development of rhodophyceans is suggestive of warm-temperate conditions, although the water at Pacific Grove is fairly cold.

But what of the heavy covering of fucoids that might be expected at such a place, a growth so commonly found on the coasts of the North Atlantic and on the more northern coasts of the Pacific? Both the general dearth of fucoids and the relative peculiarity of those forms that are present are most interesting: There is a Pacific coast *Pelvetia* (lacking on the Atlantic coast of North America), and to anyone familiar with European coasts this immediately suggests a short plant forming a restricted belt at a high level. Although there is such a plant at Pacific Grove, it is not the California *Pelvetia,* as one might expect, but *Pelvetiopsis limitata,* a close ecological counterpart of the European *Pelvetia canaliculata.* The California *Pelvetia,* on the other hand, is a large plant (sometimes exceeding 3 feet) reminiscent of the European *Ascophyllum nodosum,* but less coarse; it also resembles *A. nodosum* in its distribution, occurring in variable amounts among the rhodophycean turf in relatively sheltered places, more freely in some kinds of shelter than in others. Where it makes a really dense and dominant growth it overcomes the turf, but there remains an undergrowth of such rhodophyceans as can tolerate shade, interspersed with limpets, barnacles, lithothamnia, and corallines. *Fucus* is present along with the *Pelvetia,* but it is never an important item; apparently, all of it belongs to a single subspecies, *Fucus distichus evanescens.* Another fucoid, *Hesperophycus harveyanus,* may be seen occasionally in the upper fringe of the turf.

The commonest rhodophycean turf-formers, many of them dull red but most brown to blackish, include four species of *Gigartina* (*G. agardhii, G. canaliculata, G. leptorhynchos,* and *G. papillata*), a small form of *Rhodoglossum affine,* the blackish-purplish *Porphyra perforata,* and a short brown mosslike plant, *Endocladia muricata.* Constituents of the turf that are found only locally are *Ceramium eatonianum, Chondria decipiens, Cryptopleura lobulifera, Gastroclonium coulteri, Gelidium coulteri, G. sinicola, Microcladia borealis, Pikea californica, Rhodomela larix,* and various short corallines (including *Corallina gracilis*). Although these are all typical inhabitants of the lower midlittoral, they showed a distinct zonation among themselves. The turf is not necessarily continuous. In some places, it is patchy or reduced, while in others, it alternates with areas of rock inhabited by encrusting algae or animals or both.

In addition to *Pelvetia,* other larger plants occur commonly among the turf, varying independently in development, and forming leafy growths. These are various species of *Iridaea,* which may be dull green to brown or dull red, and may reach a considerable size (up to 2 feet in diameter). The primary species are *I. flaccida* and *I. splendens,* with some *I. coriacea,* and the growth, abundant in many places, is sensitive to changes in physical conditions. Although in sheltered places the *Iridaea* and the lower midlittoral turf are roughly coextensive, the *Iridaea* in exposed places typically forms a belt that is distinct from that of *Endocladia,* because the *Iridaea* retains its usual position while the *Endocladia* is displaced upwards by the increased wave action. A comparable growth of *Iridaea* (sometimes with such similar plants as *Aeodes*) is found on many South African shores, and is indeed a widespread feature of tidal waters.

The turf just described does not continue steadily down to below low-water levels. It appears to be a distinctive feature of the lower midlittoral, and at a certain level is replaced by a different, rather coarser turf, composed on the whole of other species.

TABLE **10.3**

Some algal species characteristic of the lower,
coarser turf of moderately sheltered parts
of the shore in the Pacific Grove region

Agardhiella coulteri	*Gigartina harveyana*
Botryoglossum farlowianum	*Gigartina spinosa*
Callithamnion pikeanum	*Halosaccion glandiforme*
Callophyllis megalocarpa	*Laurencia spectabilis*
Chondria decipiens	*Microcladia borealis*
Cryptopleura lobulifera	*Microcladia coulteri*
Gastroclonium coulteri	*Plocamium violaceum*
Gelidium cartilagineum	*Prionitis andersoniana*
Gigartina boryi	*Prionitis lanceolata*
Gigartina corymbifera	*Rhodymenia pacifica*

Some of the rhodophyceans of this coarser turf are brown or blackish, but most are of various shades of red and pink; they continue, in limited numbers, down among the laminarians and below low-water level. The species of the coarser turf, like those of the turf above, show a distinct zonation among themselves. Table 10.3 lists some of the species characteristic of the coarser turf.

Also present are various corallines, such as *Corallina gracilis, C. chilensis, Bossea dichotoma,* and *Calliarthron setchelliae.* On the whole, corallines are larger at low levels than higher up. They may be purple to scarlet, as well as duller shades. A further complication of the zonation is provided by one of the laminarians, the long, trailing *Egregia menziesii,* which invades the lower part of the lower midlittoral, though its holdfasts are never found attached as high as the top of that zone.

So far, we have been considering the low-level turfs of relatively sheltered places. What grows on those shores exposed to stronger wave action? The lower midlittoral turf, as a whole, disappears or becomes reduced in such places. The *Endocladia,* however, may remain as a scattered growth displaced to higher levels by the splash, extending up as high as the *Pelvetiopsis*—that is, far above the lower midlittoral. But a most interesting and typically Pacific-coast feature supervenes along the uppermost strip of the lower midlittoral and sometimes overlaps into the zone above. This is the appearance of a very distinct and rather narrow zone formed by the peculiarly attractive and characteristically Pacific-coast laminarian *Postelsia palmaeformis* (Figures 10.21 and 10.22). This plant differs in appearance and habit from most other laminarians. Instead of being dark brown, it is a light yellow-brown, darkening with age; and instead of collapsing when out of water, it stands erect, its strong trunklike stalk surmounted by a tuft of palmlike "leaves." It is an annual, flourishing in summer and disappearing before spring, and not only does it live higher up the rocks than any other laminarian, it also withstands continual and tremendous battering from the waves in very exposed positions. In places of maximum exposure, however, even *Postelsia* seemed to find little footing, especially where the slight ledges that favour it are absent.

In places too steep and exposed for a favourable growth of the lower midlittoral turf, the upper strip of the lower midlittoral may have a distinct pink belt (Plates 16 and

FIGURE **10.21**

A typical growth of *Postelsia* on the coast of the Monterey Peninsula, California. [Photograph by J. W. Hedgpeth.]

17), coloured by lithothamnia and corallines, above the dark brown drapery of the laminarian zone, or it may support a population of animals. Elsewhere, the laminarians continue right up to the *Postelsia*. In most places, an undergrowth of lithothamnia, corallines, and noncalcareous rhodophyceans continues down among the laminarians to below low-water level.

Mussels and Stalked Barnacles

Although mussels and barnacles are very widespread on the shores of the world, the prevalent barnacles between tidemarks are of the sessile (acorn) type. The occurence of abundant pedunculate barnacles on the California coast, therefore, is something exceptional. The plentiful stalked barnacle is *Pollicipes polymerus,* and the common Californian mussel is the ribbed *Mytilus californianus,* a rough-water species disappearing in sheltered places. Because these two animals commonly occur together (and the barnacle, like the mussel, typically forms clumps), they are sometimes referred to as the *Mytilus-Pollicipes* community; however, they do not occupy exactly the same zone, though their ranges do overlap, and it is doubtful whether their association is more than fortuitous. The distribution of the large, dense mussel beds of the district (Figure 10.15) is related rather precisely to wave action. In exposed positions, the beds extend

FIGURE **10.22**

The zonation of some common species on a rather steep rocky face exposed to fairly strong wave action (though somewhat protected by other rocks to seaward) at Mission Point in Carmel Bay. Here we see an interrupted belt of *Postelsia*, with barnacles, mussels, *Pollicipes*, *Endocladia*, and *Pelvetiopsis* above it, and a coralline turf with *Iridaea* and *Egregia* below it.

from the lower parts of the upper midlittoral right down to the red turf prevalent in such places, or even as far as the laminarians. Where there is more shelter, the beds become narrower (their upper limit remains in the same place, but their lower limit lies higher up), so that the brown turf of the lower midlittoral is well developed below them. With increasing shelter, they form a narrow band limited chiefly to the lower part of the upper midlittoral, become reduced in number, and finally disappear. In the wide beds, the individual mussels at the lower levels are typically very large, and most of them support sessile barnacles. Despite their tolerance of rough conditions, there seems to be a limit to what the mussels can withstand. In some of the roughest and steepest places, there are no large mussel beds, but merely small groups of stragglers. *Pollicipes* also dies out in really sheltered places. It also extends a little higher up the rocks than do most of the mussels, and is more plentiful among the higher rocks than among the lower ones.

Animals of the Lower Levels

A number of the animals of the lower levels of the shore are worthy of special mention.

Sea Anemones. Three common California anemones, whose taxonomic positions have only recently been established with reasonable accuracy, are spectacular in one way or another, and have attracted considerable attention. Two of the three become very large, while the third is relatively small. This small anemone, *Anthopleura elegantissima*, is remarkable because it occurs in mass formation: it may cover the rock surface in a continuous sheet, so that the rock is hidden and little else can coexist with the anemones. They are most extensively developed in the lower midlittoral in sheltered places, with some extension above and below this level; in places with greater wave action, they may become abundant in the infralittoral fringe as well. The amount of gravel attached to the verrucae (suctorial warts) on their bodies varies. Their tendency to cover themselves with gravel is reduced in wave-washed places, and most marked in sheltered ones (where, in any case, the proximity of gravel seems to encourage the anemones to flourish), probably because they are more exposed to sun and wind, in such regions, when the sea no longer wets them during low water, and the coat of gravel helps to prevent them from drying up. (A master's thesis including experimental notes on this subject was submitted by Bruce J. Roberts to Stanford University in 1941.) The colour variation of this anemone is relatively slight, and its gregarious habit is doubtless the result of its ability to reproduce asexually by fission. It is related to the British *Anthopleura thallia*, which, however, is a rare species.

Of the two large anemones, both of which attain a diameter of 10 inches or more, one is usually green in colour. This is *Anthopleura xanthogrammica*, and although the colours vary from one specimen to another, a common combination is a mossy-green body and oral disc, and dove-grey tentacles with pale tips, but suffused with a brilliant pale green that spreads over them from their bases. The body is heavily warted, but although the warts can and do attach gravel to themselves, many specimens do not have this coating. The animal is not as gregarious as *A. elegantissima*, but it can be surprisingly abundant despite its large size, especially in channels, pools, and crevices at low levels and in exposed positions.

The third anemone, *Anthopleura artemisia*, resembles *A. xanthogrammica*, though the two are distinguishable upon careful examination. The coloration may resemble that of *A. xanthogrammica*, but it is more varied, this anemone often possessing a strongly marked radial pattern on the disc, and spotted tentacles. The lower part of its body is less warty, and the upper warts are commonly covered with gravel. It is a more common anemone, and most individuals live buried in gravel in the lower midlittoral and below in places more sheltered than those characteristic for *A. xanthogrammica*. The habitats of the two species may overlap, however, and many individuals of *A. artemisia* are not buried.

Abalones (*Haliotis*). Except for *Haliotis tuberculata*, which appears on the coast of France and southward, one does not expect to see *Haliotis* on north Atlantic shores. In some parts of South Africa, species of the genus are so common that they may be collected for human food; but even where they are thus plentiful, they are typically concealed and inconspicuous. The Pacific coast of North America is noted for its abalones, the meat of which appears regularly on the menus of California restaurants. In the region of Pacific Grove, the very large *Haliotis rufescens*, with its heavy red shell, was still common at low levels in 1947. This is the chief species used commercially in these latitudes; the commercial catch is taken offshore, rather than in the intertidal, the maximal abundance lying at depths of 20–50 feet. A conspicuous and remarkable feature of the coast round Pacific Grove is the astonishing abundance of the smaller *H. cracherodii*, with a blue-black shell. This animal, which may occur singly on open rock, is most frequently found in large numbers in crevices; because most of these crevices are heavily wave-washed, its range extends from the lowest levels up through the lower midlittoral, and even as high as the upper midlittoral. This species is less often taken by sportsmen than *H. rufescens*, and is of much less commercial importance.

Echinoderms. The common large starfish of Pacific Grove, *Pisaster ochraceus*, was abundant in 1947, and though it exhibited several shades of colour (including gold, orange, brown, lilac, dull purple, and reddish, many of the individuals having conspicuous white spines), it was never of the vivid violet so predominant near Nanaimo. Although these starfish move about freely, so that they are not necessarily found at the same levels at all states of the tide, they are most plentiful during the period of low water in the lower midlittoral, or below the mussel beds in exposed places. Although abundant where strong surf occurs, they are by no means confined to such areas. Another common starfish is *Patiria miniata*, of which some individuals are bright red and others are of dull colours, either plain or patterned (for example, mottled with grey and brown). This starfish is found from the lower midlittoral downward, mainly in places less exposed than those occupied by *Pisaster ochraceus*.

A spectacular feature of this coast is the abundance of two species of sea urchins. The smaller one, the purple *Strongylocentrotus purpuratus*, is astonishingly abundant at the lowest levels, especially in the pools and crevices of wave-washed places. In many places, it inhabits hollows in the richly pink lithothamnia-covered rock, or forms pavements made up of thousands of individuals in pools. In some places, countless numbers of this animal and of *Anthopleura elegantissima* carpet the rock in the infralittoral fringe as far down as one can see into the water. The larger urchin, *S. franciscanus*, a species with blackish-purple or scarlet spines, is locally common at the lowest levels or below low water, but is less abundant than *S. purpuratus*. Some specimens of *S. franciscanus* are as much as 8 inches across.

COMPARISON OF PACIFIC GROVE WITH SOUTH AFRICA

Many times, in the course of this account, we have mentioned similarities between the coasts of California and South Africa. In Table 10.4, a small selection of California species is listed in the first column, and their South African counterparts are named in the second. It should not be assumed that the California species correspond *exactly*

TABLE **10.4**
Some intertidal plants and animals of the Pacific
Grove region of California and their ecological
counterparts in South Africa

California	South Africa
PLANTS	
Endocladia muricata	*Caulacanthus ustulatus*
Gastroclonium coulteri	*Champia lumbricalis*
Gigartina agardhii	*Gelidium pristoides*
Gigartina corymbifera	*Gigartina radula*
Gigartina leptorhynchos	*Grateloupia filicina*
Gigartina spinosa	*Gigartina stiriata*
Nereocystis luetkeana	*Ecklonia maxima*
Phyllospadix scouleri	*Caulerpa ligulata*
*Rhodoglossum affine**	*Gymnogongrus glomeratus*
ANIMALS	
Anthopleura artemisia	*Anthopleura michaelseni*
Dodecaceria pacifica	*Dodecaceria fistulicola*
Eudistylia polymorpha	*Sabellastarte longa*
Mytilus californianus	*Mytilus crenatus*
Tegula funebralis	*Oxystele tigrina*

*The short form of the species.

in appearance or distribution (much less in taxonomic position) to the South African ones; but one member of each pair does recall the other, and they occupy similar ecological niches.

Perhaps, as we are emphasizing similarities, certain contrasts should also be mentioned. For example, there is nothing in California resembling the *Patella cochlear* zone of South Africa, nor does the *Pyura* belt of that country seem to have a California counterpart; and although the beds of *Mytilus crenatus* in South Africa resemble those of *M. californianus* at Pacific Grove, the characteristic barnacle of similar habitats in South Africa is the sessile *Octomeris angulosa,* not a stalked form, such as the *Pollicipes polymerus* of California.

FURTHER NOTES ON ZONATION AND THE DISTRIBUTION OF SPECIES

The intertidal zonation of the Pacific Grove region, briefly outlined on page 228, will be considered here in some detail.

The Supralittoral Fringe

Two periwinkles, *Littorina planaxis* and *L. scutulata,* occur in vast numbers in the supralittoral fringe, both on open rock and in crevices. *L. planaxis,* though commonest here, continues in fair numbers down among the barnacles of the upper midlittoral zone, more in some places than in others. *L. scutulata,* on the other hand—although the species is plentiful in the supralittoral fringe, extending (though sparsely) to fairly high levels—is found in myriads among the barnacles of the upper midlittoral, and may even continue down into the lower midlittoral. Another animal characteristic of the fringe is the rapid isopod *Ligia occidentalis;* and a lively crab, *Pachygrapsus crassipes,* with a dark striped carapace and heavy pale claws, is abundant from here downward over most of the shore. The limpet *Acmaea digitalis* is also characteristic of this zone, in some places extending very high on the shore (though not as high as *Littorina planaxis*), and forming, on some nearly vertical faces, a densely populated band in the lower part of the fringe and along the upper edge of the midlittoral. Another limpet, *Acmaea scabra,* although characteristic of the upper midlittoral, also extends into the fringe, being commoner and going higher in some places than in others.

The algae of the supralittoral fringe are, for the most part, encrusting forms. *Hildenbrandia occidentalis* is a persistent feature of the zone in cracks, but not usually on open rock. (This alga, which forms a dark red crust that may, in some places, be fairly thick, is most abundant on open rock from the top of the lower midlittoral downwards in sheltered areas; but on splashed or wave-washed rock, it may extend upward into or beyond the upper midlittoral.) The irregular areas of blackening in the supralittoral fringe are probably caused by different species in different places: our samples from Cabrillo Point included *Peyssonelia pacifica, Entophysalis granulosa,* and a young green algae (*Prasiola* or *Enteromorpha*), all in various stages of parasitisation by fungi; and our samples from Point Lobos included *Entophysalis granulosa, Calothrix crustacea,* and *Plectonema battersii.* In some places (for example, at Headland Cove on Point Lobos), an irregular but definite zone of blackening develops at high levels, separated from the upper limit of the midlittoral by a gap.

The Midlittoral Zone

The Upper Midlittoral. The barnacle zone of the upper midlittoral has a sharp upper limit in some places, a vague one in others. It is characterised by *Balanus glandula,* in many places large, well developed, and occurring in dense sheets, but elsewhere smaller and more scanty. A smaller barnacle, *Chthamalus fissus,* is scattered (in some places plentifully) among the *Balanus,* but does not extend as high as the latter. A third barnacle, larger than either, is the pinkish-red *Tetraclita squamosa rubescens,* which is at its best in a belt overlapping the junction between the upper and lower midlittoral, favouring cracks, overhangs, and steep faces. It is abundant on some wave-washed promontories, where it has about the same upper limit as the mussels. The prickly barnacle *Balanus cariosus* is also seen occasionally, but is nowhere abundant. Several other animals are characteristic of the upper midlittoral, though not limited to it. These include two small whelks, the spotted *Acanthina lapilloides,* and the blackish or ringed *Thais emarginata,* which extends into more wave-washed places than the former; a short, wide chiton, *Lepidochitona hartwegii,* which occurs even on the open rock; and a black trochid snail, *Tegula funebralis,* which is common both in and out of pools. The two commonest limpets in this zone are *Acmaea scabra* and *A. digitalis.*

Algae are more varied in this zone than in the supralittoral fringe, but fewer and more scanty than in the zones below. The algal growth consists mainly of irregular patches or tufts of some of the species that build up the brown turf of the lower midlittoral. The commonest of these turf algae that extend as high as the upper midlittoral are *Gigartina papillata, Endocladia muricata,* a few green cushions of *Cladophora trichotoma,* sporelings of *Fucus,* and small plants of *Pelvetia, Porphyra,* and *Rhodoglossum affine.* Where *Pelvetiopsis* is present, it is situated mostly high up in the upper midlittoral. In the more wave-washed places, a curious change in distribution affects the *Endocladia.* Although in sheltered areas it occurs chiefly in the uppermost strip of the lower midlittoral, in rougher places it occurs almost entirely in the upper midlittoral, making a decided belt there with about the same upper limit as that of *Pelvetia,* but much wider. The crust of lithothamnia varies according to the amount of wave action: in sheltered places, it is limited to clefts and angles, but where there is strong wave action, it may also cover areas of open rock.

The Lower Midlittoral. Earlier in this chapter, we provided an account of the algal turf so prevalent in this zone (p. 230), and an account of some of its animals (p. 235).

All three of the barnacles that characterise the upper midlittoral extend somewhat into the lower, but their distribution varies greatly from place to place, in accordance with differences in wave action, angle of slope, and other variables. They are in competition with the turf of algae, and they flourish only in clearings in the turf or on surfaces little colonised by it. *Balanus glandula* may be quite plentiful at the top of the zone, but is generally diminished below. *Tetraclita* is common in many places, especially above; and *Chthamalus dalli* may form dense sheets almost to the bottom of the zone. It is interesting that all three of these barnacles are plentiful on the shells of mussels at levels lower than those that they colonised in quantity on the rock itself. This is true also of some of the limpets. In many places in the lower midlittoral, a fourth form, *Tetraclita squamosa* var. *elegans,* is the most characteristic barnacle. It differs quite markedly in appearance from the coarse, reddish *T. s. rubescens,* being smaller, neater, much more markedly ribbed, and brownish to whitish in colour, because its cuticle and outer white layer are commonly not rubbed off to show the red below. It occupies rather different habitats than *T. s. rubescens,* being particularly common in damp and shaded places, even on vertical and overhanging rocks, in the lower midlittoral. On the other hand, this barnacle is certainly connected with *T. s. rubescens* by intermediates.

Tegula funebralis may occur in vast quantities in this zone, both in and out of pools, and particularly in certain small bays. The small black limpet *Acmaea asmi* and the slipper shell *Crepidula adunea* are frequently found attached to the shells of *T. funebralis.* *Thais emarginata* is found throughout the zone, and a small chiton, *Nuttallina californica,* is locally common in its upper part. Other common chitons are *Tonicella lineata, Lepidochitona hartwegii, Mopalia muscosa,* and *Katharina tunicata.* A gregarious tube-making worm, *Dodecaceria fistulicola,* is frequently seen also, becoming abundant in suitable spots, and sponges are common in clefts. A noteworthy snail of low levels is the brown *Tegula brunnea,* with a black and orange foot. Compound ascidians—such as the thick, fleshy, fawn-coloured sheets of *Amaroucium californicum,* and the pale orange *Sigillinaria pulchra* with its small groups of polyps on thick stalks—may be found under overhangs. *Pugettia producta,* a brown or olive-green spider crab, is common in some places among the *Pelvetia.*

The Infralittoral Fringe

Like the other zones of the Pacific Grove region, the infralittoral fringe varies in appearance from place to place. In a sheltered bay with a shelving floor, the growth of laminarians looks very much like an ordinary European laminarian zone; but on steep slopes where wave action is strong, several species of laminarians with long straplike fronds make a spectacular dark curtain of lashing straps against a pink background of lithothamnia, corallines, and other red algae, and (contrary to all expectations for places so wet) the upper limit of this growth may be sharply marked.

Egregia menziesii forms tufts of thongs—many of them 15 feet long—covered with tags, which give them the appearance of dark brown strips of Turkish towelling. It is one of the dominant species in areas of fairly strong to moderate wave action, but it also continues into the sheltered regions, where (if plentiful) it makes a mat of tangled straps when left by the tide. So far as we could determine, it does not extend far below the level of lower low water. A common European straplike alga, *Himanthalia elongata*, occupies a similar ecological niche on the coasts of Britain.

Alaria marginata forms broad straps up to 12 feet long. It occurs with *Egregia* in places of strong to moderate wave action, but does not extend so far into shelter as the latter. In some places, the *Alaria* is restricted to a definite belt, with a rich growth of corallines and other red algae below it.

Lessoniopsis littoralis, another spectacular form, makes dense bushes of dark brown, trailing whips. One plant may comprise as many as 500 of these whips, which attain lengths up to 6 feet. In 1947, this species made dense growths on rocky slopes (some of them very steep) exposed to maximal wave action, and little of it was to be seen far away from these turbulent areas.

From the cliffs, one can see the tips of the giant laminarians *Nereocystis* and *Macrocystis* floating in the water offshore. These algae form great sublittoral beds of very large plants. Some of these giant laminarians grow inshore in places, especially in bays, where they can be seen floating quietly—even behind the turmoil of Point Lobos—and can sometimes be collected from the rocks of the shore. Although *Macrocystis pyrifera* and *M. integrifolia* both occur offshore, the latter appears to be the more coastal species, and is found in channels and on ledges at Cabrillo Point and elsewhere.

A number of other less abundant laminarians are also interesting. *Costaria costata* is a broad five-ribbed strap up to 9 feet long. *Laminaria setchellii* (a form recalling the British *L. digitata*) attains a length of nearly 5 feet and is locally common, typically in exposed places. The broad, ruffled blades of *L. farlowii*, which we saw only in relatively sheltered places, recall those of the British *L. saccharina*.

Also common in the infralittoral fringe is the leathery *Cystoseira osmundacea*, a member of the Sargassaceae whose blades resemble oak leaves. Further, the rock in this zone is very commonly carpeted with lithothamnia, typically of species thicker and more richly coloured than those higher up. These calcareous crusts, along with various corallines, continue up through the lower midlittoral, and even on to the open rock of the upper midlittoral in sufficiently wet places; although in the lower zones the crusts tend to be continuous, they become more patchy higher up. Finally, we must mention the brilliant green eelgrass (*Phyllospadix*), a flowering plant, belonging to the family

Zosteraceae, with long whiplike leaves. It occurs from the lower midlittoral downward, and is abundant both in sheltered places and in some of the rougher ones. The undergrowth of fairly short red algae in the infralittoral fringe has been described earlier in this chapter (pp. 231-232).

On some steep slopes in this area, one can see to advantage the transition in the population of a given zone from the end of the slope subjected to the strongest wave action to the other, much more sheltered end. The infralittoral fringe, for example, passes from a dominance of *Lessoniopsis* in the roughest water, through one of *Alaria* and *Egregia* in the intermediate parts, to a turf of mixed small algae accompanied by *Egregia* at its sheltered end. Although the upper limit of the fringe, in the rough places, is clearly enough marked by the effective upper limit of laminarians (other than *Postelsia*), it is much less distinctly perceptible in the most sheltered regions, possibly corresponding to the transition from the brown midlittoral turf to the red turf below it.

Some of the animals of the fringe have been mentioned earlier (see Animals of the Lower Levels, p. 235). A noteworthy barnacle is the large *Balanus nubilis*, locally common on mussels and rocks. We found specimens of this barnacle that were nearly two inches across the base, but saw none as large as the five-inch giants of Nanaimo, although five-inch specimens are common on the wharf pilings at Monterey, according to Bolim (1949). Other common animals are sponges, polyzoans (*Costazia incrassata*, for example), and compound ascidians, such as the thick, lilac-grey sheets of *Eudistoma psammion*.

RELATION OF THE ZONES TO TIDAL LEVELS

We attempted to obtain some idea of the relation of the zones to tidal levels by erecting a temporary tide gauge in a small, sheltered, gravelly bay behind Cabrillo Point. It was situated in front of the window of the laboratory in which we worked, and could be read with the help of field glasses at any state of the tide. Because no convenient pier or wharf was available to which the gauge might be attached, the graduated scale was built into a heavy metal frame to anchor it in the sea. Although the whole frame could have shifted under strong wave action, it apparently did not do so during our visit: we painted numbered marks on the rocks close to it, so that the levels on the gauge could be checked against something stable. Although the site was relatively sheltered, the sea surface was often far from flat (the waves sometimes showed an amplitude of 2 feet or more), so that the degree of precision with which the gauge could be read was not always very close. Nonetheless, readings taken over a whole tidal period on 14 October 1947, during the early part of which the sea was by no means calm, yielded so smooth a curve that our confidence in the approximate accuracy of our observations was increased.

The port of reference for the tides at Pacific Grove is San Francisco (Golden Gate). The tides near Pacific Grove follow the San Francisco pattern closely, with small local differences for which corrections can be made. One low water in each day is commonly much lower than the other; the two high waters differ also, but to a lesser degree. The highest predicted tide for 1947, corrected for Monterey, was 6.2 feet above datum, and the lowest was 1.8 feet below it; the tidal range for that year, therefore, was 8.0 feet. During our visit, unfortunately, we did not experience the whole of this range: the highest tides rose to 5.6 feet above datum, and the lowest fell to 0.9 feet below. Therefore, we did not see the infralittoral fringe exposed to the greatest possible extent, but this was offset by the fact that the waves often withdraw so low on the outer slopes that we were able to see far down the rocks before the next crest overwhelmed them.

We do not propose to present our tidal observations in detail, but the following points seem worth noting, and many of them are included in the Pacific Grove column in the tidal diagram, Figure 10.23 (see also Chapter 2, pp. 17–19). They all apply to Cabrillo Point, and all predicted tidal levels referred to have been corrected for Monterey.

1) We fixed the average upper limit of the lower midlittoral turf in sheltered places at about 3.5 feet above datum. This figure agrees with the result arrived at by Hewatt (1937) for the heavy growth of *Pelvetia* and *Fucus* present in the early 1930s, and coincides with line *4* of Figure 10.23 (mean minimal lower high water), which also lies at 3.5 feet.

2) The upper limit of barnacles in quantity in sheltered places lay near 5.8 feet (Hewatt put it nearer to 5.0 feet, probably over a more limited area). This seems to be related to line *1* of Figure 10.23 (mean maximal higher high water), which lies at 5.6 feet.

3) The transition from the brown turf of the lower midlittoral in sheltered places to the red turf below it lay just below 1.0 feet, which is very near to lines *6* and *7* of Figure 10.23 (mean maximal higher low water and mean minimal lower low water), which lie at 1.1 and 0.8 feet, respectively.

4) The average depth of the upper midlittoral in sheltered areas worked out at about 4.7 feet (3.5–5.8 feet), and that of the lower midlittoral in sheltered areas, at about 2.2 feet (0.9–3.5 feet). The upper limit of *Egregia* holdfasts in such places lay near 2.3 feet. In rougher areas, on the other hand, even the mussels extended above 6.3 feet, and *Balanus glandula* extended much higher: the barnacles thus continued above the average level of the highest tides (line *1* in Figure 10.23, 5.6 feet), but were wetted, nonetheless, as much as they needed.

One of the puzzles of the zonation problem is that, although one expects (and indeed finds) a relation between the boundaries of zones and tidal levels in sheltered places, this relation is usually considered to be a function of the amount of time that the rock, at any particular level, is submerged in the sea or left uncovered at the mercy of sun and wind. But in such a region as Pacific Grove, one finds places exposed to very strong wave action, where the rock on many days is always wet throughout the tidal belt, and often above it, even at low water; and yet the zonation there is as well marked as it is anywhere else, and may even be unusually neat. Several examples of this are given in this chapter. Plate 16 shows a steep slope flanking the Devil's Cauldron at Point Lobos where, even at low water in moderate weather, the waves surge up to and above the *Endocladia* zone, and the splash goes far higher, the whole supralittoral fringe being thoroughly wetted. Plate 17 shows a slightly less exposed area at Mission Point where

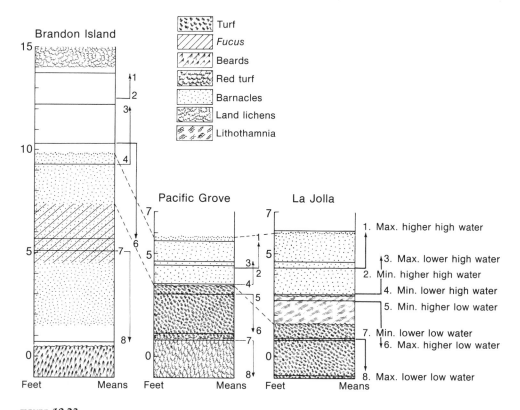

Turf
Fucus
Beards
Red turf
Barnacles
Land lichens
Lithothamnia

1. Max. higher high water
3. Max. lower high water
2. Min. higher high water
4. Min. lower high water
5. Min. higher low water
7. Min. lower low water
6. Max. higher low water
8. Max. lower low water

FIGURE **10.23**

The relationship between some zone boundaries and the eight mean tide levels at three points on the Pacific coast. The information is based both on direct measurements made on the shore and averages worked out from tide tables. For Brandon Island, the Sand Heads predictions were considered approximately applicable. For Pacific Grove and La Jolla, corrections were made on the predictions for San Francisco and San Diego. Zone boundaries believed to be comparable are connected by *broken diagonals* between the columns. The locality at Brandon Island was a very sheltered one with a wide tidal range. Those at Pacific Grove and La Jolla were moderately sheltered with smaller tidal ranges.

much the same applies, the waves washing with great force, even when the tide is low on a calm day with heavy swells, at least as high as the *Postelsia*. Figure 10.19 shows the zonation on a steep face in Sand Hill Cove (in a position rather more sheltered than the Devil's Cauldron) where even the heavy swells of a smooth sea move up and down over the rock ceaselessly, so that most of the intertidal rock is wet at any state of the tide; yet the zonation at Sand Hill Cove is unusually sharp. Finally, Plate 17 and Figure 10.18 illustrate a slope facing Sea Lion Cove, at Point Lobos, on which, again, the zones are very neatly delimited. Although the wave shock here is less than that on the rocky faces previously mentioned, the waves sweep sideways over the slope and, even at low water, often wash over the rock up to the supralittoral fringe.

THE LA JOLLA REGION

Our study of the rocky coasts of the region round La Jolla, California, was carried out between November of 1947 and January of 1948. We studied seven stretches of rocky beach between a point north of the pier of the Scripps Institution of Oceanography and the reef south of La Jolla known as Bird Rock (Figure 10.24).

INTRODUCTION

The coast near La Jolla is unlike those near Pacific Grove and Nanaimo (see Figure 3.5, p. 37). It is open to the Pacific, with low cliffs above sandy or rocky beaches, but the wave action in ordinary weather is not as powerful as that of the Pacific Grove region. Although large waves develop that are strong enough to cause heavy explosions and massive spray against the more exposed cliffs, they are less impressive than the colossal turmoil of the surf at such a place as the Devil's Cauldron on Point Lobos.

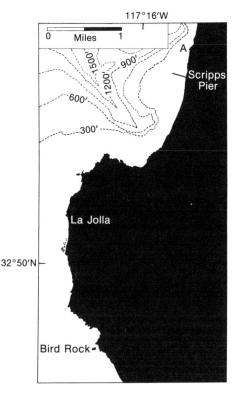

FIGURE **10.24**

Map of the La Jolla district, California. Submarine contours are shown as broken lines. Lat. 32°50′N and long. 117°16′W intersect in La Jolla. The localities described in the text lie between the point *A* in the north and Bird Rock in the south.

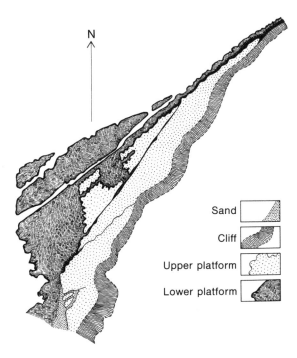

FIGURE **10.25**

A rocky area at La Jolla, California, showing the typical arrangement of the intertidal rocks in the form of two broken platforms—an upper one and a lower one—below a cliff.

As in the Pacific Grove region, a transition in the population of a given zone occurs between those rocks on which the wave action is strongest and those on which it is weakest.

The configuration and structure of the seashore rocks are particularly interesting and important in this area, because they affect the occurrence of some of the organisms quite markedly. There is a general tendency towards the following arrangement, realised in slightly different ways on different beaches: First, there is a steep or overhanging cliff with a rocky platform at its foot that is followed, to seaward, by a second, lower rocky platform (Figure 10.25 and Plate 19). Both platforms are much broken and dissected, and complicated in shape by headlands, fissures, and inlets; their seaward edges may be precipitous or overhanging. The upper platform may be reduced to isolated ridges, stacks, or low eminences; where this reduction is most marked, there appear to be two structural features only, the cliff and the lower platform. The lower platform is submerged even by the minimal high and low waters, whereas the highest parts of the upper platform are not completely submerged even at the higher high waters, though they are commonly wetted completely by surge or spray. The foot of the cliff may be submerged at high water, and the face of the cliff, where it is most exposed to the waves, may be thoroughly drenched by spray. Apart from the main sandy beaches, which are typically free from rocks, there may be small stretches of sand or pebbles behind the reef at the foot of the cliff. Boulders are plentiful, either scattered on the reef, in pools, and in creeks, or in definite concentrations, perhaps at the foot of the cliff or sloping shore, or at the seaward edge of the rocks.

The rock colonised by the intertidal organisms is of a curious type. Most of it is sandstone, and of this, the greater part is relatively soft. At its softest, one can rub down the surface of the rock with a fingernail or remove it easily with a knife. But from one small area to another, the consistency varies, and in some places, there are tracts of sandstone that are much harder. In many places, small concretions, boulders, or eminences of hard rock are inset in the softer sandstone, so that it is possible to make a direct comparison of the population of one of the hard insets with that of the surrounding softer rock at the same level. The soft rock, being too friable, forms a poor surface for the attachment of barnacles; but at the same time, it is porous and absorbs sea water, so that it retains moisture in its superficial layers when the tide has receded, which enables lithothamnia to encrust it to levels at which they would otherwise die from desiccation. Conversely, the hard rock, which dries off readily and discourages lithothamnia at such levels, is heavily colonised by barnacles. That these facts have a curious effect on the zonation will become apparent in the following pages. Apart from the sandstone, there are seams and areas of shale, which also vary in hardness. In some places, the soft shale is riddled by boring bivalves, such as *Petricola californiensis* and *Pholadidea penita*. Conglomerate also occurs locally.

Although many plants and animals prevalent at Pacific Grove occur also at La Jolla, certain southern species, absent or less conspicuous at Pacific Grove, have become important here, and certain northern ones have disappeared. The effects of these changes, taken together with those of the alternation of harder and softer rock and those of the less violent wave action, produce a picture decidedly different from that seen farther north. Here again the California coast bears a considerable resemblance to South Africa, but whereas the Pacific Grove region recalls the colder parts of the South African coast, the La Jolla region is strongly reminiscent of the neighbourhood of East London on the southeastern coast of South Africa (a warm-temperate area that includes rock formations—Bat's Cave Rocks—not unlike those at La Jolla). Here, as at San Francisco, the lowest tides of summer occur in the early morning. Coastal fog is not uncommon at La Jolla, especially in autumn and winter, but is less prevalent than at Pacific Grove.

During our visit, there were days of unusually cold winds, with frost inland. On December 11 and 31, 1947, the minimum air temperatures at Lindburgh Field, San Diego, were 39° and 38°F, respectively, and it is more than likely that the temperature on the shore, in certain areas, fell to near freezing on these and perhaps other nights. From December 11 onwards, we could see the effect of the cold on the algae of the lower midlittoral turf. Many of the algae were injured or killed, and the effects were marked by a change of colour. When healthy, the algae varied from brown through blackish and dark reds to dull greyish pinks, but as they were dying, they became dull purple, lurid pink, greenish white, and finally pure white. So many of the algae died at that time that the turf appeared to have been whitened with a light fall of snow. The algae most affected were the short, high-level corallines of the inshore turf, and such noncalcareous species as *Laurencia pacifica*, *Centroceras clavulatum*, *Pterocladia pyramidale,* and *Gigartina canaliculata.* Larger forms, such as *Egregia* and *Phyllospadix,* were somewhat affected, but more of them were damaged than killed. This makes an interesting demonstration of the risks run by algae in a warm-temperate area occasionally subjected to colder conditions.

When examining the biota of a California shore, one must bear in mind that holiday-makers and week-end fishermen frequent many parts of the coast, and that many of them collect, either for food or amusement, various of the plants and animals. Should a shortage of something occur, it is difficult to know whether it is natural or artificial. Collecting of intertidal organisms is very common at La Jolla, but because people's interests are very restricted, it seems to change the general picture very little. Collecting does, however, decidedly reduce the numbers of *Haliotis* at La Jolla.

Summary of the Zonation

The zonation at La Jolla is normal for a warm-temperate region, except insofar as it is modified by the nature of the rock (Figure 10.26 and Plate 18). The supralittoral fringe has the usual population of periwinkles, limpets, and high-level isopods. The black zone, discontinuous in some places, is heavily marked in others.

The midlittoral zone is unusually complex. The lower midlittoral is normal—as at Pacific Grove, it is carpeted by a brownish turf of rather small algae, typically with a sharp upper limit. In the upper midlittoral, however, the barnacles are so patchily developed that they are much less conspicuous than they are on most shores; this tends to make the limpets (especially *Acmaea digitalis*) unusually conspicuous. Moreover, there

CHTHAMALUS ☐ TURF ▨ LITHOTHAMNIA ▦

ACMAEA DIGITALIS ⦙∴⦙ PHYLLOSPADIX ▨ HALIDRYS, ETC. ▟

FIGURE **10.26**

The zonation of a rocky area south of La Jolla and a little north of Bird Rock. This figure supplements Plate 18.

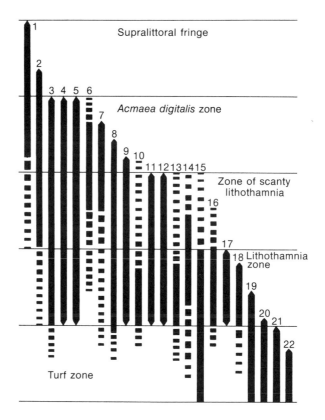

FIGURE **10.27**

The distribution of some common species in relation to the zones in the La Jolla district: 1, *Littorina planaxis;* 2, *Acmaea digitalis;* 3, *Chthamalus;* 4, *Acmaea scabra;* 5, *Acmaea conus;* 6, *Scytosiphon;* 7, *Endocladia;* 8, *Petrospongium;* 9, *Lottia;* 10, *Nuttallina;* 11, *Balanus glandula;* 12, *Pollicipes,* 13, *Tetraclita squamosa rubescens;* 14, *Mytilus;* 15, lithothamnia; 16, *Anthopleura elegantissima;* 17, *Spiroglyphus;* 18, *Acmaea limatula;* 19, *Phragmatopoma;* 20, *Pseudochama;* 21, turf; 22, *Halidrys.* The diagram refers to the La Jolla district in general, though exceptions to it are found in specific areas, especially where the lithothamnia zone is suppressed.

is a belt immediately above the lower midlittoral turf in which, on the soft rock (but not on the hard), a thin crust of lithothamnia was more or less continuous—a very unusual feature. Above this, near the middle of the zone, there is a belt of intermittent lithothamnia.

The infralittoral fringe is populated by a continuation of the algal turf of the lower midlittoral (with a change in the constituent species), among which larger plants (especially *Egregia, Eisenia, Halidrys,* and *Phyllospadix*) are common. This constitutes a laminarian zone, but a weak or shadowy one, compared with that of Pacific Grove or with those of cold-temperate regions in general (Figure 10.27).

DISTINCTIVE FEATURES OF THE REGION

Limpets

The genus *Acmaea* is very strongly represented at La Jolla, as at Pacific Grove, and its species show a similar zonation. Except for *A. scutum* and *A. mitra*, all of the *Acmaea* species that we found at Pacific Grove (see p. 230) we found also at La Jolla. *A. pelta* is by no means as common, however, and is generally smaller and less typical in appearance. *A. conus*, a small southern species with a dark centre and a spotted edge and resembling a *Siphonaria*, is not found at Pacific Grove, but is abundant in the upper midlittoral at La Jolla. *Lottia gigantea* and the hybrid *A. pelta* × *A. digitalis* are both common.

Turfs of the Lower Levels

A dense turf of short algae, primarily rhodophyceans, is as characteristic of La Jolla as it is of Pacific Grove, and the turfs of both places have similar general features (Figure 10.26 and Plate 18). At La Jolla, the turf typically covers the lower platforms extensively and has a sharply marked upper limit. It occupies the lower midlittoral and the infralittoral fringe; its colour is different at different places, according to the level and the amount of surge and splash. Its higher and more inshore parts tend to be a pale, dingy, pinkish brown, owing to a high proportion of short, dull-coloured corallines and a considerable amount of silt. Lower down, patches of olive, dark brown, blackish, and reddish prevail (Plate 19). Lower still, and in splashed seaward positions (such as along the steep edges of lower platforms) the prevailing colours are dark or bright reds and purples, due in part to vividly coloured corallines. These turfs comprise a great many species, some of the commonest of which are listed in Table 10.5. Although many of these species are found in both the lower midlittoral and the infralittoral fringe, some (including *Centroceras clavulatum*, *Gigartina leptorhynchos*, and some of the corallines) appear to be more characteristic of the former, and others (including *Gigartina eatoniana*, *G. armata*, *Rhodymenia californica*, *R. palmettiformis*, *Nienburgia andersoniana*, *Laurencia diegoensis*, *Gelidium cartilagineum*, and *G. pulchrum*), more characteristic of the latter. There is no doubt that these turf-forming species show a subsidiary zonation among themselves.

Some of the species tend to produce different forms at different levels. For example, *Laurencia pacifica* is a prevalent species in the higher turf, where it typically forms low, dark-coloured cushions. It is also common lower down, where it tends to grow larger, with branches like slender bunches of grapes. In some inshore positions, the turf is much thinner and finer than elsewhere, being similar in texture to velvet: it consists of very small species, or of dwarf plants of generally larger ones (including *Laurencia scrippsensis*, *Polysiphonia hendryi*, *P. simplex*, *Gelidium coulteri*, *Centroceras clavulatum*, and *Corallina* sp.).

Several larger species of algae can be found among the turf. Most of these belong primarily to the infralittoral fringe, but tend to penetrate the lower midlittoral in many places; they are described in the section on the infralittoral fringe (p. 259). One of

TABLE 10.5

Some of the commonest constituents of the low-level algal turfs of the La Jolla region

Chlorophyceans	Rhodophyceans (contd.)
Ulva californica	NONCALCAREOUS SPECIES (contd.)
	Gelidium cartilagineum var.
Phaeophyceans	*robustum*
Colpomenia sinuosa	*Gelidium coulteri*
Dictyopteris zonarioides	*Gelidium pulchrum*
Dictyota binghamiae	*Gigartina armata*
Dictyota flabellata	*Gigartina canaliculata*
Endarachne binghamiae	*Gigartina eatoniana*
	Gigartina leptorhynchos
Rhodophyceans	*Gigartina serrata*
	Gracilariopsis andersonii
CORALLINES	*Laurencia diegoensis*
Bossea orbignyana	*Laurencia pacifica*
Corallina chilensis	*Nienburgia andersoniana*
Corallina gracilis	*Plocamium pacificum*
Corallina gracillis f. *densa*	*Pterocladia pyramidale*
Corallina sp.	*Rhodoglossum affine*
Jania capillacea	*Rhodymenia californica*
Jania mexicana	*Rhodymenia palmettiformis*
Lithothrix aspergillum	*Ricardia saccata*
	Zanardinula linearis
NONCALCAREOUS SPECIES	
Centroceras clavulatum	
Chondria californica	
Cryptopleura crispa	

them, *Pelvetia fastigiata,* is characteristic of the upper parts of the lower midlittoral. It is found locally; in some places—especially on crests or other high places, typically inshore in somewhat sheltered positions among the upper turf—it is plentiful and vigorous. Where it is especially dense, it has a characteristic undergrowth of lithothamnia, short corallines, and small noncalcareous rhodophyceans (including *Pterocladia pyramidale, Gigartina canaliculata,* and *Laurencia pacifica*) that is inhabited by such animals as the spider crab *Pugettia producta,* the keyhole limpet *Fissurella volcano,* the cemented bivalve *Pseudochama exogyra,* the anemone *Anthopleura artemisia,* the vermetid *Aletes squamigerus,* and the barnacle *Tetraclita elegans.* Another alga often seen among the turf was *Codium fragile,* with its thick, clubbed, green branches.

Acorn barnacles

The distribution of acorn barnacles in the La Jolla region is quite unusual. In the upper midlittoral, which is the main habitat for these animals, the prevalent species are the small *Chthamalus fissus* and *Balanus glandula,* and the large red *Tetraclita squamosa rubescens.* The distribution of these species is strongly affected by the hardness of the rock: on the softest rock, few if any barnacles were found; on the slightly firmer (but still soft) rock that makes up most of the reefs, their distribution is erratic and patchy—in some places, there are none, but in others, there are dense patches of *Chthamalus*. Although the patchiness is doubtless correlated with the hardness of the rock, it is also affected by other local variations in the environment. On the hard rock, there is an immediate change. *T. s. rubescens* becomes plentiful (especially just above the turf), and *C. fissus* and *B. glandula* flourish also.

Although *Balanus glandula* is common enough near La Jolla, and is found in dense formation in some places, it is never a conspicuous and dominant feature of the zonation, as it is at Pacific Grove; the leading upper midlittoral barnacle (apart from *Tetraclita*) is *Chthamalus fissus*. *Tetraclita squamosa rubescens* (described on p. 239 in the section on Pacific Grove) is common in suitable habitats on the La Jolla reefs. A larger barnacle than any of those yet mentioned, *Balanus tintinabulum californicus* is much less plentiful than the others, but not uncommon in seaward positions at low levels.

Lithothamnia

The lithothamnia zone at La Jolla is unusual, and we can recall no exact parallel to it in other parts of the world. On most shores, the crust of lithothamnia ceases to make a continuous coating on the open rock somewhere in the lower midlittoral (which means, at La Jolla, in the upper part of the turf); above this, it is found only discontinuously on open rock; at the higher levels, it is restricted to pools, damp clefts, and similar special habitats. At La Jolla, however, there is a belt situated immediately above the turf (thus occupying the lowest strip of the upper midlittoral) in which a thin greyish-pink crust of lithothamnia covers the open rock more or less continuously (see Plates 18 and 19 and Figures 10.26 and 10.27). This growth is able to survive at such a high level purely because of the porous nature of the surface of the soft sandstone, which can retain moisture for a long time while the rock is exposed to the air. On the harder rock, which retains less moisture, the coating of lithothamnia is unable to develop, or is developed only slightly, unless some local circumstance makes a particular patch of rock damper than the rest. It is replaced on hard rock by a dense growth of *Tetraclita* and other organisms, typically including *Pterospongium*, *Anthopleura elegantissima*, *Balanus glandula*, *Chthamalus*, *Acmaea*, *Mytilus*, *Pollicipes*, *Lottia*, and *Phragmatopoma*. There are, of course, transitional populations on rocks of intermediate hardness. The shale, even though soft, does not seem to retain superficial moisture as well as the soft sandstone, and lithothamnia cannot colonise it extensively.

The population of the lithothamnia belt is interesting. Knowledge of the lithothamnia themselves is generally lacking, so that it is not now possible to say how many species formed the encrusting growth, but it may be that *Lithophyllum decipiens* is primarily responsible for it. In any case, the lithothamnia belt is typically a thinner, greyer crust than the more highly coloured lithothamnia crusts of lower levels. In the lower part of the lithothamnia belt, there may be a considerable growth of short, dull-coloured corallines (including *Corallina gracilis* and *C. gracilis* f. *densa*). A very different characteristic alga is the fleshy, brainlike, brown or blackish *Petrospongium rugosum*. An exceptional feature is the presence, in some parts of the lower lithothamnia belt, of innumerable closely set boat-shaped pits in the rock, each occupied by a small chiton, usually *Nuttallina fluxa.* Higher up, the pits become fewer. Elsewhere, this chiton is common but does not inhabit pits. It is often found on unpitted rock, both hard and soft.

Mussels and Stalked Barnacles

As at Pacific Grove, *Mytilus californianus* and *Pollicipes polymerus* are plentiful. The mussels, however, are by no means as impressively abundant as they are farther north, and *Mytilus edulis* is often present among the *M. californianus.* The zonation of these forms is similar to that at Pacific Grove, and the mussels and *Pollicipes* varied independently; in other words, one is favoured particularly by one type of situation, while the other responds especially well to another—in spite of the fact that they usually occur together—and the maximal development of the *Pollicipes* is a little higher on the shore than that of the mussels. The mussels in this district do not form vast sheets as they do at Cabrillo Point (see Figure 10.15), but flourish in small groups on rocky surfaces particularly favourable to them. In some places, they are abundant on hard rock at levels that would be occupied by the lithothamnia belt, were the rock softer; and they are characteristic of the lithothamnia belt also, extending above it on a more limited scale. They also commonly form caps on boulders (Figure 10.28).

Animals of the Lower Levels

A number of animals of the lower levels are worthy of special mention.

Sea Anemones. It was interesting to find that two of the common *Anthopleura* species of Pacific Grove were again abundant near La Jolla. The small, gregarious *A. elegantissima,* as usual, forms gravel-covered sheets of various sizes, or occurs in

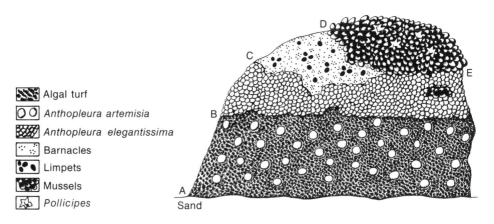

Algal turf
Anthopleura artemisia
Anthopleura elegantissima
Barnacles
Limpets
Mussels
Pollicipes

FIGURE **10.28**

A boulder from the area north of Scripps pier, La Jolla. *A–B,* a crust of *Pseudochama* and other organisms overgrown by algal turf, with large specimens of *Anthopleura artemisia* in the hollows; *B–C,* a dense sheet of *Anthopleura elegantissima; C–D,* an area of barnacles and limpets; *D–E,* mussels and *Pollicipes.*

more scattered formation, its region of maximal abundance being the lower part of the upper midlittoral. It is particularly conspicuous where the rocks run down to the sand and the wave action is less than maximal. The larger *A. artemisia* is more conspicuous than at Pacific Grove, and is particularly plentiful in small hollows in the turf or buried in (and coated with) sand or gravel. In some places, it forms a fairly regular mosaic in the turf on the sides of boulders (Figure 10.28), though it is never as dense as *A. elegantissima*. It also produces some striking colour varieties, including one with a pattern of violet and brilliant green rays on the disc, with the tentacles shading to violet distally. In dark places, or when they have perhaps been recently buried in sand, some individuals show very pallid shades. During our visit to La Jolla, we did not see a single specimen of the giant green anemone (*Anthopleura xanthogrammica*) that occurs by the thousands at Pacific Grove.

Abalones (*Haliotis*). It was impossible, in 1947, to estimate the natural abundance of these animals between tidemarks. On the whole, they are more accessible to collectors here than they are at Pacific Grove, and many of the boulders under which they live, in low-level pools or shallows, are turned over by collectors and left upside down. *Haliotis fulgens* and *H. corrugata* were nonetheless common in 1947—the former was surprisingly so, in spite of the human interference. In past times, the populations were no doubt much greater. *H. corrugata* is still commercially important, but it is most plentiful about 20–80 feet below tidemarks. *H. fulgens* was more important to the fishery in the past than it is now.

Echinoderms. The large starfish *Pisaster ochraceus* occurs in countless thousands near Pacific Grove. We were told by biologists at the Scripps Institution of Oceanography that it is common near La Jolla also, and always to be seen on the piles of the Scripps pier. During our visit, however, we found only one specimen on the rocks of the shore; because we looked for *P. ochraceous* over a period of ten weeks and found no further specimens, and because the species is far too conspicuous to be overlooked accidentally, we are convinced that it was virtually absent from the rocky intertidal region round La Jolla from November of 1947 until January of the following year.

The purple urchin *Strongylocentrotus purpuratus*, although locally common in small cavities from the lithothamnia belt downwards, is by no means as plentiful as it is near Pacific Grove. On the other hand, small specimens of the starfish *Patiria miniata* are fairly common, and so are the larger prickly *Astrometis sertulifera* (many of them mottled with grey and reddish) and small examples of *Linckia columbiae*, all of which are found most frequently under boulders at low levels. This habitat, as usual, also favours the ophiuroids, many of which (including *Ophioderma panamensis*, *Ophioplocus esmarki*, *Ophiothrix spiculata*, and *Ophiopteris papillosa*) are plentiful. Associated with the ophiuroids, in many places, is the stout, warted holothurian *Parastichopus parvimensis*.

The Cryptofauna under Boulders. In addition to some of the echinoderms mentioned in the preceeding paragraph, a number of other animals live under boulders that are situated with their bases under water in pools or shallows. This fauna is particularly rich at La Jolla, as it is on many shores, and Table 10.6 gives some idea of this variety, though the echinoderms already named have been omitted.

TABLE 10.6
Some common constituents of the cryptofauna under boulders in the La Jolla region

Molluscs

Aletes squamigerus—a vermetid gastropod with large tubes
Astrea (*Pomaulex*) *undosa*—a large turbinid gastropod with a curious operculum that is white, pointed at one end, and strongly marked with three raised ridges
Crepidula lingulata—a small, flat, irregular slipper shell
Fissurella volcano—a small keyhole limpet
Haliotis corrugata—the pink abalone
Haliotis fulgens—the green abalone
Ischnochiton conspicuus—a large chiton, typically pink and grey.
Megathura crenulata—a very large fissurellid with the shell nearly covered by the black mantle
Tegula ligulata—a spotted or chequered trochid snail
Aplysia californica (also known as *Tethys californica*)—the black sea hare (it may be brown or mottled when young), a tectibranch that may attain a length of 12 inches at maturity (also often seen in pools)

Crustaceans

Herbstia parvifrons—a small spider crab
Hippolysmata californica—a highly coloured shrimp
Paraxanthias taylori—a small crab
Pelia tumida—a small spider crab
Petrolisthes eriomerus—a small, flat anomuran crab
Tetraclita squamosa elegans—a barnacle, rather unlike its relative *T. s. rubescens*

Sponges

Leucosolenia macleayi—a small, creamy-white, stalked species
Verongia thiona—a bright yellow species (also common under overhangs)
Encrusting sponges of various species

Ascidians

Euherdmania claviformis—a highly gregarious, translucent, stalked species, reminiscent of the European *Clavelina*
Metandrocarpa michaelseni—a small, orange-red species, occurring in sheets resembling those of small zoanthids, it is perhaps the most common animal in this habitat

Apparent Shortages

The distribution of whelks in the La Jolla region is distinctly odd. Although one species or another is common nearly everywhere in the Pacific Grove region, the picture at La Jolla is quite different: the small *Acanthina spirata* is plentiful north of Scripps Pier, but apparently nowhere else; *A. paucilirata* is locally common under landward boulders; but in many areas, there seem to be few if any whelks, even among mussels, where they might be expected to be plentiful. It is possible, for example, to spend 20 minutes removing mussels from the rock and separating them from one another and—although this procedure would produce crowds of small crabs, chitons, and polychaets—to find only a single small whelk, such as *Pterorytis nuttalli*. The local shortage of whelks might possibly be due to human interference.

We saw no *Fucus* in the district (its southern limit is believed to lie farther north, in San Luis Obispo County) and no *Postelsia*. A very striking difference from Pacific Grove was that, although the shores there were heavily draped by species of *Iridaea* in September and October of 1947, in the following months, not a single plant belonging to this genus, large or small, could be found near La Jolla. The occurrence of *Ulva*, too, is different. Near Pacific Grove, both large and dwarf forms are abundant; near La Jolla, a dwarf cabbagelike species half an inch long (*U. californica*) is common among the turf, but the only forms larger than this are 2–3 inches long, and are seen only locally on inshore rocks and boulders. This restriction of *Ulva* may be seasonal, but that of *Iridaea* seems less likely to be so.

COMPARISON OF LA JOLLA WITH SOUTH AFRICA

A resemblance between the La Jolla region and the South African coast near East London has already been mentioned (p. 246). This resemblance is due not only to the fact that some of the La Jolla species have approximate ecological counterparts near East London, but also to the extraordinarily close resemblance between the population of the infralittoral fringe in these two areas. Near East London, the fringe has a turf of short algae, primarily rhodophyceans, among which *Hypnea spicifera* (brilliant green in colour) is conspicuous among the red species; there are also areas of another bright green plant, *Caulerpa ligulata*. Among these shorter plants, there are some larger brown algae, notably the small prickly laminarian *Ecklonia radiata*. It is a low-level turf of more or less short algae with a reduced growth of small laminarians—not a full-fledged laminarian zone, but rather the shadow of one. Near La Jolla, there is a strikingly similar arrangement: Here, the turf lacks a green rhodophycean; it is mostly red, but has vivid green areas of *Phyllospadix*, recalling those of *Caulerpa ligulata* at East London. Moreover, it has the small laminarian *Eisenia arborea*, which forms, like the very similar *Ecklonia radiata* at East London, the merest indication of a laminarian zone. On the other hand, East London has a conspicuous development of large, gregarious, simple ascidians in certain parts of the fringe, and this community apparently has no counterpart at La Jolla.

FURTHER NOTES ON ZONATION AND THE DISTRIBUTION OF SPECIES

The Supralittoral Fringe

The supralittoral fringe is well developed, and extends very high on heavily splashed cliffs. The characteristic littorinids are *Littorina planaxis* and *L. scutulata*. The former extends to higher levels than any other mollusc, occurring on open rock as well as in pools, hollows, and crevices. *L. scutulata* is not uncommon, but seems to be less plentiful than *L. planaxis,* and less plentiful, too, than it was at Pacific Grove. Both species extend into the upper midlittoral, though more plentifully into its higher parts above the lithothamnia zone. *Acmaea digitalis,* which continues higher than most of the *Chthamalus fissus,* is common in the lower part of the fringe. *Ligia occidentalis* is plentiful. The black zone is well marked, in most places, and is either patchy or fairly continuous. It is best developed in the lower part of the zone and overlaps into the upper midlittoral; the blackening is due mainly to such algae as *Microcoleus tenerrimus, Calothrix scopulorum, Entophysalis granulosa,* and *E. conferta.* In pools, there is a limited selection of other algae, including black patches of *Isactis plana, Enteromorpha prolifera,* and the furlike *Polysiphonia simplex.*

The Midlittoral Zone

The Upper Midlittoral.   The upper midlittoral zone at La Jolla is divisible into two parts, the *Acmaea digitalis* zone and the lithothomnia zone.

1) The *Acmaea digitalis* zone is the upper part of the midlittoral above the belt of continuous lithothamnia. In its lower part, a crust of lithothamnia occurs discontinuously on open rock in some places, but higher up it is limited to special habitats, such as pools and crevices. We call this belt the *Acmaea digitalis* zone simply because it contains the main population of *Acmaea digitalis.* This species is incredibly abundant, in some places, and is conspicuous wherever the barnacles are scanty or lacking, and where there is no clearly visible upper limit to the midlittoral (in such places, *A. digitalis* continues into the zone above; see Figure 10.29). It also congregates particularly densely in certain places at high levels—under slight overhangs, for example (Figure 10.30). On the harder rock, where barnacles become more abundant, *A. digitalis* becomes correspondingly less noticeable. (There is an interesting parallel here with the northern parts of the west coast of South Africa, where barnacles are reduced on open rock surfaces and a common high-level limpet, *Patella granularis,* becomes unusually conspicuous. Thus, the *Patella granularis* zone of South Africa is comparable to the *Acmaea digitalis* zone of La Jolla.) Other limpets of this zone are *A. conus, A. scabra, A. pelta,* and the hybrid *A. digitalis* × *A. pelta;* none of them extended as high up as *A. digitalis,* however. *Lottia gigantea* is also common, but first appears lower down than the others, and occupies a narrower zone. The crab *Pachygrapsus crassipes* is plentiful in this belt, as well as above and below it. The small, smooth, broad chiton *Lepidochitona hartwegii* is often found in this zone (and in the lithothamnia zone), even on the open rock; another small chiton, *Nuttallina fluxa,* also occurs, but does not extend as high up as *L. hartwegii.* Although the algae of this region are not numerous, they are most interesting. The small, brown or blackish, mosslike *Endocladia muricata* (an important species near Pacific Grove) occurs widely in this zone, but ceases a little below the *Chthamalus.* It extends downwards into the top of the lower midlittoral turf but, so far as we were able to determine, is much less important at these lower levels than it is near Pacific Grove. Another alga widespread in the zone—becoming maximal at different levels in different areas, and particularly characteristic of steep slopes—is *Scytosiphon lomentaria,* with its tufts of brown filaments. A local feature that becomes very marked on some well-splashed cliffs (perhaps especially in shaded areas) is a broad band darkened by brown encrusting algae, notably the thin *Ralfsia californica* and the thicker, more fleshy *Petrospongium rugosum.*

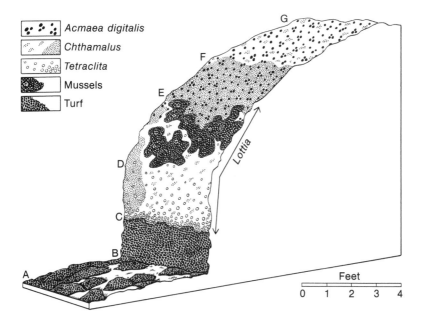

FIGURE **10.29**

A block diagram, based on a measured level section, of a strip of hard rock situated north of Scripps pier, La Jolla. *A-B*, a gentle slope of sand with turf-covered boulders and various algae; *B-C*, the upper part of the zone of algal turf; *C-D*, a zone of *Tetraclita*, with both dense and sparse growths of *Chthamalus*; *D-E*, the zone of maximal growth of *Mytilus*, with both dense and sparse growths of *Chthamalus*; *E-F*, a zone of dense *Chthamalus* with *Acmaea digitalis*; *F-G*, a zone of *Acmaea digitalis* with sparse *Chthamalus*. This area contrasts with the one illustrated in Figure 10.30 in that it has no lithothamnia zone and *Tetraclita* and *Chthamalus* are much more abundant. Vertical and horizontal scales are the same.

2) The lithothamnia zone and its population have already been described, in part, on page 251. Where there is a heavy growth of corallines both in its lower part and in the upper part of the turf below, the boundary between the two (very sharp, in most places) becomes blurred. Several animals are plentiful in the zone. The gregarious *Anthopleura elegantissima* tends to be sparse or only moderately dense where the lithothamnia is well developed, but it forms very solid sheets in some sheltered areas, and sometimes makes a conspicuous band on the boulders just above the turf (see Figure 10.28); it occurs both on hard and soft rock. *Acmaea digitalis*, *A. conus*, and *A. scabra* are still common in this zone. The small colonial vermetid gastropod *Spiroglyphus lituellus*, which makes groups of tubes that recall those of polychaets and resemble those of its South African counterpart *Vermetus corallinaceus*, is also common at these levels. Where it is most abundant, it forms extensive sheets from turf-level to the top of the lithothamnia zone both on hard and soft rock, making a narrow band just above the turf, in some places. Another vermetid with larger tubes, *Aletes squamigerus*, is also common here, and is even commoner below this zone. The red barnacle *Tetraclita squamosa rubescens* is typically sparse on the soft rock. Boulders at the level of the

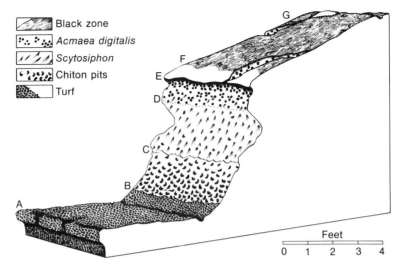

FIGURE **10.30**

A block diagram, based on a measured level section, of a strip of soft sandstone north of Bird Rock, La Jolla. *A–B*, a zone of algal turf; *B–C*, a zone of continuous lithothamnia with chiton pits; *C–D*, bare rock with patchy lithothamnia, a few chiton pits, and *Scytosiphon; D–E*, the zone with the greatest concentration of *Acmaea digitalis; E–F*, the marginal part of upper slope (bare rock); *F–G*, the main upper slope, blackened by myxophyceans, with *Acmaea digitalis* in clefts. Although *Chthamalus* was in fact reduced in this area, the strip was a "potential" *Chthamalus* zone right to the top: harder rock in this locality would have been colonised heavily by this barnacle. Vertical and horizontal scales are the same.

lithothamnia zone or of the upper turf are commonly well populated by *Acmaea limatula; Pachygrapsus crassipes, Fissurella volcano, Petrolisthes eriomerus,* and the spotted snail *Tegula ligulata* are common beneath such boulders, though they are not confined to this habitat.

The Lower Midlittoral. The turfs of this zone have been described on page 249. Beneath them is an extraordinary abundance of chamid bivalves, probably *Pseudochama exogyra*. In some places, they form a continuous crust over the rock, with the turf growing on their shells. Those we collected were *Pseudochama exogyra*, but we cannot certify that this was the only species represented. Snails commonly seen include the large turbinid *Astrea undosa* (especially in pools) and the magnificently coloured trochid *Norrisia norrisii*, whose chestnut-and-black shell has a vivid green area underneath, and whose body is black and scarlet. Another trochid, *Tegula funebralis*, can be found on and between boulders among the turf, and in pools both here and higher up; it is plentiful locally. A third trochid, the spotted *T. ligulata*, is found particularly in pools and under stones. *Acmaea limatula* is locally common in the upper turf zone and just above it, especially in clearings and on boulders. *Anthopleura artemisia* occurs freely in pools and hollows (some of them among the crust of *Pseudochama*); in some places, such small hollows are occupied by two large chitons, *Ischnochiton conspicuus* (commonly pink and grey) and the hairy *Mopalia muscosa*. The polychaet *Dodecaceria fistulicola*, which makes masses of small tubes, may be seen occasionally at low levels. Although animals are plentiful enough among the turf, especially at higher levels, the algae tend to dominate at the lower levels, and animals are not as abundant there as they might otherwise be, except under overhangs or under boulders, where the turf is suppressed. Thus, the lower limit of some animal may, as a rule, lie where the turf begins; but where the turf is absent, the animal may continue to a lower level.

The Infralittoral Fringe

The chamid bivalves described in the preceeding paragraph may continue down into the turf of the infralittoral fringe. The turf, however, has been described on page 249, so we may now refer to the larger plants characteristic of the fringe. Most of the larger plants of the upper part of the fringe extend more or less into the lower midlittoral. Consequently, while the lower plants are perpetually awash or wetted at low water of the lower tides, the upper ones are exposed to the air for a time—although, owing to the splash, they do not necessarily dry out. The growth of fair-sized brown algae (both laminarians and fucoids) makes only a weak "laminarian zone." There are, however, strong beds of very large, truly sublittoral laminarians (such as *Macrocystis* and *Pelagophycus*) offshore, but these have nothing to do with the tidal zone. Some species that occur both intertidally and below low water become larger when permanently submerged.

Phyllospadix scouleri, an eelgrass (a flowering plant of the family Zosteraceae), with long, narrow, bright green, straplike leaves, is abundant in seaward positions, including pools and creeks. In many places, it covers large areas and may extend inshore, in smaller amounts, in suitable situations (see Plate 19 and Figure 10.26).

Eisenia arborea is a small laminarian with serrated "crêpe-rubber" blades. A southern species found only south of Point Conception, it is characteristic of the infralittoral fringe, and common there. *Egregia laevigata* is a more southerly species than the *E. menziesii* so prevalent at Pacific Grove. It is a much longer laminarian than *Eisenia*, and forms tagged straps as long as 15 feet. It was common among the other low-level brown species.

Halidrys dioica and *Cystoseira osmundacea* are both coarse, brown fucoids with blades shaped rather like oak leaves. They occur at low levels along with *Eisenia*, the *Halidrys* being generally more abundant. In some places, there is a short, coarse turf consisting of the holdfasts and basal parts of *Halidrys* plants whose distal portions have been torn away by waves. *Sargassum agardhianum* is a medium-sized species with small blades and slender spikes of small bladders. On some reefs, it is common on the tops of low-level boulders and among the turf, both low down among the sea grass and higher up in more seaward places. It is also commonly reduced to short tufts like those of *Halidrys dioica*.

Zonaria farlowii is a smaller brown alga belonging to the Dictyotales. It has flat, fan-shaped blades, and is common in channels and in pools at various levels. It also occurs out of water among the turf, in some places.

Many of the animals of the infralittoral fringe have been mentioned in the section Animals of Lower Levels (p. 252). They congregate particularly under boulders and overhangs, although some inhabit the rock or turf between the larger plants. The sabellarian worm *Phragmatopoma californica,* which builds masses of coarse sandy tubes, is common in this zone, and extends also into the lower midlittoral and the lithothamnia zone of the upper midlittoral.

RELATION OF THE ZONES TO TIDAL LEVELS

It was not feasible for us to use a tide gauge at La Jolla as we had done at Nanaimo and Pacific Grove, so we had to rely upon direct observation of the behaviour of the tides in relation to the shore as a whole, to measured traverses, and to the tide tables for San Diego.

San Diego is the reference station for La Jolla, and the tidal range there for 1947 was from 7.8 feet above datum to 2.0 feet below it, that is, a range of 9.8 feet. The ratio between the tidal range at La Jolla and that at San Diego is given in the tables as 0.9 feet, so the range for La Jolla for 1947 was roughly 8.8 feet. The difference between the mean of the higher high waters and the mean of the lower low waters is given as 5.8 feet for San Diego, and 5.2 feet for La Jolla.

So far as we could determine, the following are valid relations between the zonation of the La Jolla region and the tidal levels predicted for San Diego (corrected for La Jolla):

1) In less exposed areas, the upper limit of the lower midlittoral turf lies between 1.0 and 2.0 feet above datum, and that of the lithothamnia zone near 3.0 feet above datum. The upper limits of the *Chthamalus* zone and supralittoral fringe are more difficult to fix, but can hardly be below 6.0 and 9.0 feet above datum, respectively, and may be even higher.

2) On cliffs exposed to strong wave action, all of the upper limits are higher: that of the turf zone, between 3.0 and 4.0 feet above datum; that of the lithothamnia zone, near 6.0 feet above datum; that of the locally occurring dark belt of *Ralfsia* and *Petrospongium,* near 13.0 feet above datum; and that of *Littorina planaxis* in quantity, between 17.0 and 18.0 feet above datum.

3) On fairly exposed faces of hard rock lacking a lithothamnia zone, the upper limit of the turf is unusually low down (near 1.0 foot above datum); the upper limit of mussels in quantity (abundant here in a rather restricted belt) is about 6.0 feet above datum, and that of *Chthamalus* is near 9.0 feet above datum.

4) In Figure 10.23, the upper boundaries of the zones in less exposed places are related to average tide levels corrected for La Jolla from the San Diego tables. From this, it can be seen that the upper limit of the turf lies above lines *6* and *7* (representing the means of the maximal higher low waters and minimal lower low waters, which almost coincide). The upper limit of the lithothamnia zone lies near line *4* (mean minimal lower high water); the upper limit of *Chthamalus* is always above lines *2* and *3* (which lie close together), but seems typically to vary from near line *1* (mean maximal higher high water) to well above it.

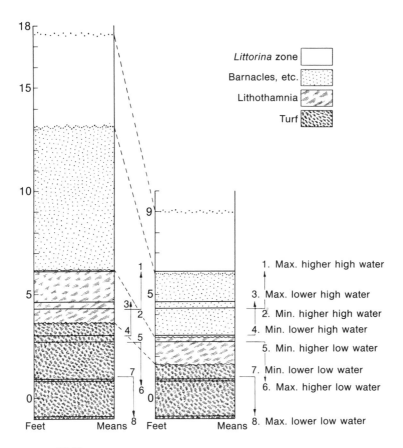

FIGURE 10.31

A comparison of the zonation of two rocky areas at La Jolla: *left*, a cliffed face subject to very strong wave action; *right*, a more gradual slope exposed to less wave action. Although the zones present are essentially the same in both columns, the upper limits are much higher up in the first column than in the second. Consequently, there is a much closer relationship between upper limits and tidal means (numbered *1–8*) in the second column than in the first. The upper midlittoral (*stippled*) in the first column was more conspicuously populated by *Ralfsia* and *Petrospongium* than by barnacles. Data are from direct measurements on the shore, correlated with calculations from tidal predictions.

5) On cliffs exposed to strong wave action (Figure 10.31), everything is higher up: both the dark zone of *Ralfsia* (where present) and the population of *Littorina planaxis* extend far above line *1*, the mean of the maximal higher high waters (the upper limit of *Littorina* in quantity reaches to nearly 17.7 feet above datum, more than 10 feet above the highest high water of the year). Under these conditions, there is no longer any close relation between zone boundaries and tide levels.

6) At those low waters that recede the least, the turf remains submerged, except in those places where the wave action is sufficient to keep it constantly wet. Such low waters may lie just below, or may approximately coincide with, the upper limit of the lithothamnia zone where it is narrower; where the lithothamnia zone is wider (that is,

where its upper limit is higher), such low waters are usually sufficient to keep it wet, except in unusually calm weather. At such low tides, the upper *Chthamalus* are emerged and dry in the quieter areas. The minimal high waters have a relation to the zones that is similar to that of the minimal low waters: the means for both lie near 3.0 feet above datum.

CHILE

A detailed ecological study of the intertidal region of the coast of Chile from Arica to Chiloé Island was undertaken by Guiler (1959a, 1959b), who visited the country as a member of an expedition in 1955. The following account is based on his work.

Access to the coast is by no means easy, and in southern areas, a stout vessel of some sort is an essential. The turbulence of the seas frequently makes studies of the lower zones extremely hazardous. Therefore, it is not possible to give a detailed analysis of the biota, but enough information has been obtained to enable broad outlines to be drawn and useful comparisons to be made.

INTRODUCTION

The Chilean coastline is about 2600 miles long, extending from Cape Horn (lat. 55°58'S) in the subantarctic to Arica (lat. 18°20'S) in the tropics (Figure 10.32). The air temperatures at the coast are slightly warmer than the sea temperatures, and the difference increases from 0°C near Valdivia to about 6°C near Arica. North of Chiloé Island, the chain of offshore islands ceases, and the coast becomes fully exposed, with rocky headlands, open beaches, and fewer and fewer sheltered bays. At about lat. 30°S, desert shores appear, and these continue to Arica near the border of Peru. The continental shelf is very narrow, and the wave action is exceptionally heavy, backed as it is by prevailing onshore winds.

Although the long coast of Chile extends well into the tropics, cool coastal currents, together with intense upwelling, reduce the temperatures of all coastal waters. Indeed, cold surface water comes closer to the equator along the west coast of South America than along any other continental mass in the southern hemisphere. Although Arica, the most northerly Chilean station, is well within the tropics, there are no mangrove swamps, nor do there appear to be any other tropical elements in the biota.

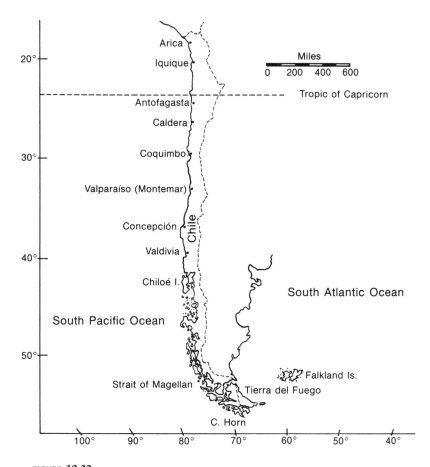

FIGURE **10.32**

A map of the coast of Chile, showing the position of the stations referred to in the text.

The cold Humboldt Current (a northerly branch of the West Wind Drift), reinforced by upwellings of subsurface water rich in nutrients, reaches the South American shore near lat. 38°S; it then passes north along the coasts of Chile and Peru to about lat. 4°S, where it turns west-northwest across the Pacific Ocean towards the Galápagos Islands.

Sometimes, for various reasons, the southeast trade winds that blow on to the west coast of South America may fail, and be replaced by a north wind that brings torrential rains and a flow of warm water of low salinity from the South Equatorial Current. This phenomenon—*El Niño*, as it is called locally—has a catastrophic effect on fish and intertidal life. It occurs most frequently in Peru, but in exceptionally severe cases, it can extend well into northern Chile.

Westerly winds—shifting a little north or south in winter and summer seasons—

prevail south of lat. 30°S. North of this latitude to the Peruvian boundary, the coast is dominated by cool, southerly winds. A southwesterly sea breeze is a coastal peculiarity that can whip up short, steep waves on top of the usual oceanic swell and produce such confused and violent seas that many coastal activities are curtailed. This extreme wave action is a major factor controlling the width of the intertidal zones.

The tides are semidiurnal. The highest high waters occur at noon and midnight, which, combined with fog and cool seas, reduces desiccation of the intertidal life in the summer, and protects it from cold in the winter.

Air temperatures, particularly in the north, are reduced by persistent cold winds along the coast. On hot shores, fog is very common, protecting intertidal organisms from the sun. The absence of rain on desert shores, however, has its own effects on the distribution of intertidal biota.

Sea temperatures are controlled by the prevailing cold currents from the south, which effectively inhibit the spread of tropical elements southwards. These currents also produce the rather startling result that the mean sea temperatures between Arica and Valparaíso show a difference of only about 4°C in summer and 5°C in winter—trivial difference on a shoreline of this length (Figure 10.33).

Therefore, because of the winds, currents, and temperatures peculiar to this area, there emerge only two main regions, which overlap near Chiloé Island: a warm-temperate region in the northern and central parts, and a cold-temperate region in the south, which becomes almost subantarctic at lower latitudes.

THE COAST FROM ARICA TO MONTEMAR

Guiler (1959a) gave detailed accounts of the zonation at the following stations: Arica (lat. 18°20'S), Iquique (lat. 20°15'S), Antofagasta (lat. 23°40'S), Coquimbo (lat. 29°59'S), and Montemar, a beach near Valparaíso (lat. 32°57'S).

Figure 10.34, with its explanatory legend, should give an adequate overall idea of the arrangement of the common species of this coast, whose zonation is exceptionally uniform. As there is a definite limit to the number of symbols that can be intelligible in a text figure, only the main zone-forming elements of the biota are represented; other animals that are widely distributed but do not form definite belts (including various crabs and gastropods) are listed in Table 10.7.

It should be noted that zonation at Antofagasta is exceptional when compared with that of the other stations along the coast. The abundance there of the ascidian *Pyura chilensis,* which swamps all other species from the midlittoral to low-water level (see Figure 10.35), would seem to indicate that conditions at Antofagasta are optimal for this animal's growth. Elsewhere, from Peru to Chiloé Island, *P. chilensis* is recorded only as scattered specimens in the infralittoral fringe. Excessive growth of *Pyura* is also a feature of both the south coast of South Africa and the coast of New South Wales in Australia. Figure 10.33 shows that the maximum sea temperatures at Antofagasta are higher than elsewhere on the Chilean coast; members of the genus *Pyura,* for the most part, inhabit warmer seas.

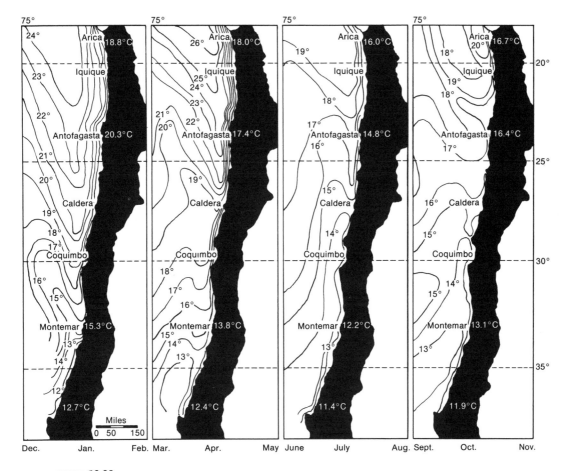

FIGURE **10.33**

Mean seasonal surface-water temperatures around the Chilean coast. Values are averages for the years 1945–1955. [Adapted from Chile, Departamento de Navegacion y Hidrografia de la Armada (1959, fig. 18).]

The Supralittoral Fringe

Except at Iquique, which has a definite littorina belt, the supralittoral fringe consists of bare rock; lichens, however, may grow among the maritime plants at its upper limit. In some places, the barrenness of the fringe may be due to the trampling of numerous sea birds and to the physical and chemical effects of their guano, and in others, to the violence of the sea.

The Midlittoral Zone

The biota of the midlittoral zone is plentiful, and both the number of species and the number of individuals increase to seawards. The barnacle *Chthamalus cirratus* is ubiquitous. Mixed with it, in the upper midlittoral, there is usually a belt of *Littorina peruviana,* whose exact relation to *Chthamalus* varies with local conditions, as littorinids

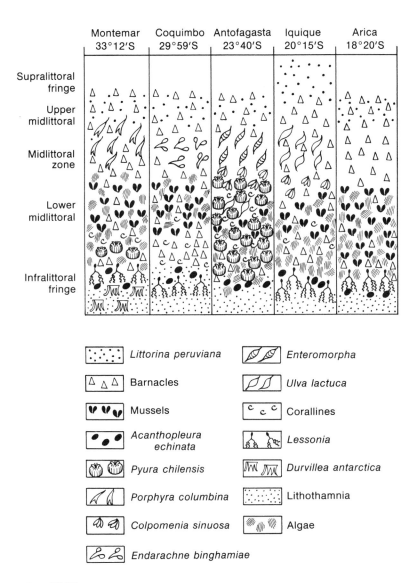

	Montemar 33°12′S	Coquimbo 29°59′S	Antofagasta 23°40′S	Iquique 20°15′S	Arica 18°20′S
Supralittoral fringe					
Upper midlittoral					
Midlittoral zone					
Lower midlittoral					
Infralittoral fringe					

Littorina peruviana   Enteromorpha

Barnacles   Ulva lactuca

Mussels   Corallines

Acanthopleura echinata   Lessonia

Pyura chilensis   Durvillea antarctica

Porphyra columbina   Lithothamnia

Colpomenia sinuosa   Algae

Endarachne binghamiae

FIGURE **10.34**

A comparison of some of the principal species that characterise the zones at the various stations studied on the Chilean coast. The vertical scale has little significance, because the actual widths of zones are variable from place to place.

prefer shaded and damp situations. Lower down, *Balanus laevis* and *B. flosculus* appear. They may mingle with, or replace, *Chthamalus*. At Montemar, for example, *B. laevis* replaces *Chthamalus*, but at other places, *B. flosculus* becomes dominant at lower levels, and can extend into the infralittoral fringe. Another barnacle, the large, conspicuous *B. psittacus*, is recorded from the infralittoral fringe of all stations, and at Coquimbo, it forms a definite belt (see Figure 10.36).

TABLE **10.7**
Some common intertidal species of the coast of Chile between Arica and Montemar

Characteristic species of the midlittoral zone	Characteristic species of the lower midlittoral and the infralittoral fringe (contd.)
ANIMALS	ANIMALS (contd.)
Brachyodontes granulatus	*Fissurella limbata*
Chaetopleura peruviana	*Leptograpsus variegatus*
Chiton cumingsi	*Pachycheles grossimanus*
Chiton granosus	*Petrolisthes* spp.
Chiton latus	*Pilumnoides perlatus*
Littorina auracana	*Pseudonereis galapagoensis*
Nacella magellanica	*Sargartia chilensis*
Siphonaria laeviuscula	*Stichaster striatus*
Siphonaria lessoni	*Tegula atra*
TURF-FORMING ALGAE	*Tetrapygus niger*
Centroceras clavulatum	*Thais chocolata*
Colpomenia sinuosa	*Turbo niger*
Gelidium filicinum	
Iridaea laminarioides	ALGAE
Characteristic species of the lower midlittoral and the infralittoral fringe	*Chaetomorpha linum*
	Codium dimorphium
ANIMALS	*Ectocarpus* sp.
Acanthocyclus gayi	*Gigartina lessonia*
Concholepas concholepas	*Glossophora kunthii*
Enoplochiton niger	*Halopteris hordacea*
Fissurella concinna	*Padina* sp.
Fissurella costata	*Plocamium pacificum*
Fissurella crassa	*Rhodymenia* sp.

All stations, with the exception of Antofagasta, have a definite belt of mussels reaching well into the lower midlittoral. It is usually composed of two species—*Brachyodontes purpuratus* and the smaller, edible *Mytilus chorus*—both of which prefer moderately sheltered places. At Antofagasta, most of the mussels are replaced by *Pyura chilensis*, with an overgrowth of corallines and *Ulva*; only a few small beds of *B. purpuratus* persist.

Belts of algae are conspicuous in the midlittoral. At Montemar, usually below the littorinids, there is a well-developed band of *Porphyra columbina* that is conspicuous in spring and summer. At Coquimbo, *Endarachne binghamiae* forms a strip just below the middle of the *Chthamalus* belt; in the same position at Antofagasta, a species of *Enteromorpha* is very common, yielding to *Colpomenia* lower down. At Primeras Rocas, near Iquique, a place of normal exposure for this wild coast, *Ulva lactuca* grows in considerable quantities among barnacles, and two species of limpets, *Acmaea viridula* and *Collisella arausana*, are sufficiently numerous to form a belt. The starfish *Heliaster helianthus* is common. Except at Arica, *Corallina chilensis* has been recorded from the lower midlittoral to the infralittoral fringe at every station. Other common inhabitants of the midlittoral zone are listed in Table 10.7.

FIGURE **10.35**

The belt of *Pyura chilensis* at Antofagasta, Chile. There is no *Lessonia nigrescens.* [Photograph by E. R. Guiler.]

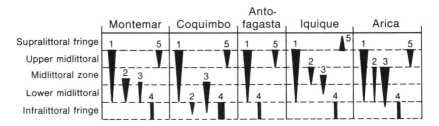

FIGURE **10.36**

The distribution of four species of barnacles and one littorinid in the zones at the stations studied by Guiler on the Chilean coast: 1, *Chthamalus cirratus;* 2, *Balanus laevis;* 3, *Balanus flosculus;* 4, *Balanus psittacus;* 5, *Littorina peruviana.* Only at Iquique does *L. peruviana* form a distinct belt in the supralittoral fringe; elsewhere, it is mixed with *C. cirratus* in the upper midlittoral. *C. cirratus* appears at all stations, ranging throughout the midlittoral, being most numerous in the upper part of the zone. *B. laevis* and *B. flosculus* have a distribution within the middle and lower midlittoral, with interesting variations in their exact levels, and are present at all stations except Antofagasta, where they seem to have been "ousted" by *Pyura.* *B. psittacus,* the largest species on this coast, is ubiquitous in the infralittoral fringe, but only at Coquimbo does it form a recognisable belt.

The Infralittoral Fringe

In certain places, owing to weather conditions, Guiler's investigations of the infralittoral fringe were much restricted; this was particularly true at Arica and Coquimbo.

The most characteristic alga of the infralittoral fringe is the dark brown *Lessonia nigrescens* with its many branched holdfasts, in many places almost forming a boundary line for the top of the fringe (Figure 10.37). It is recorded from all stations except Antofagasta. At Montemar, the southernmost station, scattered specimens of *Durvillea antarctica* appear, and in more sheltered positions, beds of *Macrocystis pyrifera* (Figure 10.38). Both species become enormously abundant in the Cape Horn region. There is a dense mass of algae in the upper part of the fringe, the most typical constituents of which, along with the associated fauna, are listed in Table 10.7. At every station, a crust of lithothamnia forms a pink background.

FIGURE **10.37**

The shore south of Iquique, Chile. Dried salt spray whitens the upper rocks; in the middle distance, barnacles cover the upper midlittoral zone; and the rocks in the foreground are overgrown with lithothamnia, below which there is a belt of *Lessonia nigrescens*. [Photograph by E. R. Guiler.]

FIGURE **10.38**

A view across dense beds of *Macrocystis pyrifera* at Ventanas, Chile, a little north of Montemar. [Photograph by E. R. Guiler.]

The most obvious isolated animal that extends below this turf of algae at Montemar is the large chiton *Acanthochiton echinata*. As at Antofagasta, however, *Pyura* dominates. Another notable feature of the infralittoral fringe at Montemar is the quantity of the brilliantly coloured anemone *Phymactis clematis*, reminiscent of the carpets of *Bunodactis reynaudi* found on the coasts of the Cape Peninsula in South Africa. Also plentiful are the anemone *Sargartia chilensis* on barnacles and the edible urchin *Loxechinus albus*. Sponges are poorly represented, and are usually found attached to the fronds of seaweed.

THE COAST SOUTH OF CHILOÉ ISLAND

From Chiloé Island southwards, the coast of Chile becomes more and more precipitous and wild (see Figure 10.39). In these regions, the weather is almost always bad— particularly so in the summer—and the winds blow almost continually from west to east. The mountains along the shore rise to 9000 feet, and great glaciers pour down their ravines into the sea. In good shelter, the antarctic beech trees (*Nothofagus*) can grow to 200 feet tall. Elsewhere, the terrific winds flatten all the vegetation, and trees crawl along the ground forming layers of horizontal trunks. Above these grow dwarf cypress trees so small and thick that they form a "turf" resembling long heather.

FIGURE **10.39**

The algal mat of the infralittoral fringe, with *Durvillea antarctica* and *Lessonia nigrescens,* at Chepu on the west coast of Chiloé Island, Chile. [Photograph by E. R. Guiler.]

Between this coastline and the open sea lie a string of offshore islands. Most of the islands are high enough to give shelter from the storms outside, and thus provide protected water for inshore navigation (Figure 10.40). The thick kelp on the rocks, which grows down to 40 feet or so below water, is so strong that a small vessel can be moored to it safely.

The following account is based on unpublished ecological notes made by Knox (1960).

The Supralittoral Fringe

The supralittoral fringe has a line of black lichens, and irregular patches of the dark red algae *Hildenbrandia lecannellieri,* which can extend to the zone below. There are no littorinids here, and there are none in any of the other zones.

The Midlittoral Zone

The upper and central parts of the midlittoral zone have typical inhabitants. The dominant animals are the barnacle *Chthamalus scabrosus,* the small mussel *Brachyodontes purpuratus,* and the siphonarid *Kerguelenella lateralis.* A very conspicuous feature in exposed situations is a close carpet of *Bostrychia mixta,* which may spread into the fringe

FIGURE **10.40**

Isla Bertrand and the Wollaston Islands of Cape Horn, the southernmost area of farm and forest clearance in the world. A typical landscape in a protected southern coastal channel. [Photograph by M. W. Holdgate.]

FIGURE **10.41**

A view along the shore of Isla Hermite, Cape Horn. Note the prominent lichen zone, *Macrocystis pyrifera*, and *Durvillea antarctica*. [Photograph by E. R. Guiler.]

and may be mixed with *B. scorpioides* in more sheltered areas. Associated with these algae are a number of subsidiary species, notably *Chaetangium fastigiatum* and *Catenella fusiformis*. In many localities, species of *Cladophora*, *Cladophoropsis*, and *Rhizoclonium* give a bright green colour to the zone.

Although *Chthamalus* is still present in the lower midlittoral, species of *Elminius* and *Balanus* prevail. *Mytilus edulis chilensis* takes the place of *Brachyodontes*, and *Siphonaria* spp. are the dominant limpets. Bands of algae occur, the chief constituents of which are species of *Iridaea*, with *Adenocystis utricularis*, *Caepidium antarcticum*, *Ceramium rubrum*, *Porphyra umbilicalis*, *Scytosiphon lomentaria*, *Scytothamnus fasciculatus*, *Ulva lactuca*, and *Utriculidium durvillei*.

In the lower midlittoral, the rocks are covered by corallines and lithothamnia, which form a thick pink covering extending down into the sea. Locally, the mussel *Aulacomya ater* may form dense beds; other notable animals are chitons and species of *Fissurella* and *Nacella*.

The Infralittoral Fringe

In the infralittoral fringe of wave-swept shores, the bull kelp *Durvillea antarctica* is the most impressive alga; *Lessonia* and *Macrocystis pyrifera* grow just below it (Figure 10.41). Among the holdfasts of these large brown algae, and in association with lithothamnia, is a wealth of mixed red algae; representative species include *Ahnfeltia* spp., *Ballia callitricha*, *Ceramium rubrum*, *Desmarestia lingulata*, *D.rossi*, *D. willii*, *Plocamium* sp., and *Rhodymenia cunefolia*.

SELECTED REFERENCES

Aleem, A. A., 1955 (1956). A quantitative study of the benthic communities inhabiting the kelp beds off the Californian coast, with a self-contained diving apparatus. *Int. Seaweed Symp. 2nd Trondheim 1955,* pp. 149-152.

Bolin, R. L., 1949. The linear distribution of intertidal organisms and its effect on their evolutionary potential. *Proc. 13th Int. Congr. Zool.* 6: 459-60.

Brattström, H., and E. Dahl, 1951. Reports of the Lund University Chile expedition, 1948-49. *Lunds Univ. Årssk. Avd. 2* 46(8).

Chile, Departamento de Navegacion y Hidrografia de la Armada, 1959. *Perfiles Batitermográficos Frente a la Costa de Chile.* Valparaíso: Dep. Naveg. Hidrogr. Armada.

Clarke, R., 1962. Research on marine resources in Chile, Ecuador and Peru. *Fish. News Int.* 1(5).

Cox, K. W., 1960. Review of the abalone in California. *Calif. Fish Game* 46(4): 381-406.

Dawson, E. Y., M. Neushul, and R. D. Wildman, 1960. Seaweeds associated with kelp beds along southern California and northwestern Mexico. *Pac. Natur.* 1(14): 2-81.

Garrett, H., 1958. Animal life in wave-beaten rocks. *Nature* 184: 1652.

Guiler, E. R., 1959a. Intertidal belt-forming species on the rocky coasts of northern Chile. *Pap. Proc. Roy. Soc. Tasmania* 93: 33-58.

————, 1959b. The intertidal ecology of the Montemar area, Chile. *Pap. Proc. Roy. Soc. Tasmania* 93: 164-183.

Hewatt, W. G., 1937. Ecological studies on selected marine intertidal communities of Monterey Bay, California. *Amer. Midland Natur.* 18: 161-206.

Knox, G. A., 1960. Personal communication.

Mason, M., 1968. The Cape Horn region. Chapter 19 *in* Sir Alec Rose, *My Lively Lady.* Lymington, England: Nautical Publishing Co.

Paul, A. R., 1958. Eastbourne marine shore fauna. *Proc. Zool. Soc. (London)* 131(4): 527.

Stephenson, T. A., and Anne Stephenson, 1961a. Life between tide-marks in North America, IV A. Vancouver Island, I. *J. Ecol.* 49: 1-29.

————, and ————, 1961b. Life between tide-marks in North America, IVB. Vancouver Island, II. *J. Ecol.* 49: 227-243.

Womersley, H. B. S., 1954. The species of *Macrocystis* with special reference to those on southern Australian coasts. *Univ. Calif. Publ. Bot.* 27(2): 109-132.

11

AUSTRALASIA

Properly speaking, Australasia includes all of the islands of Oceania in the South Pacific, along with New Guinea, Australia, New Zealand, and associated islands. Our discussion of Australasian shores, however, will be limited to those of Australia and New Zealand and those of their adjacent, smaller islands.

INTRODUCTION

A broad overview of the physical and biogeographic factors influencing the shores of Australasia is important to a clear understanding of the ecology of those shores. In this introductory section, therefore, we will present certain basic information about ocean currents and sea temperatures in Australasia, and will discuss the biogeographic elements of the region.

CURRENTS

There is a relatively uninterrupted Southern Ocean encircling the globe in the Southern Hemisphere. The dominant feature is the West Wind Drift, which influences the whole of the temperate coasts of Australia. The tropical and warm-temperate shores of the region are affected by the anticlockwise current systems of the Indian Ocean, the Tasman Sea, and the South Pacific Ocean.

The Coral Sea, the Eastern Coast of Australia, and the Tasman Sea

The dominant current along the eastern seaboard of Australia is the warm, southward-flowing East Australian Current. It is derived from the westward-flowing Trade Wind Drift, which lies to the south of the South Equatorial Current. The East Australian Current flows strongest in the north, and decreases south of latitude 34°S. The spread of the current and its southward penetration vary with the seasons and the winds. A weaker branch of the Trade Wind Drift enters the West Coral Sea, where it turns and becomes integrated with the East Australian Current.

In the South Tasman Sea, the general flow of the Tasman Current—a mixture of subtropical and subantarctic water—is from west to east, with an increasing deviation to the north as it approaches New Zealand (Figure 11.1).

The Northern and Western Coasts of Australia

The coastal waters of northern and western Australia are influenced by water flowing from the South China Sea, from the North Pacific Ocean through the East Indian Archipelago, and from the tropical waters of the Indian Ocean. The rather scanty information available suggests that there is an inshore current along the western and northwestern coasts that flows north in the summer (December to May), and in the reverse direction in the winter (April to November). The general circulation within the Indian Ocean tends to be anticlockwise, with a general eastward flow between South Africa and Australia.

Tasmania and the Southern Coast of Australia

The main current south of Australia is the broad West Wind Drift; published charts indicate that it is variable across the Great Australian Bight.

A warm current from the Indian Ocean passes eastward round Cape Leeuwin in winter; it seems to be a surface current lying above the cold West Wind Drift. The coastal current off Cape Leeuwin in the summer, however, flows in the opposite direction.

In Bass Strait, the circulation of the water is very complex. In winter, predominantly cool water passes eastward through Bass Strait, along the Victoria coast, towards the south coast of New South Wales, and along the west and south coasts of Tasmania. In summer, the axis of this drift moves south; a branch of it still washes the west Tasmanian coast, but some of it turns towards the Great Australian Bight.

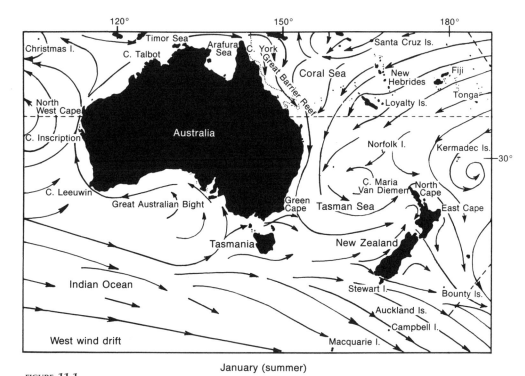

January (summer)

FIGURE **11.1**

The direction of the main currents around Australasia in summer (*above*) and winter (*below*). [From "Stream Drift Chart of the World," U.S. Naval Oceanographic Office, revised March 25, 1963.]

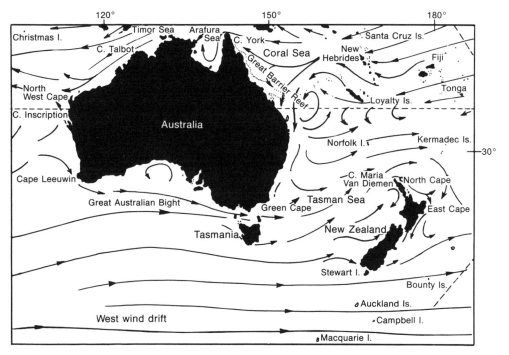

July (winter)

New Zealand

The details of the system of coastal currents around New Zealand are better known than those of Australia. Very simply, there are three principal surface-water movements: the cold West Wind Drift to the south of South Island, the warmer Trade Wind Drift to the north of North Island, and the Tasman Currents of mixed cool-temperate and subtropical water, which flow broadly east and northeast across the Tasman Sea and, in the winter, influence the region north of North Cape. On the northern part of the east coast, from North Cape to East Cape, the current flows southeast throughout the year, although occasionally, especially in summer, it may continue southwards as far as Cook Strait. On the east coast of South Island, the prevailing current seems to be northward throughout the year; it generally continues along the east coast of North Island as far as East Cape, which is the most common meeting place for the currents that flow round the north and south of New Zealand.

SEA TEMPERATURES AND THE DISPOSITION OF COASTAL WATER MASSES

T. A. Stephenson (1947) provisionally classified the intertidal shallow-water faunas into types according to their respective temperature ranges (tropical, warm- and cold-temperate, and so forth). This classification was extended by Knox (1960) to apply to temperate and antarctic shores of the southern hemisphere. The rough table below gives approximate temperature and salinity variations for the different coastal water masses of Australasia.

Type	Temperature (°C)		Salinity (‰)
	Winter (Aug)	Summer (Dec)	
Tropical	20 (min)	25+ (max)	35.5+
Warm-temperate and subtropical	12–20 (mean)	16–25 (avg max)	34.9–35.5
Transitional warm-temperate [for Australasia (Knox, 1960)]	12 (mean avg)	20 (mean avg not exceeded)	—
Mixed cold-temperate [for Australasia (Knox, 1960)]	7 (mean avg)	11–18 (mean avg)	34.5–34.8 (approx)
Subantarctic and cold-temperate	3–11 (mean avg)	4.5–5.5 (mean avg)	34.5
Antarctic	0 (mean)	10	—

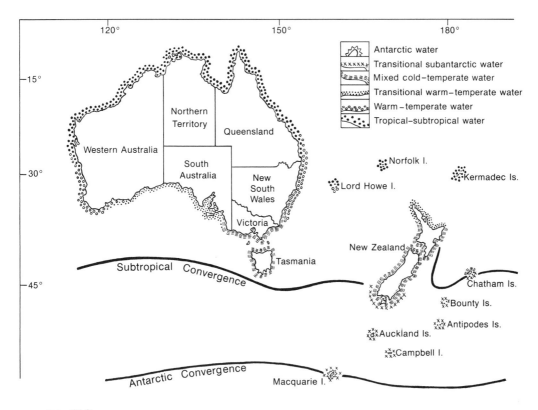

Legend:

- Antarctic water
- Transitional subantarctic water
- Mixed cold–temperate water
- Transitional warm–temperate water
- Warm–temperate water
- Tropical–subtropical water

FIGURE **11.2**

A map of Australasia showing the disposition of the principal coastal waters according to their temperatures, and the mean positions of the Antarctic and Subtropical convergences. [After Knox (1963).]

Figure 11.2 shows the disposition of coastal water masses of different temperatures around Australasian coasts. At the Antarctic Convergence, somewhere between lat. 54°S and lat. 62°S, there is a sudden change in the temperature of the surface water; the colder antarctic water lies south of this convergence. Between lat. 35°S and lat. 47°S, at the Subtropical Convergence, the cool subantarctic water sinks below the warmer subtropical water to the north. The whole of Australia and Tasmania lies to the north of the Subtropical Convergence, but the southern part of New Zealand straddles it.

A short note concerning the nomenclature for the temperature ranges of inshore waters may be suitably introduced here. The simple terminology of the *Discovery* Expedition Reports (*antarctic, subantarctic, subtropical,* and *tropical*), which was worked out primarily for the pelagic regions of the southern parts of the southern hemisphere, does not apply satisfactorily to inshore waters, where a greater number of distinctions are required. *Antarctic, subantarctic, cold-temperate, warm-temperate, subtropical,* and *tropical* provide a better series of terms for describing the temperature variations of inshore waters. This matter is discussed at length by T. A. Stephenson (1947, pp. 214–217).

BIOGEOGRAPHICAL ELEMENTS
OF THE AUSTRALASIAN BIOTA

In both Australia and New Zealand, the Indo-Pacific element dominates the biota.

Since Upper Cretaceous times, New Zealand has been a solitary land mass in a wide expanse of ocean, which has restricted the migration of species. The large number of Australian species in New Zealand is an indication of its long isolation. Australia, on the other hand, because of its great size and its extension well into the tropics, gives far greater opportunities for coastal migrations; consequently, it is much richer than New Zealand in numbers of genera and species. Australia's long southern coast provides chances for southerly migrations to it from western and eastern Australian shores, and has an unusually high percentage of endemics. The northwestern coast has a South African and Indian Ocean element; the biota of the northeastern coast is mainly Malayo–Indo-Pacific. There are small groups of species confined to the midnorthern section that appear to be of purely Malayan origin.

The fluctuations of mixed cold- and warm-water boundaries around the Subtropical Convergence are important factors in the distribution of the constituent species of the biota. All evidence suggests that migration runs with the prevailing westerly currents from Australia to New Zealand. In the temperate-sea areas of these countries, many groups have closely related or identical species. The biota shows a cold, circumpolar, West Wind Drift (Austral) element. This is more pronounced in New Zealand, of course, as it extends further south, and it is most pronounced on the shores of Chatham Island and other islands lying still farther south.

It is an interesting fact that, although the northern parts of New Zealand and the central and western coasts of South Australia are considered to be in the warm-temperate zone, their biotas retain many features of the cold-temperate area to the south.

Fleming (1962) proposed a classification of biogeographic elements for New Zealand that can be easily adapted for Australia. The original scheme has been followed here for the New Zealand elements, and the adapted scheme has been used for the Australian ones.

New Zealand Elements

Malayo-Pacific Species. Plants and animals of tropical and subtropical derivation, other than those that come from Australia.

Australian Species. Forms that cross the Tasman Sea from Australia.

West Wind Drift or Austral Species. These can be divided into Neoaustral and Paleoaustral species. The Neoaustral species are originally tropical forms that came into the Southern Ocean from Australia, South Africa, and South America. Some animals, such as *Mytilus edulis*, are bipolar, and their origin is unknown. The Paleoaustral species comprise an older element with fossil records and southern land connections; they are no longer being dispersed.

Endemic Species. These species have no close relations in other countries to indicate their place of origin, either because they developed in New Zealand (primary endemics) or because they became extinct elsewhere (secondary endemics). These must not be confused with species now endemic to New Zealand whose relationships and origins are known.

Cosmopolitan Species. These organisms are so widespread that their particular route of colonisation cannot yet be determined.

If a statistical analysis could be made of New Zealand biota, it would undoubtedly show the Malayo-Pacific group to be the dominant one, and the Australian and West Wind Drift groups to be next in importance.

Australian Elements

Indo-West-Pacific Species. These are plants and animals of tropical and sub-tropical derivation; they form the greater part of the shallow-water biotas along the coasts of the northern half of the continent. Indian Ocean representatives are prominent on the north and northwest shores.

Australasian Species. These are species also found in New Zealand, mostly transferred there from Australia across the Tasman Sea.

West Wind Drift or Austral Species. As is to be expected, this element is less conspicuous in Australia than it is in New Zealand, which extends farther into the Southern Ocean.

South African Species. Australia has many species derived from South Africa that have travelled directly across the Indian Ocean; naturally, a very small number of these reach New Zealand.

BIOGEOGRAPHICAL REGIONS OF AUSTRALASIA

Australian ecologists have divided the biogeographical regions of Australasia into provinces, each with a distinctive name (some are named after prominent persons in the history of their country, and some after geographical features—there is no uniformity). The differentiation of these provinces seems to be based mainly on the distribution and affinities of certain of their recognized inhabitants. The following table lists the various regions with their provincial names.

Careful consideration of the table will show that this nomenclature is not satisfactory in this context. The approach of this book is primarily ecological, and the coastal biota of major world areas is classified and analysed with relation to the main environmental factors. A profusion of terms that are mainly meaningless to the general reader will serve to confound, rather than to clarify, his understanding of the relations of Australasia

Biogeographical region and province	Area
Tropical and subtropical region	
Damperian	Australia
Great Barrier Reef	Australia
Solanderian	Australia
Kermadecian	New Zealand islands
Phillipian	Lord Howe Island
Norfolkian	Norfolk Island
Warm-temperate region	
West Australian	Australia
Peronian	Australia
Transitional warm-temperate region and overlaps	
Flindersian	Australia
Auporian	New Zealand
Cold-temperate region	
Maugean	Australia
Tasmanian	Tasmania
Cookian	New Zealand
Moriorian	New Zealand islands
Transitional subantarctic region	
Foresterian	New Zealand
Antipodian	Antipodes Islands

to ecologically similar regions. Arguments can go on indefinitely over the minutiae of many an animal's affinities and distribution; this can lead both to an unnecessary proliferation of provincial terms and to constant, irritating alterations of the definitions of existing ones as more knowledge is obtained and as the specialists further disagree.

The biogeographical regions of Australasia (Figure 11.3) are defined simply in the following paragraphs.

The Tropical and Subtropical Region

In the tropical and subtropical region, the sea temperature does not fall below 20°C and commonly exceeds 25°C. Reef-building corals need a mean temperature in the summer of at least 23.5°C for their best development, and are limited in growth if temperatures fall significantly below 20°C.

This region includes a tremendous length of coast. It extends along the whole northern shore of Australia, and stretches for considerable distances down the eastern and western coasts—approximately as far as Houtman on the west, and to near Point Vernon on the east. The temperature is higher on the northwest coast (24°–29°C) and slightly lower on the east (20°–28°C). There are, of course, areas where this region overlaps the warm-temperate region; these will be discussed later.

In New Zealand proper, there is no tropical and subtropical region, but certain islands in the vicinity fall into this category, including the Kermadec Islands, Norfolk Island,

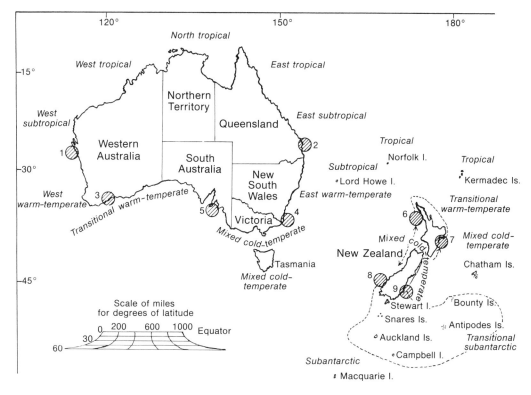

FIGURE **11.3**

A map of the biogeographical regions of Australasia and their overlaps (numbered *1–9*), which are discussed in the text.

and Lord Howe Island. The last is the most southerly locality at which reef-building corals are found. Subtropical areas have temperatures roughly between 20°C and 25°C, and such temperatures exist in parts of some warm-temperate areas.

The Warm-Temperate Region

The warm-temperate region has a mean winter temperature of 12°–20°C; in summer, one average maximum is about 16°–25°C, or a little higher.

Warm-temperate conditions prevail along the southwestern and eastern shores of Australia. Along the former, this temperature range extends from about lat. 30°S, round Cape Leeuwin, to the vicinity of Albany. Figure 11.1 shows a considerable confusion of water in this area caused by the West Australian Current, which can reverse its flow at different seasons of the year, but no really cold West Wind Drift water penetrates here. On the eastern shores, the picture is different. The warm East Australian Current leaves the coast near Twofold Bay, but from this bay northwards to the neighbourhood of the Queensland border, warm-temperate conditions prevail.

There do not appear to be any warm-temperate areas in New Zealand, and there is only a small one along the south shore of Lord Howe Island, which lies to the northwest.

The Transitional Warm-Temperate Region

The transitional warm-temperate region was defined by Knox (1963). The mean temperature is between 12°C and 20°C, and the latter temperature is never exceeded. This region extends all along the concave coast of the Great Australian Bight, tailing off into the mixed cold-temperate region in the vicinity of Adelaide and Kangaroo Island. Transitional warm-temperate waters are also found around the northern coasts of North Island, New Zealand.

The Mixed Cold-Temperate Region

In the mixed cold-temperate region, the water temperatures lie between a winter mean of 7°C and a summer mean of 18°C. This is the region that embraces the shores of Victoria. The complex and variable currents in and around Bass Strait produce sea temperatures 2.3°–3.5°C lower than would be expected for a shore at that latitude. In warm months, the coldest stretch of water lies between Cape Otway and the border of South Australia; the temperature rises markedly west of Kangaroo Island, where this region overlaps the transitional warm-temperate region. Tasmania is surrounded by mixed cold-temperate water, but its northeast coast tends to be slightly warmer than the others.

The coasts of the central section of New Zealand also lie in a mixed cold-temperate area, as does the northern coast of Chatham Island.

The Cold-Temperate and Subantarctic Region

The waters of the cold-temperate and subantarctic region have a range of mean temperatures from 3°C to 14.5°C. This region surrounds the extreme southern end of South Island, the south coast of Chatham Island, and the Antipodes. Macquarie Island lies close to the Antarctic Convergence, and has temperatures ranging from 2.8°C to 7.2°C.

Areas of Overlapping

Naturally, there must be considerable stretches along the coasts of any country where one type of environment merges with another. The lengths and positions of these overlaps, and the abruptness with which the changes take place, are dependent on many factors (topography, currents, river outlets, and so on). In Australia and New Zealand, there are nine obvious overlaps; the numbered areas in Figure 11.3, which correspond to the numbers in parentheses in the following two paragraphs, indicate their locations. Figures 11.4 and 11.37 (p. 329) will aid in locating the place names mentioned in the text.

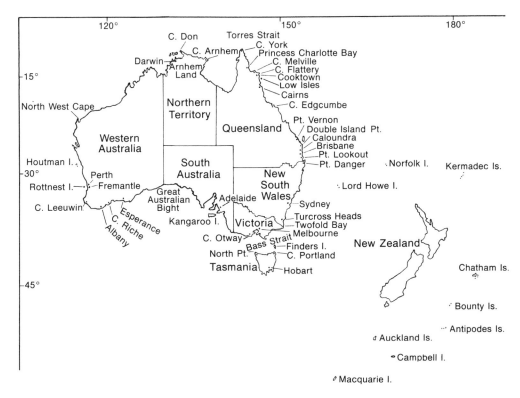

FIGURE **11.4**

A map of Australia and adjacent islands, showing the localities named in the text.

In Australia, the tropical and subtropical region intermingles with the warm-temperate one around Houtman Island (*1*) on the west coast, and in the neighbourhood of Double Island Point (*2*), near lat. 26°S, on the east. In the region between Cape Riche and Esperance (*3*), the warm-temperate waters of the west coast gradually yield to the cooler transitional warm-temperate seas of the Great Australian Bight, with its very complicated system of currents. On the east, the circumstances are different: owing to the cold water flowing through Bass Strait, there is a fairly sharp transition from warm-temperate conditions in the area of Turcross Heads and Twofold Bay (*4*) to mixed cold-temperate ones. These latter conditions continue all along the coast of Victoria to beyond Kangaroo Island (*5*), where water temperatures rise considerably, owing to the admixture of the transitional warm-temperate waters of the Bight.

Because New Zealand is a narrow, isolated country that runs almost north and south and is surrounded by deep ocean, the situation there is relatively simple. There are four overlaps: two between the transitional warm-temperate and the mixed cold-temperate regions around North Island—near Auckland (*6*), and near Poverty Bay (*7*)—and two more on South Island, where mixed cold-temperate seas give way to cold-temperature and subantarctic ones—just north of West Cape (*8*), and near Dunedin (*9*).

AUSTRALIA

WESTERN AUSTRALIA–WEST WARM-TEMPERATE

New South Wales, Victoria, and Tasmania have been studied intensively, and detailed accounts have been given of South Australia and parts of Queensland, but the published information about Western Australia is meagre (Hodgkin, Marsh, and Smith, 1959). However, it serves to provide some comparison between warm-temperate regions of the east and west coasts of Australia (Figures 11.5 and 11.6).

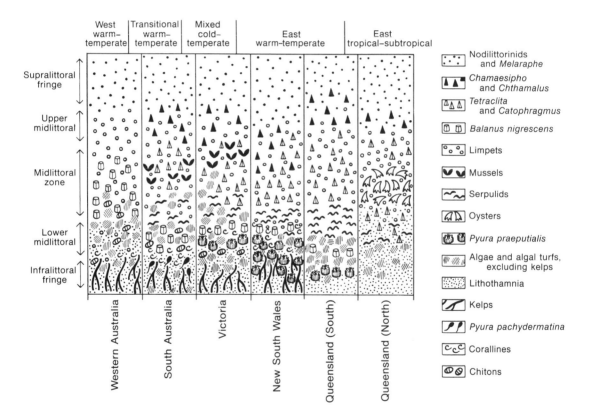

FIGURE **11.5**

The basic types of zonation of characteristic coastal regions around Australia. The vertical scale is only roughly approximate, as the widths of zones depend entirely on environmental conditions. Only a selected number of typical species has been used in order to keep the figure uncomplicated and intelligible. Actively moving species are omitted because, although they wander between certain definite boundaries, they do not form permanent belts. See also Figure 11.6.

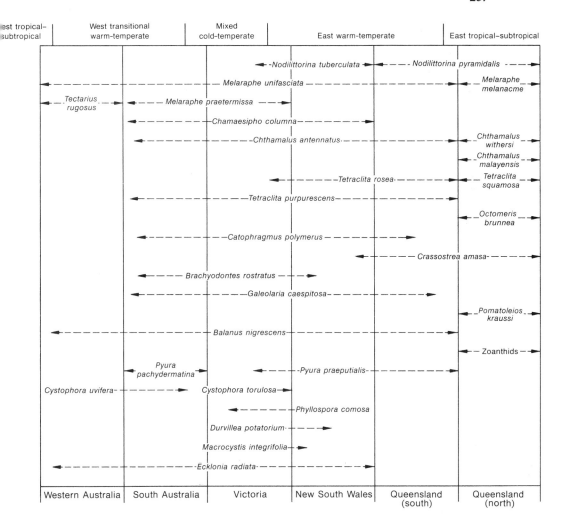

FIGURE 11.6

The extent of the distribution of various elements of the fauna and flora of the coasts of Australia.

Characteristic formations of the rocky shores of the southwest coast of Australia are limestone cliffs 6–8 feet high, typically overhanging. Stretching below these are flat intertidal platforms of various widths. Many of them are deeply undercut at the seaward edge, but some of them form sloping ramps to the sea.

Two localities in Western Australia will be considered here—Cape Leeuwin and Rottnest Island. Both localities are in the warm-temperate region.

Cape Leeuwin

The Supralittoral Fringe. The supralittoral fringe at Cape Leeuwin has an upper band of blue-green algae that is typically very pronounced. Below are two littorinids—*Melaraphe unifasciata*, which is dominant, and *Tectarius rugosus*, which is subtropical and replaces the *M. praetermissa* of colder conditions.

The Midlittoral Zone. There are no barnacles in the upper part of this zone, which is mainly occupied by the endemic limpets *Notoacmea onchyitis* and *Siphonaria luzonica*. Below the limpets, there is often a dense formation of *Balanus nigrescens*, within which the limpets *N. onchyitis* and *Patellanax peroni* and the chiton *Clavarizona hirtosa* are common. This chiton penetrates somewhat into the lower midlittoral, an area dominated by the limpet *P. laticostata* and the chiton *Poneroplax costata*. The rock surface is coated with lithothamnia, *Laurencia* spp., and other small algae. The polychaete *Galeolaria caespitosa* and the alga *Hormosira banksii*, so common on platforms elsewhere in Australia, are not present.

The Infralittoral Fringe. This zone has no *Pyura praeputialis* belt. It is dominated instead by the kelps *Ecklonia radiata* and *Cystophora* spp., but these latter are far less important here than they are in South Australia. A number of subtropical algae (such as *Caulerpa ramosa*, *Halimeda cuneata*, and *Penicillus nodulosus*) are prominent. There is the usual growth of corallines and lithothamnia. *Haliotis roei* appears to be the dominant animal.

Rottnest Island

Rottnest Island is similar to Cape Leeuwin, but has the following additional species: In the upper part of the midlittoral zone, there is a seasonal growth of *Enteromorpha* spp., *Ulva lactuca*, *Chaetomorpha aerea*, *Cladophora* spp., and *Porphyra umbilicalis*. On the wide platforms of the midlittoral, sandy tufts contain *Caulerpa cylindracea*, *Centroceras clavulatum*, *Laurencia heteroclada*, and the coralline *Jania fastigiata*; among the turf, a number of animals abound, including small crustaceans and polychaetes. Other algae in the infralittoral fringe at Rottnest Island are *Sargassum* spp., *Cystoseira abrotanifolia*, *Pterocladia capillacea*, and *Hypnea musciformis*; among these occur browsing gastropods (such as *Euplica bidentata*, *Pyrene* spp., and *Senectus intercostalis*) and certain anemones, including *Isanemonia australis*. In some exposed places, the sea urchin *Echinometra matthaei* is very abundant.

The west warm-temperate coast is conspicuously different from the east warm-temperate one, in that the dominant species are not the same. Many of the western species are endemic, or perhaps from South Africa, and all cold-temperate molluscs and echinoderms are lacking. The biota of Western Australia slowly grades into the South Australian biota from around Albany to Esperance.

SOUTH AUSTRALIA–TRANSITIONAL WARM-TEMPERATE

The rocks of the coast of South Australia, which mainly consist of granite, schist, and quartzite, descend rather steeply into deep water. There are also cliffs of calcareous sandstone with wide wave-cut platforms stretching out into the sea at low-water level (Figure 11.7).

The tides are semidiurnal and of relatively small amplitude, but they vary considerably along the coast. The lowest low waters occur in the summer (January and February), usually during the middle of the day.

Womersley and Edmonds (1958) proposed a terminology for intertidal zones that is theoretically applicable to rocky, sandy, and muddy shores alike, but because the dynamics of sandy and muddy shores are totally different from those of rocky ones, we find their system unsuited to our purpose. In this work, only rocky shores are discussed, and the Stephenson scheme is adhered to throughout for the naming of the various zones.

The shores of South Australia, from Kangaroo Island to Western Australia, have

FIGURE **11.7**

The coast at Point Sinclair, a little east of the western border of South Australia, at very low water. The smooth granite point in the background shows the zonation very clearly. The very dark upper band is a growth of *Calothrix* in the supralittoral fringe; the light grey bands below are *Chamaesipho* and *Catophragmus*, barnacles of the midlittoral; and the dark belt just above water level is the coralline mat of the infralittoral fringe with its population of *Balanus nigrescens*. In the foreground is a reef of sandstone consolidated by lime, typical of many parts of these coasts, with the surface mainly dominated by algae and various molluscs. [Photograph by H. B. S. Womersley and S. J. Edmonds, CSIRO, Melbourne, Australia.]

a biota very similar to that of Victoria, although sea temperatures are generally higher. There are a number of interesting variations in the biota in westerly districts. The small section of the coast from Kangaroo Island east to Victoria is in the mixed cold-temperate region, and will be discussed with the coasts of Victoria and Tasmania.

The Supralittoral Fringe

The lichens *Lichina difformis* and *Verrucaria* sp. are common in many places in the supralittoral fringe. In areas of strong wave action, the blue-green alga *Calothrix fasciculata* forms an extensive and conspicuous black band. *Melaraphe unifasciata* is the commonest littorinid, and *M. praetermissa* is common also; both follow their usual habits, and may extend downward, in some places, among the barnacles of the upper midlittoral. *Ligia australiensis* scuttles about in the higher parts of the fringe.

The Midlittoral Zone

The midlittoral zone, as usual, is dominated by barnacles. *Chamaesipho columna* and *Chthamalus antennatus* are common in the upper midlittoral; they are particularly prolific in exposed areas, and decrease in importance in more sheltered ones. *Tetraclita purpurascens* occurs in shaded places.

Somewhat lower on the shore, on steep, wave-swept slopes, *Catophragmus polymerus* prevails; and on gentler slopes, among or just below the *C. polymerus,* the mussel *Brachyodontes rostratus* occurs, very much as it does in Victoria. Scattered among both mussels and barnacles are isolated gastropods, including *Patelloida latistrigata, Siphonaria dimenensis,* and *Neothais textilosa.* Blue-green algae spread widely here, but they vary greatly in abundance; common species are *Rivularia firma* and *Isactis plana,* in some places with species of the algae *Polysiphonia* and *Nemalion.*

In moderately exposed places on the western part of the South Australian coast, *Splachnidium rugosum* may grow over and among the *Catophragmus,* either as scattered plants or as a community (this has been observed also on parts of the coast of Victoria).

On less exposed shores, a number of additional gastropods occur, including *Notoacmea septiformis, Patelloida alticostata,* and *Cellana tramoserica.* The last two species may become codominant. The serpulid *Galeolaria caespitosa* is also found on such shores, usually as scattered tubes on the rock surface.

Just above the infralittoral fringe on unprotected shores, *Balanus nigrescens* and various corallines are the chief features. Associated with the mat of corallines are a number of short or stunted algae, such as *Laurencia heteroclada, Wrangelia plumosa, Dasyopsis clavigera, Nemasto feredayae, Centroceras clavulatum,* and *Gigartina* spp. Other than the usual chitons *Poneroplax albida* and *P. costata,* there are few macroscopic animals.

The Infralittoral Fringe

The infralittoral fringe is dominated by the alga *Cystophora intermedia.* The rocks under its fronds are typically covered with a dense mat of corallines composed mainly of

Corallina cuvieri, and bare areas of such rock are covered with encrusting lithothamnia. In slightly sheltered places, *Ecklonia radiata* may be found, and *Cystophora intermedia* gives way to *C. spartioides, C. subfarcinata, C. paniculata, C. siliquosa,* and smaller forms of such algae as species of *Caulerpa* and *Xiphophora.* Animals are not conspicuous, but *Haliotis roei* and the stalked *Pyura pachydermatina* are notable. Starfish and numerous small crustaceans may be found in the algal mat.

Cliffs Behind Rocky Platforms

Cliffs behind rocky platforms are protected from violent wave action, and this produces variations in their zonation. The supralittoral fringe is the normal one for such situations. The midlittoral, typically, has a "mollusc-*Galeolaria*-blue-green algae" zonation (see Figure 11.8). Barnacles are diminished here, and molluscs are either equally plentiful or dominant. Many additional molluscs occur, including *Austrocochlea odontis, Melanerita melanotragus,* and *Bembicium melanostoma. Galeolaria caespitosa* is prominent at low levels among the well-developed blue-green algae (mainly *Rivularia firma* and *Isactis plana*). The black mussel *Brachyodontes rostratus* is variably abundant. There are good communities of various anemones in damp places.

FIGURE **11.8**

The midlittoral of a sheltered rock at Cape Westall, a little east of Point Sinclair (see Figure 11.7), showing a distinct zone of *Galeolaria caespitosa* with a few *Cellana* and *Melanerita* scattered above; below, *Hormosira banksii* marks the upper limit of the infralittoral fringe. [Photograph by H. B. S. Wormersley and S. J. Edmonds, CSIRO, Melbourne, Australia.]

Horizontal Reefs

Horizontal reefs extending from the cliffs are usually dominated by algae, and the ecological variations of the biota of these reefs in the western and central parts of South Australia are relatively minor. The most prominent algal species is *Hormosira banksii,* and *Enteromorpha clathrata, Ulva lactuca, Ectocarpus confervoides,* and *Codium capitulatum* are also present. Well-developed communities of *Brachyodontes rostratus* occur in some places. The various areas on such reefs that are never uncovered, and are therefore not strictly intertidal, support many additional animals and algae.

VICTORIA INTO SOUTH AUSTRALIA
AND TASMANIA—MIXED COLD-TEMPERATE

There is direct evidence of the existence of a body of colder water between Victoria and Tasmania; its existence has been verified by the records of the Danish *Galathea* Deep Sea Expedition of 1950–52, as well as by other sources. The coastal waters of Victoria and Tasmania are consistently at least 2.3°C lower in temperature than those bathing the adjacent coast of New South Wales and the central and western coast of South Australia. (The sea temperature rises markedly around Kangaroo Island, so the mixed cold-temperate region extends beyond Victoria along the coast of South Australia only as far as that neighbourhood.) The average summer sea temperatures off Victoria lie between 16°C and 19°C, and the winter ones average between 12°C and 13°C. These temperatures parallel those of Tasmania.

It follows, naturally, that there must be many affinities between the biota of Victoria and the biota of Tasmania. Further, there are many affinities between these biotas and those of New South Wales to the east and of South Australia to the west. In a sense, the coasts of Victoria and Tasmania are vast overlaps of cold-temperate and warm-temperate biotas, but their very particular conditions justify making them a region of their own (see Figures 11.5 and 11.6). Victoria and Tasmania have no tropical component, such as appears in certain areas on the coast of New South Wales. The merging of the latter's warm-temperate waters with the mixed cold-temperate ones of Victoria is gradual and extends over a very long stretch of the coast.

Victoria

The Supralittoral Fringe. In the supralittoral fringe in Victoria, the isopod *Ligia australiensis* abounds (in New South Wales, it only occurs in bays and inlets). The littorinids *Melaraphe unifasciata* and *M. praetermissa* are present, but *Nodilittorina tuberculata* is missing. A marked feature, particularly towards the west, is a very dark and definite band of lichens, consisting of *Lichina confinis* and *Verrucaria* sp.

The Midlittoral Zone. This has the usual barnacle population—*Chamaesipho columna, Chthamalus antennatus,* and *Catophragmus polymerus,* in that order, from higher to lower levels. These species vary in density of formation according to environ-

FIGURE **11.9**

Intertidal zonation of a very exposed area at Wilson's Promontory, the most southerly point in Victoria, as seen on an exceptionally calm day. The barnacle zone, which stretches upward as much as 40 feet, can be seen clearly on a slope in the background, where there is also a very obvious black belt of the mussel *Brachyodontes rostratus;* in the foreground, only a few can be seen. Below the feet of the woman there are, in descending order, a band of large *Balanus nigrescens;* one of *Pyura praeputialis,* with a rather bare strip below; and *Durvillea potatorum,* of the infralittoral fringe, extending into the sea. [Photograph by E. Pope.].

mental conditions. On the whole, barnacles are fewer here than they are farther west, and *Tetraclita rosea* is petering out in this area. In the upper regions, limpets are more varied in species and more numerous than they are in New South Wales. In many places, there are two bands of mussels—*Modiolus pulex* and *Brachyodontes rostratus.* The latter is dense enough, in some places, to exclude other animals (Figure 11.9). Associated with these mussels are belts of algae, which include the seasonal species *Splachnidium rugosum* and *Ilea fascia. Rivularia firma* only occurs in the extreme east of this area.

In the lower midlittoral, there is a belt of *Galeolaria caespitosa,* which, in very exposed places, is replaced by a mixed algal community that includes species of *Pocockiella, Jania, Padina, Pterocladia, Peyssonnelia, Caulerpa, Platysiphonia,* and *Laurencia*—a picture very reminiscent of South Africa. The mussels, serpulids, and algae provide shelter for numerous animals—polychaetes, amphipods, limpets, and so on (a list of them is given by Bennett and Pope, 1953). East of Cape Otway, these populations vary little from those of New South Wales. West of that point, however, a number of species disappear, including the barnacle *Ibla quadrivalvis*, the spider *Desis crosslandi*, and the pulmonate *Onchidiella patelloides*. The last two species apparently reappear in South Australia. Particular mention must be made of *Balanus nigrescens*, which can occur, as a definite narrow fringe of conical individuals $1\frac{1}{2}$ to 2 inches tall, above the *Pyura praeputialis*. In sheltered places, it is largely replaced by the alga *Hormosira banksii*, which also characteristically carpets fairly level rock platforms.

The Infralittoral Fringe. This shows a striking change from that of New South Wales, for the dense growth of *Pyura praeputialis* is petering out. Isolated groups may be found here and there, but it no longer exists as a zone. Where *P. praeputialis* is absent, its level of the shore is characteristically bare, or covered by algal growths of *Cystophora* spp. (*C. intermedia* in exposed places and *C. torulosa* in sheltered places). It is noteworthy that some species common here and in New South Wales flourish better in one place than the other. For instance, *C. columna* is two or three times taller here than it is in New South Wales, the chiton *Poneroplax* is considerably larger, and *Hormosira banksii* thrives more and grows lower on the shore.

The large brown algae *Ecklonia radiata* and *Phyllospora comosa* are found in sheltered places and deep rock pools, but they play a secondary role to *Durvillea potatorum*, which is the dominant kelp in Victoria. Slightly lower on the shore, there is a stand of *Macrocystis integrifolia*. Between the holdfasts of these plants, the rocks are coated with lithothamnia, as are the shells of *Patellanax squamifera*, which inhabits this part of the shore.

Tasmania

Tasmania lies a little to the north of the Subtropical Convergence (see Figure 11.2). According to Deacon (1967):

> The currents round the coast of Tasmania are known to vary about the pattern shown in [Figure 11.1]. The West Wind Drift is the major feature, and as shown by the current arrows south of Australia, this has a slight northward trend in winter and a southward trend in summer. Along the west coast of Tasmania itself there is, however, more evidence of northward movement in summer than in winter. There is also evidence of a greater influence of the South Equatorial current and the East Australian current in the region east of Tasmania during the summer, and some differences in water conditions seem to occur between the two sides of the island.

The wave action in Tasmania is strongest and most constant on the west coast, but (according to Bennett and Pope, 1960) it is not quite as extreme as that on the western coast of Victoria. On the south and east coasts, it is slightly less strong than it is on the west (Figure 11.10). The north coast is sheltered from the severest wave action, and has no oceanic swell. The prevailing winds are westerly or northwesterly for most of the year, but in the late summer months (February and March) and during part of the winter (June), cold southerly winds prevail.

The zonation pattern is fairly uniform along the exposed shores of Tasmania, but there are differences from Cape Portland to North Point on the north coast, where the seas are consistently calmer (Figure 11.11).

FIGURE **11.10**

The zonation pattern of an open but sheltered shore on the southeast coast of Tasmania. At the top is a dense band of *Brachyodontes rostratus,* followed by stunted *Lithophyllum* and small algae, with some *Mytilus edulis* on the right and two chitons (*Peneroplax costata*) near the centre, and by a broad belt of *Pyura praeputialis.* [Photograph by E. Pope.]

The Supralittoral Fringe. Some workers have mentioned a particularly bright orange lichen that makes a frieze on the upper shore, but this really belongs to the supralittoral *zone* of maritime plants, not to the supralittoral fringe. *Lichina confinis* is a common lichen here, but it extends to a lower level than it does in Victoria. The littorinids *Melaraphe unifasciata* and *M. praetermissa* are both present, but the latter

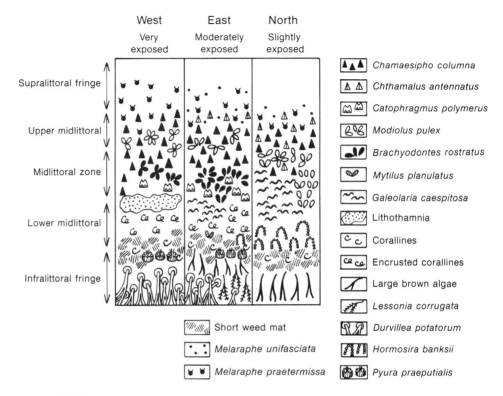

West — Very exposed
East — Moderately exposed
North — Slightly exposed

Supralittoral fringe
Upper midlittoral
Midlittoral zone
Lower midlittoral
Infralittoral fringe

▲▲▲ *Chamaesipho columna*
▲ ▲ *Chthamalus antennatus*
🔶 *Catophragmus polymerus*
Modiolus pulex
Brachyodontes rostratus
Mytilus planulatus
Galeolaria caespitosa
Lithothamnia
c c Corallines
Ce Ce Encrusted corallines
Large brown algae
Lessonia corrugata
Durvillea potatorum
Hormosira banksii
Pyura praeputialis

Short weed mat
• • *Melaraphe unifasciata*
😾 😾 *Melaraphe praetermissa*

FIGURE 11.11

The changes of some common constituents and their positions in zones on the shores of Tasmania, according to the amount of exposure to which they are subjected. It is clearly shown that the basic zones of a shore fall or rise, widen or narrow, according to the local conditions. Note that the various coasts of Tasmania have their typical exposures.

is the more important species—indeed, it is the only one on the west coast, and it is codominant with the other (which prefers warmer conditions) on the east coast. The isopod *Ligia australiensis* appears to be common and numerous everywhere but in the south.

The Midlittoral Zone. Barnacles are the dominant animals at the top of this zone along much of the exposed coast, but they are not as numerous here as they are on the adjacent mainland. Littorinids may overlap into the barnacles from the belt above. On the north and east shores, *Chthamalus antennatus* and *Chamaesipho columna* are found as they are on the mainland, but in the extreme south of Tasmania, the former species disappears. *Catophragmus polymerus* occurs sporadically below *Chamaesipho columna,* but on the east coast, it approaches the limit of its geographical range. It may be associated with the mussel *Brachyodontes rostratus,* which may form extensive sheets that eliminate the barnacles in places. Another smaller mussel, *Modiolus pulex,* also occurs in abundance in suitable areas.

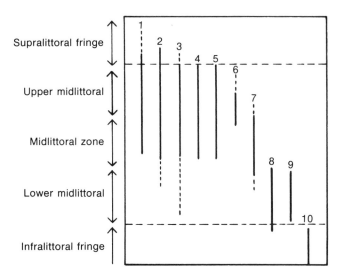

FIGURE **11.12**

The vertical range of common limpets and siphonarians on an exposed coast of Tasmania: 1, *Siphonaria diemenensis* (maximal in shelter); 2, *Patelloida latistrigata latistrigata;* 3, *Cellana solida* (range extended in shelter); 4, *Siphonaria tasmanica;* 5, *Siphonaria funiculata* (east coast only); 6, *Notoacmea petterdi* (east coast only); 7, *Notoacmea mayi* (maximal in spray); 8, *Patellanax peroni* (maximal in exposure); 9, *Patelloida alticostata* (maximal in shelter); 10, *Patelloida victoriana; unbroken lines,* a fairly even distribution; *dashed lines,* a sporadic occurrence. [After a drawing by F. J. Beeman.]

Limpets play a particularly important part in the midlittoral fauna. *Notoacmea mayi* and *Cellana solida* are much more common here than they are on the mainland; *C. tramoserica* is absent here, being replaced by *C. solida.* With *Chthamalus antennatus* and *Chamaesipho columna* are found the predatory gastropod *Lepsiella vinosa* and certain limpets, including *Patelloida latistrigata* and (in eastern localities) *Notoacmea petterdi.* Three pulmonate species, *Siphonaria diemenensis, S. tasmanica* and *S. funiculata,* commonly occur in the barnacle zone (Figure 11.12). Exclusive to Tasmania is the darker, less eroded form of the chiton *Sypharochiton pelliserpentis.*

The upper midlittoral carries only a sparse lichen and algal population similar to that of the mainland, with such familiar species as *Ulva lactuca,* stunted *Hormosira banksii,* small patches of *Gelidium pusillum,* and bits of *Splachnidium rugosum.*

A band of *Galeolaria caespitosa* may be found in the lower midlittoral in bays and other sheltered places (Figure 11.13). In more exposed places, it is replaced by the hard, mustard-yellow coralline *Lithophyllum hyperellum,* which flourishes especially on the west coast (Figure 11.14). Patches of this species are found only occasionally in Victoria. Below the *Lithophyllum,* there is an algal turf composed primarily of corallines. *Corallina officinalis* and *C. cuvieri* are the prevailing species, but *Jania fastigiata* is important in some places. Examples of algae associated with them are species of *Bryopsis, Leathesia,*

FIGURE **11.13**

The zonation of a very sheltered shore near Port Arthur, Tasmania. On the bare rock above are some black patches of *Lichina confinis* with a few *Cellana;* below this is a well marked zone of *Galeolaria*, followed by an ample growth of *Hormosira*, which becomes mixed with *Cystophora* under water. [Photograph by E. R. Guiler.]

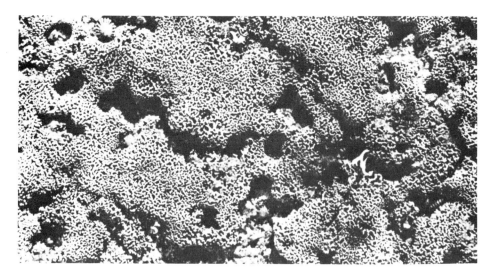

FIGURE **11.14**

A dense growth of the hard coralline *Lithophyllum*, which replaces *Galeolaria* in Tasmania. [Photograph by I. Bennet.]

Chaetomorpha, Plocamium, Centroceras, Pterocladia, and *Laurencia.* In deep pools and crevices, species of *Caulerpa, Codium,* and *Cystophora* are common. This turf is best developed on the west coast.

Molluscs are the dominant animals of the lower midlittoral, and those recorded from the west coast of Tasmania are very much the same as those recorded for Victoria. There are the familiar limpets, chitons, amphipods, polychaets, and so forth (detailed lists of the species are given by Bennett and Pope, 1953, 1960). One striking difference between Tasmania and the mainland is that the alga *Hormosira banksii,* which is dominant on platforms on the exposed coasts of southern Australia, is found in Tasmania only in very sheltered places.

The Infralittoral Fringe. The simple ascidian *Pyura praeputialis* is confined to small clumps in particularly favourable localities, typically rather higher on the shore than is usual elsewhere; in many localities, there is a bare strip of rock below the *Pyura*—perhaps the constant lashing of the fronds of giant fucoids has something to do with this. *Pyura* does not seem to extend below the upper limits of fucoid holdfasts, but still seems to mark the top of the infralittoral fringe.

The dominant organisms are the giant alga *Durvillea potatorum* and bright pink encrusting lithothamnia (Figure 11.15). Only on the north coast is the former absent;

FIGURE **11.15**

An exposed area at the end of a reef on the southeast coast of Tasmania, showing the dense growth of *Durvillea potatorum.* There is a patch of *Xiphophora* on the extreme left bottom corner and an irregular growth of *Pyura preputialis* above the *Durvillea.* [Photograph by E. R. Guiler.]

it is replaced there by other species, mainly *Cystophora torulosa* (not found in South Australia) and *C. paniculata. C. intermedia,* common along southern Australian coasts, is absent here, and *Ecklonia radiata* can only be found in extremely sheltered pools. In less exposed areas, there is a band of the brown alga *Xiphophora gladiata;* with increasing shelter, it may be associated with *Lessonia corrugata.* It occurs above the *Durvillea* and *Macrocystis,* both of which are indicative of cold-temperate water. (Extensive beds of *Macrocystis* are common off the eastern coast of Tasmania.)

The only large animals consistently found here are *Patelloida victoriana* and, less commonly, the fissurellid *Notohaliotis ruber.* The kelps of tempestuous seas do not seem to provide an environment suitable for an epiphytic biota.

The differences between the zonation of the shores of Tasmania and that of adjacent mainland shores confirm that Tasmania, particularly in the south, has a better developed cold-temperate biota.

NEW SOUTH WALES—EAST WARM-TEMPERATE

The warm-temperate coast of New South Wales presents a very close parallel to the southern and southeastern shores of South Africa (Figure 11.16). The following is a brief account of some of its main features (see Figures 11.5 and 11.6).

FIGURE **11.16**

A typical intertidal rock platform of central New South Wales at low water. [Photograph by E. Pope.]

The Supralittoral Fringe

The supralittoral fringe is populated by great numbers of the littorinids *Nodilittorina tuberculata* and *Melaraphe unifasciata*, the former reaching the highest part of the shore. In bays and inlets, there are quantities of *Ligia australiensis* (a different species of *Ligia* is characteristic of this zone in South Africa).

The Midlittoral Zone

The midlittoral zone has the typical barnacle population; *Chthamalus antennatus*, the highest barnacle on the shore, may occur in massed formation wherever spray can reach. Mingling with the *Chthamalus* at its lowest levels, in some places, and becoming dominant still lower down, is the smallest barnacle of these oceanic rocky shores, *Chamaesipho columna*, a most important zone-marking species. Below, the barnacles *Tetraclita rosea* and *Catophragmus polymerus* grow together, both preferring areas of heavy surf (Figures 11.17 and 11.18). These barnacles are associated with the dense growth of the serpulid polychaete *Galeolaria caespitosa* (represented in South Africa by *Pomatoleios crosslandi*). Another common barnacle, *Tetraclita purpurescens*, inhabits damp shady places at practically any level of the seashore from the top of the infralittoral upwards. In a limited number of localities with suitable conditions, a clear band of

FIGURE **11.17**
The vertical face at the back of a rock platform near Sydney, Australia, showing a dense growth of the surf barnacles *Tetraclita* and *Catophragmus* above the *Galeolaria* zone. [Photograph by W. J. Dakin.]

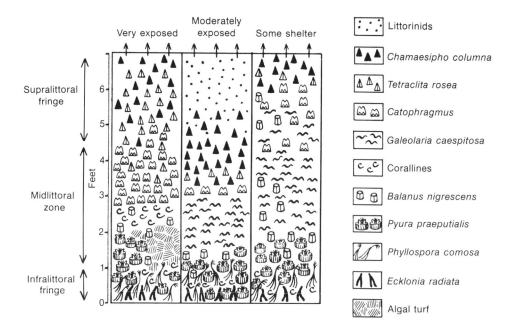

FIGURE **11.18**

The changes of some common constituents and their positions in zones on the shores of New South Wales, according to the amount of exposure to which they are subjected.

the limpet *Notoacmea petterdi* may be found just above the *Galeolaria;* in other places, this limpet occurs only in scattered colonies. (In South Africa, there are the definite belts of limpets, for which it is famous.)

The *Galeolaria* zone marks the lower part of the midlittoral, and extends seawards to the *Pyura* belt, at the top of the infralittoral fringe (Figure 11.19). The upper limit of *Galeolaria* is often extraordinarily regular. In very exposed places, it may be repressed by growths of the red alga *Pterocladia capillacea* or by corallines. At the other extreme, under conditions of exceptional shelter (especially in sandy places), it may be replaced by *Chamaesipho columna,* or by a stubby growth of algae (corallines, *Padina pavonia, Pocockiella variegata,* and stunted *Sargassum*), or by a combination of both. Among the *Galeolaria* tubes is a rich association of small animals; notable are the stalked barnacle *Ibla quadrivalvis,* the pulmonate *Onchidiella patelloides,* the tremendously common small bivalve *Lasea (Kellia) australis,* and the peculiar spider *Desis crosslandi,* represented in South Africa by *D. tubicola.*

Much of the barnacle zone of New South Wales lies on the rocky platforms so characteristic of these shores (Figure 11.20). The rich algal growth of the platforms is mainly *Hormosira banksii,* which prefers a damp habitat, and in drier areas, it may form thick fringes around pools. In many places, it is associated with close mats of corallines characteristic of the infralittoral fringe, but here, they continue upward into the midlittoral, where their growth form is short and stubby. Their colours are dull and faded, and the whole growth has a shabby appearance (this is also true of corallines in many places in South Africa). On the lower parts of these rocky platforms, there

Figure **11.19**

A maximal growth of *Galeolaria* at Newcastle, New South Wales. Compare with Figure 8.7 (p. 106), which shows a growth of *Pomatoleios* on the South African coast. [Photograph by E. Pope.]

are moderate growths of *Ulva* and—in some places—extensive ones of *Enteromorpha*. There are also other surface algae, including species of *Bangia* and *Ectocarpus*.

Large numbers of the conspicuous limpet *Cellana tramoserica* may be found in the lowest part of the midlittoral. Other conspicuous gastropods are the pulmonates *Ellsiphon scabra*, *Austrocochlea concamerata*, *A. obtusa*, and the carnivorous *Morula marginalba*. *Melanerita melanotragus* and *Bembicium melanostoma* collect in huddles in damp places as the tide recedes. Other animals of this level include various anemones, the chiton *Sypharochiton septentriones*, and scattered oysters.

The Infralittoral Fringe

The infralittoral fringe in New South Wales follows the pattern familiar in many other similar parts of the world. The most conspicuous feature is the belt of the ascidian *Pyura praeputialis*. For its best development, *P. praeputialis* requires unadulterated ocean water and considerable—but not violent—wave action. It extends for many miles along certain coasts and reaches down sporadically into the regions of the big kelps. These animals are associated with a large population of other creatures; conspicuous among them is the large *Balanus nigrescens*, which also occurs further up the shore as isolated

FIGURE 11.20

A very exposed area near Sydney, Australia. The *Galeolaria* zone lies below *Ulva*, corallines, and associated forms on the rock submerged by the backwash. The main barnacles here are *Tetraclita rosea* and *Catophragmus polymerus;* those on the top of the flat rock in the foreground are *Chthamalus antennatus*. [Photograph by H. D. Hughes, Australian Museum, Sydney.]

specimens. The chitons *Poneroplax pateliana* and *Onithochiton quercinus* are most plentiful on flat surfaces at the upper margin of the *Pyura*, as is the spectacular worm *Sabellastarte indica*, which also inhabits permanently exposed pools at higher levels. Algae among the *Pyura* include *Ulva lactuca, Colpomenia sinuosa, Dictyota dichotoma,* and *Pocockiella variegata*.

Below the *Pyura* belt lies the region of the large brown kelps *Phyllospora comosa* and *Ecklonia radiata*. Where these two fade out on the northern coast of New South Wales, corallines come into their own: *Pterocladia capillacea* occupies the most exposed places, particularly vertical faces exposed directly to the ocean. Some typical animals of this lowest intertidal zone are the sea urchin *Heliocidaris erythrogramma*, the "cart-rut" shell *Dicathais orbita*, and, in more sheltered places, *Haliotis ruber*, large tritons such as *Charonia rubicunda*, the black fissurelid *Scutus antipodes*, and the large turban *Ninella torquata*. Anemones are common in the intertidal regions, but most notable on the lowest parts of the shore is the orange coloured *Corynactis australis*, which lives in thousands in sheltered crevices, and is rarely uncovered by the sea. As in so many other parts of the world, there are countless hollows in softer rocks occupied by burrowing echinoderms—here, the species are *Heliocidaris erythrogramma* and *H. tuberculata.* Sponges and small anemones are numerous. In many places, corallines and lithothamnia give the open rock surfaces their typically pinkish colour.

Certain tropical organisms are to be found sporadically along the coasts of New South Wales: the outstanding ones are species of molluscs and echinoderms; in some places, there are also a few corals, zoanthids, and tropical algae. All these tend to be concentrated in various small circumscribed areas scattered along the coast where temperatures are higher than the usual—for example, where eddies from the warm Tasman currents reach the shore.

QUEENSLAND—EAST SUBTROPICAL AND TROPICAL

Introduction

The aim of the following section is to give a general impression of the various biotas that are to be found in subtropical and tropical Queensland on the intertidal rocky shores of the mainland and on the various types of adjacent islands and reefs (that is, high wooded islands, low wooded islands or island-reefs, cays, and the inner and outer reefs of the Great Barrier Reef itself). No attempt has been made to compile detailed lists of all the fauna and flora, for such lists have been published in numerous papers (including T. A. Stephenson, Tandy, and Spender, 1931; Endean, W. Stephenson, and Kenny, 1956; W. Stephenson, Endean, and Bennett, 1958; Guiler, 1960; and Knox, 1963). Rather, our aim has been to provide a clear and general picture of the region.

The coast of Queensland can be divided into two parts—northern Queensland, with a tropical and subtropical biota, and southern Queensland, with a warm-temperate one. The overlap of these two biotas lies roughly between lat. 25°S and lat. 26°35′S, that is, between Point Vernon and Double Island Point. The extent of this overlap is very small (see Figure 11.4).

Southern Queensland has a characteristic coastline of rocky headlands separated by long sandy beaches. The fauna and its zonation are very similar to those of the north of New South Wales, with the usual modifications brought about by variations in the environment. Figures 11.21 and 11.22 show the zonation of the more plentiful animals and some of the algae in northern and southern Queensland under different conditions of exposure. Figure 11.23 shows the percentage of species of a typical northern Queensland coast and of a typical southern Queensland coast plotted against latitude. Note that the actual boundary between northern and southern species lies in the region of lat. 25°S. In the Moreton Bay-Caloundra area, there is a definite northern element with corals, alcyonarians, zoanthids, tropical molluscs, and echinoderms; but 30 miles south of Double Island Point, only one typically northern form has been recorded. This dramatic change is comparable with the sudden change from cold-temperate to warm-temperate biotas on the south coast of Cape Province, South Africa.

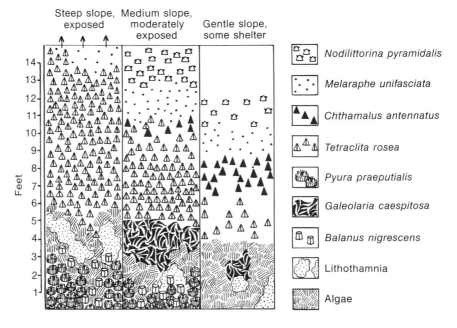

FIGURE **11.21**

A sequence of typical zone patterns on the shores of southern Queensland. [After Endean, Kenny, and Stephenson (1956).]

The northern Queensland area is extremely complicated. There are barrier reefs running parallel to the coastline with numerous islands of many types between them and the shore. The coast has been subject to submergence: in places, it consists of a succession of bays opening to the north and bounded to the east by rocky headlands. In other places, mountains plunge directly into the sea, and there are deep embayments partially filled with alluvium. Recent alluvial material covers low-lying areas; clean sandy beaches are rarer than they are farther south, and mangrove swamps are common in sheltered localities.

The name "Great Barrier Reef" has been applied to the reefs stretching from the Torres Strait to Lady Elliot Island (see Figure 11.24); as Steers (1937) has pointed out, however, this is a misleading expression, as it is not until north of Cairns that there is any true barrier. South of Cairns, a series of reefs extends southward adjacent to the shore, reaching their furthest distance offshore in the south at the Swain Reefs, which are more than 100 miles from the coast. The outer barrier reefs come closest to the coast in the north at Cape Melville, from which they are only 20 miles distant. Particularly in this region, the coast and inshore reefs are always sheltered from excessive wave action by a wide channel of protected water, and the full force of the Pacific Ocean is expended on the seaward edges of the outer reefs (Figure 11.24).

No intensive ecological survey of the coastline seems to have been carried out north of lat. 16°39′S, which leaves the shore from there to Cape York (lat. 10°41′S) virtually

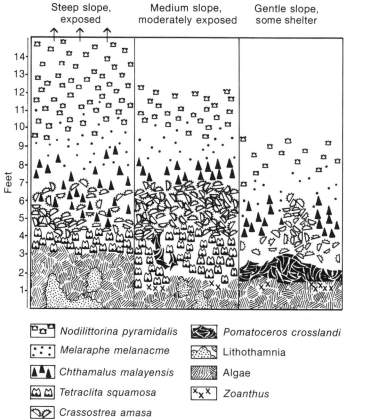

Steep slope, exposed   Medium slope, moderately exposed   Gentle slope, some shelter

Feet

Symbol	Species
Nodilittorina pyramidalis	
Melaraphe melanacme	
Chthamalus malayensis	
Tetraclita squamosa	
Crassostrea amasa	
Pomatoceros crosslandi	
Lithothamnia	
Algae	
Zoanthus	

FIGURE 11.22

A sequence of typical zone patterns on the shores of northern Queensland. [After Endean, Kenny, and Stephenson (1956).]

uninvestigated. Fairbridge (1950), in his paper on the Recent and Pleistocene reefs of Australia, comments on the paucity of fringing reefs south of Trinity Opening, but notes the existence of uninterrupted fringing reefs from Cape Flattery to Cape Melville and from Princess Charlotte Bay to Cape York. Endean, Kenny, and W. Stephenson (1956) note the lack of common reef animals and coral on the mainland coast of northern Queensland, but the most northerly point they investigated was Whitecliffs, south of Low Isles. Therefore, the most tropical part of this particular shoreline is apparently still ecologically unknown. There is, however, a great volume of work about the innumerable reefs and islands in the steamer channel. The following paragraphs give a rough classification of the various formations in this area, and brief descriptions of their topography.

High Wooded (Continental) Islands. Much of the work on islands and reefs off the Queensland coast demonstrates only too clearly the impracticability of using tidal levels, rather than the limits of the communities, as the boundaries between zones.

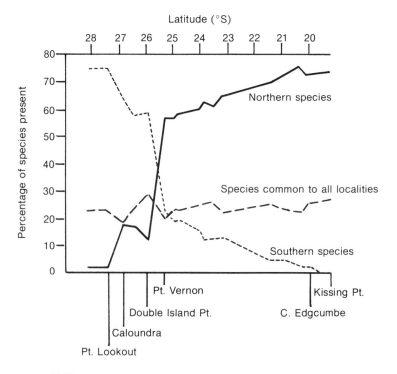

FIGURE **11.23**

The percentages of northern and southern species found at various stations on the Queensland coast. [After Endean, Kenny, and Stephenson (1956).]

It is difficult either from the diagrams or the detailed text to obtain a general picture of the zonation of animals on high wooded islands.

It would appear, however, that the zonation of high, wooded islands is very similar to that of the adjacent mainland, allowing, of course, for differences in the physical environment, such as different degrees of exposure (there is also a certain amount of variation in species). There is a supralittoral fringe with littorinids; a midlittoral with upper (*Chthamalus*) and lower (*Tetraclita*) barnacle zones, below which there is a zone of oysters (*Crassostrea amasa*), which becomes more dominant as latitude decreases. Another change as one travels north is that the carpet of algae in the infralittoral fringe is replaced by ever increasing quantities of lithothamnia, zoanthids, and corals.

Low Wooded (Coral) Islands and Cays. These are formed on patches of reef, and are confined to the area enclosed between the outer barrier reefs and the mainland coasts.

Cays, which may be composed of sand or shingle, are found on the leeward sides of reefs. In its simplest form, a cay consists merely of a mass of sand or shingle piled up on a reef. Its height is typically not more than 4 or 5 feet above high water of

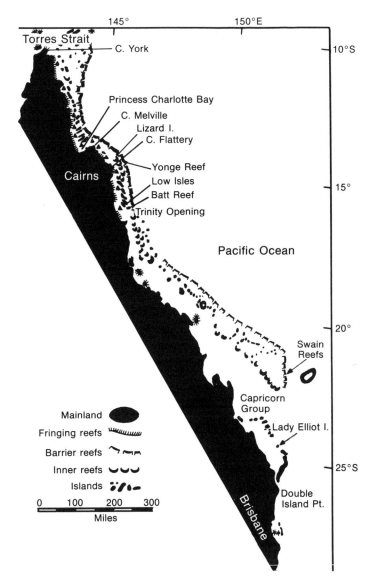

FIGURE **11.24**
The Great Barrier Reef from Brisbane to Torres Strait.

spring tides; it is flat-topped, and may be covered with vegetation. Beach rock of lithified sand may form around a cay, or coarse shingle may be consolidated into a hard coral conglomerate. A cay remains stable as long as the conditions that gave rise to it—a reef of adequate height and constant southeast trade winds—remain constant. A cay may change its shape or be almost swept away in extreme weather conditions during the summer, but it may re-form again when the trade winds are re-established.

Low wooded islands, or island reefs, are more complex and varied. They occur where the outer barrier reefs are fairly near the mainland and form a true barrier. The enclosed channel is a region of persistent southeast trade winds with a short season of summer calms and cyclones. Nearly all of the reef platforms on which these islands rest have the shape of a horse's hoof. Typically, there are two distinct islets. One is a simple sand cay crowned with vegetation and more or less surrounded by beach rock. The other, to the weather side (southeast) of the first and separated from it by shallow water, is a cay of shingle and mangroves, characteristically with a platform of shingle conglomerate on its outer side. The whole is protected by a well-developed outer rampart of coral fragments (which may be partly cemented), in most cases with boulders on its crest.

This outer rampart may almost encircle the reef platform. Enclosed within it is a more or less continuous moat, and a variable number of inner ramparts, which move slowly over the surface with ridges of coral shingle (some islands may have lesser moats, depending on local conditions) until a reef flat covered by shallow water is reached. The flat has a substratum of honeycomb rock, sand, and mud, which, but for deep pools, may be exposed at low water. It has a distinctive flora and fauna, including corals. Some corals may show above the surface, while others are permanently submerged. Scattered mangroves stray into this area. A belt of rich coral growth encircles the whole reef, varying in luxuriance, but "coral heads" (mounds of rock overgrown by living coral) and the boulder zone reach their most exaggerated forms on the leeward side.

Low Isles (lat. 16°23′S) is the most southerly island of this type in the area of the outer barrier reefs, and has been the most intensively studied; research there has continued since the Great Barrier Reef Expedition of 1928-1929. It appears that Low Isles is in an advanced stage of development (Figure 11.25). The sand cay is the dominant feature, and there is scarcely any shingle cay, whereas on most other islands of the same type, the shingle cay is the more conspicuous. There is an unusual parklike area where mangroves have invaded the reef flat. The anchorage is particularly pronounced. The biota of Low Isles is described on pages 314-322.

Inner Barrier Reefs. Batt Reef (lat. 16°23′S) is typical of inner barrier reefs. Although no detailed survey has been carried out here, enough information has been gathered about it to provide a good impression of this type of reef.

Batt Reef is extremely large, about 10 miles long and 4 miles wide, its major axis running roughly ESE to WNW. Its shape is roughly that of a distorted crescent with its convexity facing the southeast (Figure 11.26). Only a part of its eastern edge is open to the Pacific Ocean; the southern part of that edge is protected by Tongue Reef, and the western edge faces the protected steamer channel. There is no deep water inside the arms of the reef, but, instead, an immense sand flat, whose central part is exposed at extreme low water. At first appearance, the flat seems empty. The biota of this reef is described on pages 322-323.

Outer Barrier Reefs. Yonge Reef (lat. 14°36′S), named by the Hydrographic Department of the Royal Australian Navy, is representative of a great number of the reefs in the outer barrier-reef series. It is accessible only at low water of the most extreme

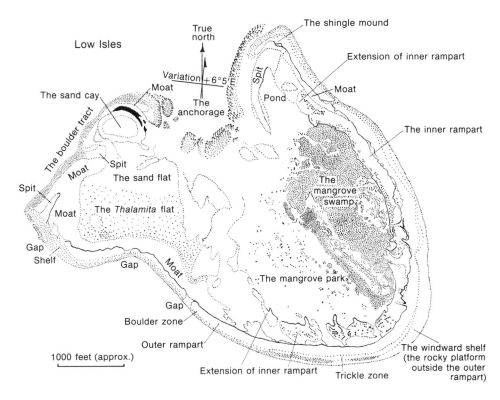

Low Isles

True north

Variation +6°5'E

The shingle mound

Extension of inner rampart

Moat

The sand cay

Moat

Pond

The anchorage

Moat

The inner rampart

The boulder tract

Moat

Spit

The mangrove swamp

Spit

The sand flat

Moat

Spit

The *Thalamita* flat

Gap

Shelf

Gap

Moat

The mangrove park

Gap

Boulder zone

Outer rampart

1000 feet (approx.)

Extension of inner rampart

Trickle zone

The windward shelf (the rocky platform outside the outer rampart)

FIGURE **11.25**

Low Isles, Australia. [After Stephenson et al. (1931).]

spring tides. It is an elongated reef about 4 miles long and 1 mile wide, with its major axis directed nearly NNW to SSE (Figure 11.27). The two extremities of the reef form two incurved horns that contain, within their crescents, coral heads and reef patches. The reef has a "backbone" of solid coral rock, swept clean of debris, that is about 3 miles long and some 160 yards wide. It is exhilarating to walk along it—a highway in the sea! On its seaward side, this crest slopes gently down to an outer moat that is practically continuous and about 100 yards across. The outer moat is contained by an outer ridge that is lower and much narrower than the reef crest; it is interrupted in various places by wedge-shaped clefts or surge channels. From this ridge, it can be seen that the reef slopes gently into the Pacific for a short distance and then appears to descend steeply into a chasm. The coral growth on this outer edge is magnificent, as are the enormous breakers that curl and crash upon it.

On the landward side of the reef crest, there is a shallow and ill-defined inner moat; beyond it, there is a rather haphazard belt of boulders and coral fragments, which gradually fades out as the depth of the water increases, and an area of rich coral growth—the anchorage coral zone. Farther to landward, the water deepens into the steamer channel, which lies between the outer barrier reefs and the mainland. The biota of this reef will be discussed on pages 323–325.

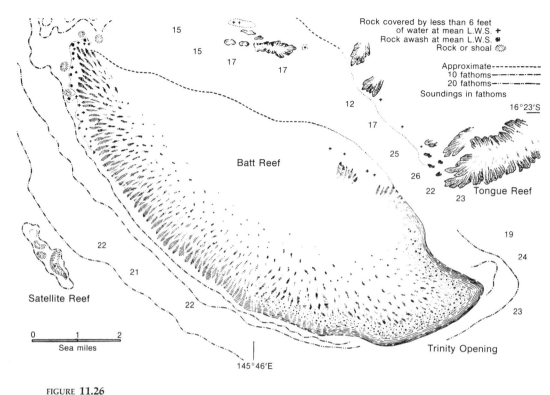

FIGURE **11.26**

Batt Reef, Australia. This chart summarizes the knowledge of the reef gained by the Barrier Reef Expedition of 1928-1929. The representation is purely conventional: the extent of the pavement indicated by *parallel shading* round the windward angle of the reef is not known; the *unshaded* part of the reef represents sand. [After Admiralty Chart No. 2924.]

Summary.   There are many distinctive types of habitat along the Queensland coast. Such factors as wave action, substratum, shore profile, salinity, turbidity, and rainfall are highly variable. The whole area, however, is subject to the same gradations of sea temperature and is not disturbed by strong ocean currents. In the widest sense, the environment is fairly constant.

It appears that the intertidal fauna of high wooded islands is much more closely related to that of the mainland than to that of the barrier reefs. There is, of course, an overlapping of the two faunas, but for practical purposes, the mainland fauna and that of the outer barrier reefs may be regarded as entities. This may not be true of the flora, however; the algae of this area have been little studied.* Australian workers recognize

---

*There are numerous papers on Queensland coasts, but, unfortunately, much of the recent work directs only the most cursory glance at the algae. This produces a restricted ecological picture. Fauna and flora react with each other in numerous ways to produce a complex ecological picture that cannot be thoroughly interpreted without due consideration being given both to plants and to animals. See page 9 in Chapter 2.

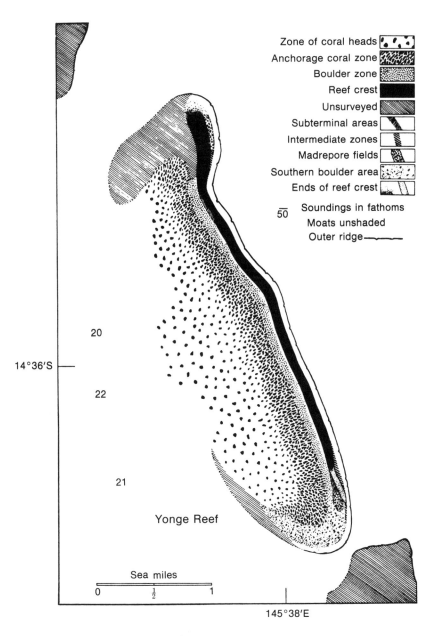

Zone of coral heads
Anchorage coral zone
Boulder zone
Reef crest
Unsurveyed
Subterminal areas
Intermediate zones
Madrepore fields
Southern boulder area
Ends of reef crest

$\overline{50}$ Soundings in fathoms
Moats unshaded
Outer ridge

20

14°36'S

22

21

Yonge Reef

Sea miles

0   ½   1

145°38'E

Moderately sheltered

FIGURE **11.27**

Yonge Reef, Australia. The relative widths of the zones along the oceanic side were obtained from pacings along a transverse section. The remaining details are sketched in, and the sizes and relations of the small areas near the ends of the reef are only approximate. [Size, outline, and orientation from Admiralty Chart No. 2923.]

a Great Barrier Reef Province and a mainland one that they have designated the Solanderian Province; the innumerable islands would then appear to constitute an assorted collection belonging variously to both provinces.

It hardly seems reasonable to break up, into separate provinces, an area with essentially the same environment containing a great variety of habitats. It is preferable to designate the whole region as East Tropical and Subtropical, and to describe briefly the different types of islands in groups—namely, high wooded (continental) islands, low wooded (coral) islands or island reefs, cays, and inner and outer barrier reefs—and to give short accounts of their respective, and predictably changing, biotas.

Low Isles

Although Low Isles is not entirely typical of island reefs, so much work has taken place there that, in a general survey such as this, it provides the best picture available of the type of biota found on such formations (see Figures 11.25, 11.28, and 11.29).

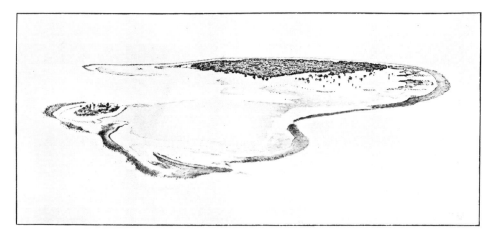

FIGURE **11.28**

An aerial view of the Low Isles reef, looking east. The lighthouse and the roofs of the huts used by the expedition may be seen on the cay. [After a photograph by F. S. Russell.]

FIGURE **11.29**

An aerial view of the Low Isles reef, showing the part of the reef concealed by the mangrove swamp in Figure 11.28. The seaward edge is approximate here, because it was blurred in the photograph from which the drawing was prepared. [After a photograph by F. S. Russell.]

The Cay.   This supports a dry land biota with deciduous trees, palms, casuarinas, various bushes of lower growth, and climbing plants, with associated insects and arachnids. Swiftly running nocturnal crabs, *Ocypode ceratophthalma* and *O. cordimana*, make numerous burrows in the sand. Well below vegetation line, there is a large population of undersurface "cockles," the mactrid *Davila plana*.

Beach sandstone (cemented sand) lies in patches on the lower slopes of the cay. The receding tide leaves this rock exposed to intense heat; in spite of this, the surfaces are extensively colonised by blue-green algae, some *Enteromorpha*, the cemented bivalves *Crassostrea amasa* and *Chama jukesii*, small periwinkles, and numerous sea anemones (*Anthopleura* sp.) in the cracks. These animals are able to withstand daytime temperatures of more than 44.8°C. On the undersides of the sandstone, chitons, boring bivalves, and sipunculids are common. Acorn barnacles and limpets are present, but not in great numbers. Crabs rush about the area during certain states of the tide.

The Reef Flat.   This consists of several parts: (1) the mangrove swamp, with its pools, glades, park, and shingle tongues; (2) various rocky areas; (3) the *Thalamita* flat; and (4) various sandy areas.

1) The mangrove swamp (Figure 11.30) is a very specialised area that shelters a diversity of organisms adapted to its peculiar conditions (for example, animals modified

FIGURE **11.30**
The outskirts of the mangrove swamp at Low Isles, showing the system of stilt roots of *Rhizophora mucronata*.

for aerial respiration, and plants adapted for life on a muddy substratum). Such are the jumping fish *Periophthalmus koelreuteri*, the whelks *Prazus* and *Telescopium,* the periwinkle *Melaraphe,* and the crustaceans *Thalamita anomala, Scylla serrata,* and *Clibanarius striolatus.* Many of these animals are more or less limited to the mangrove area, seldom straying far from it.

The dense woodland of mangroves is almost impenetrable. The trees are mainly *Rhizophora mucronata.* Snails, crabs, hermit crabs, and fish live among and on their roots and branches. Pigeons frequently roost in the upper branches.

The passages and glades into the swamp demonstrate the penetration of reef animals into woodland. The deeper glades are sinister places that might be a hundred miles from the coral reef, for all the connection they appear to have with it. There are openings in the trees, floored by foul black mud on which lie scores of whelks of unpleasant aspect, amid dead tree trunks riddled with shipworms. Here also lives the noteworthy black bivalve *Cyrena coaxans.*

The sandy pools and open glades of the swamp may be considered together. They have a soft or firm bottom of muddy sand on which cyanophyceans are ubiquitous. Where mangrove roots intrude, there are extensive carpets of the fernlike algae *Caulerpa verticillata* beneath them, and most of the floor has a covering of *Thalassia.* Sponges, compound ascidians, hydroids, and zoanthids grow on the mangrove roots. There is a large population of burrowing and surface-living organisms, including holothurians, crabs, hermit crabs, cockles and other bivalves, prawns, and tube-making polychaetes. Some pools contain species of clams, anemones, and corals. Often, on mangrove roots flanking the entrances to the glades and on dead coral slabs and shells, small living colonies of massive *Porites, Cyphastrea,* and *Leptastrea* can be found.

The mangrove park, consisting of outlying trees and saplings of *Rhizophora,* covers a large area of the whole reef flat and varies considerably in nature and level. The biota is not easily characterised. Because it is a part of the reef flat, which has other definite areas of rock and sand, many of its animals are common to the whole flat. There are, however, some fairly distinctive features: There is a widespread underwater greensward of *Thalassia hemprichii,* and with it, an abundance of *Halimeda opuntia.* Common also are *Caulerpa racemosa* and *Halophila ovalis. Cystophyllum muricatum* is notable, for it reaches a height of 4–5 feet above the substratum, the maximum for all forms on the reef; it is found in association with species of *Sargassum,* including *S. lanceolatum.*

The epiphytes and animals living on mangrove roots in the park are similar to those in open glades. Particularly interesting are *Valonia aegagropila* and an unusual form of *Turbinaria turbinata.* The animals include the hydroid *Myrionema amboinense* (the most distinctly characteristic animal of the park), *Crassostrea amasa,* and a large, velvet-green and brown zoanthid. Other special features are the large black *Pinna vexillum,* the great sprawling holothurian *Synapta maculata,* the black commercial sponge *Euspongia irregularis* var. *pertusa,* the spider shell *Pterocera croata,* and the box crab *Calappa hepatica.* There is a scattered population of living corals (they are not abundant here, nor do they flourish as they do in the moats), including *Pocillopora bulbosa,* massive species of *Porites,* and species of *Favia, Leptastrea, Cyphastrea,* and *Heliopora.* Common crabs are the mangrove hermit *Clibanarius striolatus* and the blue and green *Thalamita crenata.*

FIGURE **11.31**

A semicircular bank of loose shingle piled up on the inner side of the inner rampart at Low Isles. See also Figure 11.25.

Several shingle tongues project into the mangrove park. They have their own peculiar biota with variations according to their position on the reef, that is, whether they lie actually in the swamp area or outside it (Figure 11.31). Generally speaking, their highest levels are populated by ants, geckos, and various flowering plants typical of a mainland strand. There are the inevitable isopods, with probably more than one species of *Ligia*. Lower down, there are numerous small gastropods, typically under the loose shingle, and quantities of small, scuttling crabs. Notable is the large blue mangrove crab *Scylla serrata*, which can make great pits, more than a foot across, in the shingle.

2) The rocky areas—relatively continuous tracts of rock on the reef flat—have a biota that is naturally rather complicated, but not markedly different from that of any reef-flat area with dead coral blocks on it. One unusual feature of the rocky areas is that, in certain places, there is a superabundance of *Onchidiella*, a soft leathery pulmonate. Other common inhabitants are the hairy anomuran *Pilumnus*, various other crabs, brittle stars, chitons, holothurians, and attached bivalves. A short, mosslike turf covers the rocks. The constituents of the turf vary from place to place, but the usual ones are species of *Laurencia*, *Boodlea*, and *Dictyosphaeria*.

3) The *Thalamita* flat—another distinct stretch of reef flat—consists mainly of a sandy floor overlaid by slabs and boulders of dead coral and some broken honeycomb rock. There are a few deep pools with populations resembling those of the moats. Living coral is scarce; it is restricted to shallow fringes of living *Porites* around slabs of dead

coral, and to small colonies of other corals, chiefly *Favia* and *Cyphastrea*. On the upper surfaces of the dead coral rock are species of *Chama* and *Spondylus* together with acorn barnacles and *Siphonaria* sp., but the most concentrated population lies under and on the sides of these rocks. Brilliantly coloured encrusting sponges mixed with encrusting algae and *Melobesia* spp. are conspicuous. A characteristic animal is the abundant crab *Thalamita stimpsonii*. The clam *Hippopus hippopus* is common, the adults of similar size being evenly distributed. The short turf of encrusting algae is like that of the rocky areas, with a spring growth of *Hydroclathrus*. On the sandy floors, there is the usual growth of *Thalassia*, with intermingled *Spyridia filamentosa*.

4) Sandy areas are partly exposed at low water, but have permanent shallow pools. Some sand flats are bare, but most have a sparse growth of *Thalassia* or *Halophila* or both. The *Halophila* is typically covered by epiphytes, and beardlike growths of cyanophyceans (*Lyngbya majuscula* in the shallow places and *Hormothamnium solutum* in the deeper ones) may occur in pools and exposed areas. A stiking algae is *Arainvillea erecta*, with its 6- or 8-inch stipes buried in the sand; other common seaweeds are *Caulerpa racemosa* and species of *Sargassum* and *Cystophyllum*.

The fauna of the sand is distinctive but mostly concealed. A characteristic form is the anemone *Actinodendron plumosum*, with only its fernlike tentacles expanded on the surface when it is covered by water. It is somewhat unexpectedly that polychaetes are not the dominant animals here, but instead, species of the burrowing anemone *Edwardsia* and, next in quantity, the very large *Balanoglossus carnosus*. Also plentiful are sipunculids, polychaetes, and various burrowing gastropods, including *Natica*, *Nassarius*, and *Terebra*, with *Strombus* on the surface. An interesting animal is the giant squilla *Lysiosquilla maculata*, which makes a very deep burrow. In addition to these typical sand-dwellers, there are a number of species characteristic of the reef flat in general.

The Moat. The conditions prevailing here restrict the range of flourishing life. The biota of the moat must adapt itself to the following circumstances: (1) water that is shallow and still for long periods, often resulting in temperatures as high as 35°C; (2) salinity that may fluctuate markedly, particularly in the wet season; (3) considerable movement of water during the ebb and flow of tides and the resultant to-and-fro movement of sediment; and (4) variations in the oxygen saturation of the water—it may reach 27% after a daytime of low water, or fall as low as 20% at the end of a low water at night, according to Orr and Moorhouse (1933).

The constituent organisms of the biota of a moat vary considerably according to its position on the reef. The following two paragraphs contain a brief description of the population to be expected in a typical rich moat.

Coral platforms are common in seaward parts of moats, where they show a particular development. The coral colonies grow in a normal fashion until they project above the water, then, the exposed tips eventually die off, and the surface becomes covered with sediment and infested with microphytic algae and species of *Melobesia*. The result is a dead coral platform with living coral at the edges and down the sides. Such formations provide shelter for a multitude of inhabitants—commensal, boring, and shelter-loving—among which anemones, urchins, worms, crustaceans, and fish are common. The general population of the moat includes numerous tridacnids of various sizes, starfish, holothurians, sea urchins, highly coloured sponges, colonies of living coral (including, especially, species of massive *Porites*), and dense colonies of the tube-dwelling polychaete *Spirobranchus giganteus*, with its beautifully coloured fans expanded at the surface.

Algae in the moats are abundant and occur in various forms—turflike tangles, spongelike masses, and carpets of considerable extent. Characteristic forms are *Turbinaria turbinata*, tufts of *Padina* and the calcareous *Halimeda*, creeping species of *Caulerpa*, and *Dictyosphaeria*. The more exposed plants of *Dictyosphaeria* become bleached in the sun. In shallow places, such as those near the *Thalamita* flat, there are long trails of *Cystophyllum* and *Sargassum*. The coralline *Amphiroa fragilissima* is ubiquitous. The sandy floors of nearly all moats support some *Thalassia*, but the growth is relatively sparse.

The Ramparts and the Boulder Tract. The extent of the outer rampart is clearly shown in Figure 11.25. It consists of loose fragments of dead corals and shells. These fragments are not generally cemented by species of *Melobesia*, as those of the ramparts of the outer barrier reefs typically are, but the angulated and nodulated nature of the pieces makes for interlocking, and the rampart functions successfully as a breakwater. Because of its unconsolidated nature, however, the rampart is not a wholly stable feature: it may be encroached upon by layers of new shingle, gradually driven into different positions, or breached by the waves during cyclonic disturbances.

At first sight, the outer rampart appears to be a most unpromising habitat for colonisation by plants and animals. Although very little surface life is visible, investigation shows that the outer rampart conceals an extremely populous fauna. In the highest and driest parts, little exists but rapidly moving isopods; but elsewhere, mostly under shingle, there are abundant small gastropods (especially *Nerita* spp.), crabs, the echinoderms *Asterina exigua* and *Echinometra mathaei*, small holothurians, hermit crabs, bivalves, and anemones. Lower down, where there is considerable moisture from water trickling off the flat, the shingle may be bound together in places by turfs of algae, of which the most remarkable is a form of *Caulerpa racemosa*.

The biota of the boulder tract is naturally much more varied than that on the outer rampart, but one statement can suffice to describe both the boulders and the tract on the leeward side of Low Isles, because their characteristic animals and plants mingle with one another: over all other groups of animals, molluscs and crustaceans predominate, including species capable of withstanding arid conditions and unpromising environments like shingle (see note on shingle tongues, p. 317). Particularly notable on the larger boulders of the tract are the numbers of boring bivalves and sipunculids. *Tridacna crocea* is present in great numbers, and makes holes of considerable size in the rock. This large population of boring animals is responsible for a great deal of destruction of reef material.

The inner rampart differs from the outer one in many ways. The highest parts of the inner rampart are extensively colonised by certain flowering plants, including the creeping carpetweed *Sesuvium portulacastrum*, isolated patches of fairly dense *Avicennia officinalis*, and seedlings and saplings of *Rhizophora mucronata*, many of them in poor condition. The fauna is a mixture of outer-rampart and shingle-tongue species; small gastropods and crabs are the dominant forms. In damp places (near the mangrove swamp, for example) the fauna becomes more modified and richer. One unique feature of the inner rampart is the calling crab *Gelasimus tetragonon,* which makes its burrows in extraordinarily hard material.

The Anchorage. In the anchorage on the seaward slope, wherever there is a sufficient depth of water and where there are no unusual inhibiting conditions, corals and alcyonarians abound. The anchorage reefs consist of very irregular masses of dead coral with sand and debris between them. Because these reefs were produced by corals growing on a seaward-sloping shore, the height of those that reach the level of low water is naturally greater in deeper than in shallower water (Figure 11.32). In areas where the coral grows close to the surface of the water, the coral tips may be exposed or awash at low water of spring tides, which is enough to modify the coral fauna very definitely: The chief corals are platformlike growths of *Montipora racemosa, Acropora hebes,* and various others somewhat similar to those of the moat. Particularly notable is *Acropora pulchra,* with its slender buff-coloured branches with blue tips. It forms large, rounded bushes in a peculiarly limited zone, that is, on seaward faces exposed at low spring tides. Alcyonarians are also abundant.

In areas where the coral grows lower in the water, its growth becomes denser and more varied. The predominant genus is *Acropora,* with its many growth forms. Almost equally abundant are the foliose species of *Montipora.* Alcyonarians, no less than corals, find their optimum conditions here—colonies of *Sarcophyton* 2–4 feet across are not rare, and there are fields of *Sinularia* covering yards of rock, as well as many other species.

Table 11.1 names the animals (other than corals and alcyonarians) and the algae common to seaward slopes and the anchorage. It will be noted that there is a considerable variety, particularly among the tridacnids. Algae are well represented; they belong primarily to the higher levels, and their growth in deeper areas is negligible.

Seaward Slopes of the Reef. Although the anchorage is certainly a seaward slope of the reef, the fact that it is an inlet, and therefore subject to unusual conditions, justifies its separate treatment. The rocky slopes dealt with here descend from the outer rampart and boulder tract to the muddy floor that surrounds the reef. The configuration of these slopes may be summarised thus: On the windward arc, the slope is at first

FIGURE **11.32**

A section of a slope to leeward of Low Isles, illustrating the arrangement of coral masses upon it. The vertical and horizontal scales are the same. Compare with Figure 11.33.

TABLE **11.1**

Various animals and algae common to the anchorage
and the seaward slopes of Low Isles Reef, Australia

Animals	Algae (contd.)
Echinostrephus molare	*Dictyosphaeria favulosa*
*Lamprometra gyges**	*Dictyosphaeria sericea*
Linckia laevigata	*Dictyosphaeria versluysii*
Panulirus versicolor	*Dictyota bartayresiana*
Pinctada margaritifera	*Dictyota ciliata*
Spirobranchus giganteus	*Digenia simplex*
Stichopus chloronotus	*Galaxaura* spp.
Tridacna crocea	*Halimeda cuneata*
Tridacna derasa	*Halimeda opuntia*
Tridacna fossor	*Halimeda tuna*
Trochus niloticus	*Laurencia botryoides*
Algae	*Laurencia papillosa*
Amphiroa fragilissima	*Melobesia* spp.
Boodlea paradoxa	*Padina australis*
Bornetella nitida	*Padina commersoni*
Caulerpa racemosa	*Sargassum cristaefolium*
Chlorodesmis comosa	*Turbinaria ornata*

*Various other crinoids are also present.

gradual; then there is a steep drop (a declination of 9 inches every horizontal foot), after which the slope becomes gentle again until the floor is reached. At the foot of the slope, there is a slight trough. Some slopes are more gradual, and others may be straight all the way to the bottom, but there is a general tendency for a short, steep drop to occur about half way down.

The windward shelf is a pavement of pitted honeycomb rock to seaward of the outer rampart. It is considerably overgrown by a felt of algae, and its surface is studded with boulders and débris. Of all of the algae on the shelf and its boulder zone, the high-growing *Sargassum cristaefolium* is the most notable. With *Turbinaria ornata*, it makes a distinct "sargassum" zone, widest and most luxuriant at the weather extremity and in places where the trickle zone is well marked. The familiar *Chlorodesmis comosa,* a brilliant green alga characteristic of outer barrier reefs, is present all along this area, and a species of zoanthid of the genus *Palythoa* forms large masses that are exposed at low water.

Seaward of the windward shelf and its boulder zone, a region of alcyonarians and coral develops gradually until a rich growth is attained in water roughly 5 fathoms deep (Figure 11.33). The surface of the slope is mostly rugged and barren. At first, it is covered with broad low fixed blocks of dead coral; to seaward, these blocks may become large mounds more than 20 feet high, and as much as 25 feet across the top. Many of them are mushroom-shaped, with corals growing over their tops and down their sides. For some distance out to sea, the tops of some of these mounds may be exposed at exceptionally low tides. This belt of rich coral growth surrounds the whole reef, and a distinct zonation of the constituent species of coral can be recognized.

FIGURE **11.33**

The seaward slope of the leeward side of Low Isles, showing the type of coral growth. In the distance is part of the cay, and in front of it, a portion of the boulder tract. In the foreground are two coral heads, showing the overhanging form that they often assume and the way in which the principal growth of corals is on and a little below the top. The coral head on the right slopes gradually to the sea floor on one side, and here the rich growth of coral extends further down. The shore is seen at low water of a low spring tide. The proportions are approximate. Compare with Figure 11.32.

Exceptional animals of the seaward slopes are the stinging hydroid *Lytocarpus phoeniceus,* the urchin *Echinostrephus molare* (which burrows deep in solid rock), and the *Trochus niloticus* of commerce. The characteristic animals of the area were listed by T. A. Stephenson et al. (1931).

Mention must be made here of the innumerable brilliant and varied fishes that abound among the coral reefs. Many of the smaller ones are commensal with coral.

Batt Reef

The population of the large central sand flat of Batt Reef (see Figure 11.26) consists mainly of such sand-binding plants as trailing *Caulerpa, Thalassia,* and *Cymodocea.* There are isolated boulders on the flat, and scattered colonies of *Pocillopora bulbosa* and a few tridacnid clams and pearl oysters (*Pinctada*). The anemone *Stoichactis* sp. is common.

Between this central region and the margin of the reef, there is a mixed landscape of large boulders cast up in storms, pieces of rock with sand between, pools of varying depths, and patches of dead coral with a secondary growth of living coral on or around them (particularly common and notable is *Tubipora musica,* the red organ-pipe coral, which was not found on Low Isles). The biota here in many ways resembles that of the *Thalamita* Flat on Low Isles. In the deeper pools, the population is varied and interesting.

A small area situated on a reef patch, a square tract 6 feet 8 inches on a side, was studied in detail. The substratum consisted of irregular masses of deal coral much overgrown by short algae, with sand between them, whose tops were uncovered only at extreme low water. The animals present in this small area were considerable: there were 70 small colonies of living corals, 85 colonies of the alcyonarian *Xenia umbellata,* abundant ascidians, and a few assorted specimens of *Aplysia, Echinometra mathaei, Euspongia,* various nudibranchs, sedentary polychaetes, and *Tridacna fossor.* The algae growing in the same tract included 42 stems of *Turbinaria ornata,* abundant *Amphiroa fragilissima* and *Melobesia* spp., and a few specimens each of *Bornetella, Caulerpa, Dictyosphaeria, Galaxaura, Halimeda, Laurencia,* and various cyanophyceans.

The outer margin on the eastern edge of the reef consists of a gently sloping cemented rock pavement much overgrown with mossy algae. The pavement is inhabited by scattered solid corals and thick-shelled molluscs, and the surf breaks over it with tremendous force.

Yonge Reef

Yonge Reef (see Figure 11.27) was chosen for careful examination both because it was easily accessible from our base at Lizard Island, and because it appeared to be a reef characteristic of the outer barrier series. On parts of Yonge Reef, the biota is markedly different from that of Low Isles. On outer barrier reefs like this one, the complications of ramparts, mangroves, and sandy flats do not occur, and a series of bandlike zones on the seaward side seem peculiar to this type of formation. Of the numerous corals and alcyonarians, some were quite new to us, while others represent modified growth forms of species we had seen elsewhere. A notable feature of Yonge Reef is the tremendous encrustation of *Melobesia* spp., which forms a complete veneer over the coral rocks and gastropod shells, and produces the prevailing purplish-pink colour: "It renders the landscape to leeward incomparably beautiful for it has a pink foreground with the brilliant blues and greens of shallow water over white sand behind it" (T. A. Stephenson et al., 1931). The rather coarse sand to leeward of Yonge Reef is very clean and white, and the water is of glasslike clarity.

The Outer Ridge.   A rich growth of corals flourishes here (see Plate 20 and Figure 11.34), including massive species and *Acropora.* The various species of *Acropora* exhibit many different styles of growth: cones united to a firm foundation (*A. gemmifera*); dishlike brackets (*A. hyacinthus*); encrusting sheets yards in extent (*A. palifera*); and systems of heavy branches lying close to the substratum (*A. decipiens*). Yet another

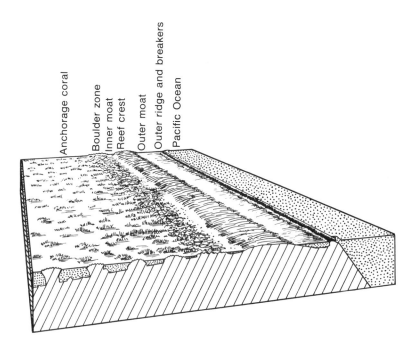

FIGURE **11.34**

The arrangement of the several zones on Yonge Reef, as seen at low water of a low spring tide. The figure has been drawn, as far as possible, to scale—the vertical scale being ten times the horizontal—but the coral heads are exaggerated in horizontal scale in order to avoid giving a false impression of their shape. Compare with Figure 11.27.

member of the genus (*A. delicatula*) produces bushes of branches so slender and brittle that they are difficult to collect; but somehow, they are able to withstand the turbulence of breakers.

The Outer Moat. This contains constantly agitated water. It has a wide range of coral growth, and numerous other animals, including tridacnids, gastropods, hermit crabs, compound ascidians, sponges, and large hydroids. As the water becomes shallower towards the reef crest and the coral becomes more and more scattered and impoverished, the boring echinoids *Echinostrephus molare* and *Echinometra mathaei* predominate, the former sometimes occurring as thickly as six individuals per square foot.

The Reef Crest. This slopes upward from the outer moat. Species of *Acropora*—generally those of shorter, stumpy growth—are dominant; *A. gemmifera* is particularly characteristic. There are also corals belonging to other genera. Alcyonarians, starfish, zoanthids, and sponges are notable features of the reef crest, and an abundant fauna may be found under boulders and in shallows. An outstanding alga is the bright green *Chlorodesmis comosa*. There are also small amounts of *Halimeda* and *Caulerpa*.

The reef crest varies somewhat from one place to another, giving rise to interesting changes in the biota. At its north, and where it fans out, there is a plentiful growth of *Favia;* to the south, where it slopes downward, the reef becomes partially bare with a poor growth of *Melobesia* spp., little coral, but an extensive turf of soft algae.

The Inner Moat. The bottom is rocky here, and such boulders as occur are mostly cemented to the floor. The coral growth is fairly strong, but in scattered clumps, the general effect being that of an impoverished version of the outer moat.

The Boulder Zone. This consists of loose boulders and other débris in some quantity, and a small amount of sand. There is a good fauna under the boulders, with thick patches of *Halimeda* and *Caulerpa*. The biota of this boulder zone is similar to that of the boulder zone at Low Isles (see p. 321).

NORTHERN TERRITORY—TROPICAL

This brief account of the shores of Arnhemland, Northern Territory, is based entirely on work done by Miss Elizabeth Pope, who visited this area in the middle 1960s. She most kindly made her notes available, for which we are extremely grateful. Her studies were carried out at Cape Don on the Coburg Peninsula, and at Port Darwin and a number of places in its vicinity.

Along this coast, which is the termination of a long, low-lying mainland far from inland mountains, there are no substantial cliffs. At first sight, the coast gives the impression of being a succession of sandy beaches, mud flats, and mangroves. Closer inspection, however, reveals a few rocky reefs, headlands, and outcrops made of sandstone and shales with bands of laterite (Figure 11.35). These formations extend down the shore for varying distances and are often discontinuous. As a rule, the shore has a very gentle slope to seaward, and an extensive fringing reef is typical. The rocks make a gradual descent in a series of shallow steps, with scattered boulders marking their edges; under the boulders, there is the characteristic fauna, including ascidians, serpulids, and so forth.

The rocks of these platforms are quite hard, but at the top of the beach, where slight cliffing has taken place, the rock is a granular sandstone that is soft enough to be scratched with a fingernail. Typically, the sandstone is streaked with colours, mainly yellow ochre and rusty red. In many places, the shores are sheltered by outlying islands, and so are not normally subject to violent wave action.

The tides are semidiurnal, and the spring tides may have an extreme range of about 26 feet. This, combined with the long, gentle slope of the shore, causes great areas of the flat beach, from one-half to three-quarters of a mile in width, to be left uncovered at low water.

There appears to be a mixture of Pacific and Indian Ocean water masses in the vicinity of Port Darwin. There are strong tidal streams and various upwellings and sinkings of water masses. In spite of this, mud is able to settle in sufficient quantities to form a well-defined muddy zone, but the customary divisions of the shore are clearly recognizable.

The Supralittoral Fringe

The supralittoral fringe is indicated by such typical inhabitants as *Nodilittorina pyramidalis*, *Melaraphe scabra*, and *Nerita striata*. A species of *Ligia* scuttles about on the

FIGURE **11.35**

Silt-covered rocks at Fannie Bay, near Darwin, Northern Territory. Light-coloured cliffs can be seen in the background, and the "top step" of the midlittoral shows clearly below them in the centre of the picture. This area of the shore is comparatively free from the silt that covers the rocks in the foreground. The receding tide has not yet uncovered the area where brain corals flourish. [Photograph by E. Pope.]

highest rocks, where there is some shade and shelter. Below this, just above the midlittoral zone, there is a marked line of the cyanophyceans so characteristic of many shores. Astride this black line, and extending well above it into the fringe, are quantities of *Chthamalus withersii.*

The Midlittoral Zone

The upper boundary of the midlittoral zone is marked by the usual concentration of barnacles—in this case, *Chthamalus malayensis,* which is mixed with *C. withersii* from the zone above. Under suitable conditions, there may also be a few scattered limpets. Below the barnacles, there is a series of bands, of which *Crassostrea amasa* forms the most conspicuous one; this is followed by a wide belt of *Tetraclita squamosa,* and then by a narrow one of chitons. On Queensland's eastern coast, the dominant chitons seem to be *Acanthozostera gemmata* and *Liolophura gaimardi;* but towards the extreme north of that coast, and at least as far west as Port Darwin, the latter species is replaced by *Acanthopleura spinosa.* Two quite different species are found on the tropical and subtropical shores of Western Australia, namely *Clavarizona hirtosa* and *Onithochiton occidentalis.*

Below the chitons, a muddy zone begins. The shore is covered by a fine, gelatinous, brown sediment. Where this is fairly thick, it rises up as a fine silt when it is disturbed, and entirely obscures the surroundings. The mud is deepest at its upper levels, and is directly comparable with the one on the Carolina coast of the USA (see pp. 177–187). It supports a varied population of species.

Under overhangs of rock above the mud there are various simple ascidians, such as *Microcosmus australis,* together with small colonial species. The barnacles *Chthamalus caudatus* and *Tetraclita costata* occur sporadically. The upper muddy zone is conspicuous for its population of sponges, including the black *Iotrochota baculifera,* the red *Spirastrella vagabunda, Xestospongia exigua,* and the rounded orange "fibreglass" sponge *Cinachyra australiensis.* The vermetid *Siliquaria ponderosa,* the hairy, mud-covered crab *Pilumnus vespertilio, Holothuria leucospilota, Macrophiothrix* spp., and *Onchidiella* spp. are present together with coated algae, including species of *Valonia* and *Padina.*

There are pools and rocks in the lower, less smothered half of the silt-covered area, especially towards its inshore side. The rocks in the blazing sun, and many of the shallow pools, support large numbers of small, oval, compound ascidians; prominent among them is *Leptoclinum (Lissoclinum) molle.* Where the muddy zone merges into the fringing reef, the fauna, which comprises animals capable of combating the mud, is extremely rich. It is dominated by brain corals, anemones, large alcyonarians, sponges, and crustaceans (Figure 11.36).

FIGURE **11.36**

The wide area that is uncovered at low tide at Dudley Point, near Darwin, Northern Territory. In spite of the silt, there is a rich and varied fauna, and brain corals dominate the midlittoral zone. [Photograph by E. Pope.]

Bivalve molluscs, such as toothed pearl shells and razor clams, are common here, but the more spectacular gastropods, such as cowries, cones, and volutes, so conspicuous in the tropical intertidal regions of the east coast of Queensland, are few in number and small in size. Apparently, asteroids and echinoids are also scarce between tidemarks, except in special, limited areas. This is not true, however, of the offshore fauna, for a great variety of gastropods and echinoderms are reasonably plentiful in collections obtained from dredgings and from material washed ashore after storms.

The Infralittoral Fringe

In the infralittoral fringe, the shore is once again clean. Coelenterates and sponges are the commonest animals. Among the notable elements of the fauna are large alcyonarians, particularly a pinkish grey one; enormous quantities of a peculiar eunicid worm, *Eunice tubifex,* which occurs sporadically; sponges of many kinds; fringes of stinging hydroids; brightly coloured compound ascidians, including *Leptoclinum virens;* a pale brown branching species of *Acropora;* and a few very large brain corals.

NEW ZEALAND

A brief general summary of the hydrography and other features of New Zealand will be found at the beginning of this chapter where Australasia is considered as a whole.

For the reasons stated on page 281, the various regions of New Zealand will be classified under the same headings as those used for Australia. As Figure 11.3 shows, New Zealand has no tropical or subtropical shores (unless we consider the Kermadec Islands, a dependency and not properly a part of the country). The area of highest temperature is the transitional warm-temperate shore around the north of North Island. On the other hand, subantarctic conditions occur in the extreme south, which is much colder than anywhere in Australia. There are extensive areas of transition between northern and southern conditions along the east and west coasts of New Zealand.

Within these transitional areas certain subdivisions can be made. On the west coast of North Island and on the east coast around Poverty Bay (see Figure 11.37), there is a definite falling off of the characteristically warm-temperate elements of the biota, which include a fair number of Australian ones. Particularly marked are changes in the algal populations of the lower midlittoral and the infralittoral fringe.

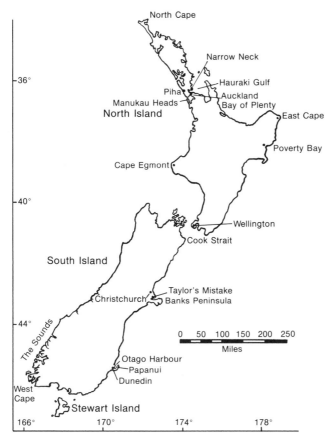

FIGURE **11.37**

A map of New Zealand, showing the localities referred to in the text.

There follows a great stretch of mixed cold-temperate waters, which extends through Cook Strait and along the east and west coasts before reaching transitional subantarctic waters in the south—near The Sounds on the west coast, and near Dunedin on the east.

Morton and Miller (1968) have carried out a very extensive survey of New Zealand shores. Other investigators have carried out extensive research on particular stretches of the coast, including: the region of the Hauraki Gulf at Auckland, North Island, in transitional warm-temperate waters (Dellow, 1955); the vicinity of Taylor's Mistake, near Bank's Peninsula, South Island, a district of mixed cold-temperate waters (Knox, 1953); and the Otago Peninsula, near Dunedin, on the edge of the transitional sub-antarctic area (Batham, 1956, 1958).

As previously stated, the most conspicuous changes in the biota of New Zealand are found among the algae of the lower zones of the shore. In order to minimise the necessity for long explanations, these variations have been shown as accurately as possible in the text figures. For details of the distribution of the common characteristic algae of New Zealand, see Moore (1961).

NORTHERN NORTH ISLAND–
TRANSITIONAL WARM-TEMPERATE

The west coast has periods of much longer-lasting and greater turbulence than the east coast. The prevailing westerly wind drives spray high onto the western shore, which raises the effective levels of high water; on the east, however, the conditions are reversed, for the tendency is for the wind to carry the sea away from the land. Because spray is not blown perpetually onto the eastern shore, the biota is subjected to longer periods of desiccation there than it is on the west. Northeasterly gales, naturally, have the opposite effect, but they occur only infrequently.

Hauraki Gulf

Dellow (1955) studied a large selection of different habitats in Hauraki Gulf on the northeast shore of North Island, and this account is taken from her general summary of the district (see Figure 11.38).

The Supralittoral Fringe. Highest of all intertidal organisms are the lichens. They are present nearly everywhere; *Lichina pygmaea*, which occurs in localised patches, tends to be the dominant species. The blue-green *Calothrix* is common on sandstone.

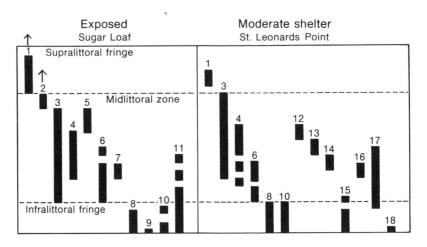

FIGURE **11.38**

A comparison of the variations of some typical species of the Hauraki Gulf, in places of shelter and exposure: 1, *Melaraphe oliveri;* 2, *Chamaesipho brunnea;* 3, *Chamaesipho columna;* 4, *Elminius plicatus;* 5, *Bostrychia arbuscula;* 6, algal turf; 7, *Xiphophora;* 8, *Carpophyllum elongatum;* 9, *Lessonia variegata;* 10, lower algal turf; 11, lithothamnia; 12, *Modiolus neozelanicus;* 13, *Crassostrea glomerata;* 14, *Pomatoceros cariniferus;* 15, *Mytilus;* 16, *Hormosira;* 17, corallines; 18, *Ecklonia; broken columns,* species present in patches; *unbroken columns,* species forming continuous belts. The vertical scale is roughly indicated, but actual depths cannot be given. In exposed places, species in the supralittoral fringe can reach heights outside the range of the figure: this is indicated by *vertical arrows.*

Verrucaria spp., and the dark brown *Hildenbrandia,* frequently form the familiar black band on cliff faces. There is the customary population of the littornid *Melaraphe oliveri.*

The Midlittoral Zone. In the upper midlittoral, the most common of all animals is the tiny white barnacle *Chamaesipho columna;* in many places, it is mixed with *Elminius modestus.* In exposed areas, *C. brunnea* becomes dominant, and may completely replace *C. columna* at the highest levels. *E. plicatus* displays a discontinuous horizontal pattern in places of swift water movement. Superimposed on the balanoid zone in different localities are a number of algae, including *Apophloea sinclairii,* reddish black *Bostrychia arbuscula,* vivid green *Enteromorpha* (seasonal), and *Gelidium pusillum.* The small, shining black mussel *Modiolus neozelanicus* occurs here, too—typically in a mixture with *Gelidium*—and it may become dominant in some places. All of the aforementioned midlittoral organisms may form distinct bands in one place or another. Small seasonal algae are common in very exposed places. Notable are *B. arbuscula* and *A. sinclairii;* the latter a leathery, encrusted form, is a widespread basic zone-former that resembles dried blood when it grows in isolated patches. Various gastropods are associated with these algae, including *Cellana ornata, C. radians, Siphonaria zelandica, Lepsiella scobina,* and *Melanerita melanotragus.*

In many places, there is a prominent belt of the oyster *Crassostrea glomerata,* which may replace *Chamaesipho columna* in shelter; its range is restricted by sediment on the one hand and by surf on the other. It provides a habitat for limpets, chitons, tube worms, and various other species, including such algae as *Caulacanthus spinellus, Centroceras clavulatum,* and *Gelidium caulacanthus.* In some places, boring gastropods (*Lepsiella scobina,* for example) cause enormous damage to beds of oysters; on the other hand, the normal growth of *Hormosira banksii* is much affected in this area by the cultivation of oysters. *Mytilus canaliculus* is also reduced by commercial collection.

Another organism that may be dominant in the lower midlittoral is *Pomatoceros cariniferus.* It is found in widely divergent situations and with great fluctuations in density; the reasons for these phenomena are still in need of investigation.

In sheltered position, *Hormosira banksii* may be associated with the cosmopolitan *Corallina officinalis* (Figure 11.39); where these species form mats in silty areas, they provide a suitable substratum for many small algae (see Table 11.2).

On the descending shore, *Corallina* and its associated turf wane and are replaced by another turf (mainly of species of *Gigartina, Polysiphonia, Caulacanthus,* and *Gelidium*) and by encrustations of pale pink lithothamnia. *Mytilus canaliculus* is scarce. Pale brown *Xiphophora* may form a *Fucus*-like band, but it is replaced, under conditions of maximal exposure, by dominating *Gigartina alveata* and *G. cranwellae,* mixed with species of *Pachymenia himantophora, Gymnogongrus nodiferus,* and *Melanthalia abscissa.*

The Infralittoral Fringe. There are several large kelps on this part of the shore (Figure 11.40). The profusely branched *Carpophyllum maschalocarpum* is almost ubiquitous, and two other species, *C. augustifolium* and *C. plumosum,* are very common. *C. maschalocarpum* is replaced by *C. augustifolium* in places of maximum exposure. *C. plumosum* has an unusual distribution being found variously in pools, in inlets, and on shifting, wave-swept boulders. *Ecklonia radiata* is abundant locally, but its holdfast

FIGURE **11.39**

A typical expansive flat reef of the sheltered inner part of the Hauraki Gulf. The rocks of the lower midlittoral are covered by an algal turf consisting mainly of corallines with some *Hormosira banksii* and small, tufted red algae. The infralittoral fringe is hardly visible; it consists of large brown algae, dominated by *Carpophyllum maschalocarpum* and *Ecklonia radiata*. [Photograph by V. Cassie.]

TABLE **11.2**

Typical turf-forming algae of Hauraki Gulf

Algae of the infralittoral turf	Algae of the coralline turf
Champia langii	*Caulerpa sedoides*
Champia novaezelandiae	*Codium adhaerens*
Cladophoropsis herpestica	*Colpomenia sinuosa*
Glossophora kunthii	*Enteromorpha procera*
Melanthalia abscissa	*Laurencia* spp.
Myriogramme denticulata	*Leathesia difformis*
Peyssonnelia rubra	*Microdictyon mutabile*
Pterocladia lucida	*Plocamium* spp.
Zonaria subarticulata	

are below tide levels. The most common algae are *Cystophora torulosa* and *Sargassum* sp. Among and under the larger algae of the fringe, there is a turf of smaller forms (see Table 11.2), among which dwell a multitude of animals, including hydroids, grey and orange sponges, polyzoans, and molluscs.

FIGURE **11.40**

A close view of the northeast tip of a very exposed rock face on Great Barrier Island, which forms the eastern boundary of Hauraki Gulf. The upper midlittoral is populated by a band of barnacles, followed by a narrow strip of lithothamnia. Below this, there is a very marked belt of brown algae, with *Carpophyllum angustatum* dominant; in some places, a few battered plants of *Durvillea antarctica* may occur towards the upper limits of this belt. [Photograph by V. Cassie.]

Piha

Piha lies on the west coast of North Island, about 7 miles to the north of the Heads at the entrance to Manukau Harbour, in or near the west-coast overlap of the transitional warm-temperate waters to the north and the mixed cold-temperate waters to the south. This very exposed shore is subject to the full force of the prevailing westerly winds of the Tasman Sea. The sea is never calm; in storms, waves and sheets of spray reach to great heights on the vertical cliffs.

The Supralittoral Fringe. This fringe is very wide because of the great heights to which waves and spray are constantly thrown. There are two species of *Melaraphe*, *M. oliveri* and *M. cincta;* as is usual in northern New Zealand, the former has the wider range. *Bostrychia* is common, except in places of extreme exposure. Species of *Enteromorpha* occur frequently on sunny faces, typically associated with cliff seepage; their best growth, therefore, is in winter and spring, when seepage is greatest.

The Midlittoral Zone. This zone is unusually wide because of the violent wave action. The familiar *Chamaesipho columna* extends throughout, but is more abundant at high levels, where it gets no competition from the frequently dominating mussel

Modiolus neozelanicus and the barnacle *Elminius plicatus. C. brunnea* is also present at upper levels. The algal mat becomes increasingly dense to seaward. Some of the characteristic constituents of this turf are *Caulacanthus spinellus, Centroceras clavulatum, Chaetangium corneum, Gelidium caulacanthus, G. pusillum,* and *Plenosporum hirtum.* There is a restricted band of the serpulid *Pomatoceros cariniferus* and the sabellariid *Sabellaria kaiparaensis* on continuous rock at the bottom of the midlittoral. Except in districts of extreme exposure, there is a better growth of these species where rocks and gullies rise from sand; in such places, the sand has a scouring action, which displaces larger algae.

The Infralittoral Fringe.   Conspicuous animals on exposed shores are *Mytilus canaliculus,* with its predator starfish *Stichaster australis,* and *Notoacmea pileopsis; Durvillea antarctica* is the dominant kelp. In less turbulent places, there is an abundance of corallines; in some places, there are belts of *Gigartina alveata, Pachymenia himantophora,* and *G. marginifera,* descending in that order. Characteristic constituents of the algal turf of the infralittoral fringe (which actually begins in the lower midlittoral above) are *Champia novaezelandiae, Laurencia* spp., *Lophurella caespitosa, Pachymenia himantophora,* and *Plocamium* sp.

It is interesting to contrast some of the chief differences between the flora on the east and west coasts of North Island, they being more striking than those of the fauna. Table 11.3 lists a few notable differences.

CENTRAL NEW ZEALAND—MIXED COLD-TEMPERATE

The region of mixed cold-temperate water extends around the southern coasts of North Island and along the northern coast and most of the eastern and western coasts of South Island. Taylor's Mistake, on the central east coast of South Island, is typical of the region.

TABLE **11.3**
Some specific differences between the flora of the
east coast of North Island and that of its west coast

East-coast species scarce or lacking on the west coast	West-coast species scarce or lacking on the east coast
Carpophyllum sp.	*Durvillea antarctica*
Ecklonia radiata	*Gigartina atropurpurea*
Glossophora kunthii	*Gigartina marginifera*
Hormosira banksii	*Pachymenia himantophora*
Lessonia variegata	*Porphyra columbina*
Xiphophora chondrophylla var. *minima*	*Stenogramma interrupta*

Taylor's Mistake

The Supralittoral Fringe. There are the usual blue-green algae in the supra-littoral fringe at Taylor's Mistake. *Melaraphe cincta* and *M. oliveri* occur here, their vertical ranges varying from place to place. (In Southern New Zealand, *M. cincta* has the wider range; in the northern districts, *M. oliveri* occurs throughout the fringe and well down into the midlittoral zone.) Here also can be found the outliers of *Chamaesipho columna.*

The Midlittoral Zone. The upper limit here, as on so many other shores, is an unmistakable line of barnacles. *Chamaesipho columna* is the dominant species, ex-tending into the lower part of the zone with *Elminius plicatus.* This latter barnacle has a shorter upward range, however, and favours shade with *Bostrychia arbuscula,* which is also common in the upper midlittoral. *E. modestus,* although nowhere common on exposed shores, can be very plentiful on sheltered ones; it can stand brackish water, mud, and sand, and therefore has a more generous distribution than the other barnacles. Its usual upper limit is about halfway up the midlittoral, and it may extend downwards to the infralittoral fringe. A common mollusc is *Lepsiella scobina;* it feeds on the mussel *Modiolus neozelanicus,* which occurs plentifully throughout a very wide range of condi-tions. Other characteristic genera are *Cellana* (*C. ornata* is ubiquitous, and *C. radians* is noteworthy because it is also common at high levels), *Sypharochiton, Notoacmea, Siphonaria,* and *Benhamina.* Larger common mussels are *Mytilus edulis* var. *aoteanus* in the lower midlittoral, and *Mytilus canaliculus,* which extends well into the fringe below. *Pomatoceros cariniferus* is widely distributed in the midlittoral, and under suitable conditions (that is, on vertical rock faces with some shelter), it may form belts of solid encrustations.

Superimposed on the barnacles in many sheltered places are bands of algae of such species as *Scytothamnus australis* and *S. fasciculatus,* with *Ralfsia verrucosa* and *Splach-nidium rugosum* a little below. Species of seasonal occurrence are *Porphyra* (which may also extend upwards to the supralittoral fringe), *Adenocystis, Ilea, Myriogloria, Scyto-siphon,* and *Ulva. Codium adhaerens* has a somewhat scattered distribution, being intolerant of sunlight and exposure. A band of *Cystophora* and *Carpophyllum* may reach up to the lower limit of the barnacles, or there may be a mixed red algal turf, some of the commonest components of which are species of *Champia, Cystophora, Glosso-phora, Halopteris, Laurencia, Lophurella, Polysiphonia,* and stunted *Xiphophora.* Under very sheltered conditions, on level platforms, there may be a belt of corallines and *Hormosira.*

The Infralittoral Fringe. *Durvillea antarctica* and *D. willana* are the kelps of exposed positions; they may intermingle, although the latter is usually found in the more seaward places. Underneath these weeds, the dominant growth is a crust of lithothamnia; in less exposed conditions, it may be replaced by corallines and mixed red algae. In quieter places, *Durvillea* may be replaced by an upper band of *Xiphophora chondrophylla* var. *maxima,* and a lower band of *Carpophyllum, Cystophora scalaris,* and

Cystophora retroflex. As shelter increases, *Cystophora torulosa* may replace *Xiphophora.* Dominant brown species in the lowest parts of the fringe are *Lessonia variegata, Ecklonia radiata,* and *Macrocystis pyrifera;* species of *Marginariella* and *Landsburgia* may also be found in suitable areas. The most obvious animals of this region are the large *Mytilus canaliculus, Haliotis iris,* and *Pyura pachydermatina.* Also notable is the ribbed mussel *Aulacomya maoriana;* it is scattered throughout the lower midlittoral and the infralittoral fringe, and is very tolerant of various conditions of exposure.

THE OTAGO PENINSULA—MIXED COLD-TEMPERATE AND TRANSITIONAL SUBANTARCTIC

The transitional subantarctic water of the south coast of South Island overlaps the mixed cold-temperate water of central New Zealand near West Cape on the west coast of South Island and near the Otago Peninsula on the east coast (see Figure 11.3).

The Otago Peninsula (Figure 11.41) lies in a district where cold south winds are the prevailing and most violent ones. Inshore sea temperatures lie between 8°C and 16°C. The tides are semidiurnal. Many parts of the area are subjected to continuous pounding by Pacific rollers, which make the tide levels difficult to define and the limits

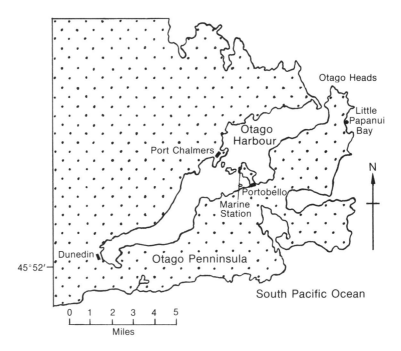

FIGURE **11.41**

A map of the Otago Peninsula of New Zealand, showing localities referred to in the text.

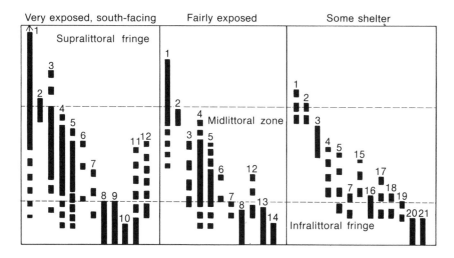

FIGURE **11.42**

The general zonation around the Otago Peninsula: 1, *Melaraphe cincta;* 2, *Melaraphe oliveri;* 3, *Bostrychia arbuscula;* 4, *Chamaesipho columna;* 5, *Elminius plicatus;* 6, *Modiolus neozelanicus;* 7, *Mytilus edulis* var. *aoteanus;* 8, *Mytilus canaliculus;* 9, *Durvillea antarctica;* 11, lithothamnia; 12, *Pachymenia lusoria;* 13, *Xiphophora chondrophylla* var. *maxima;* 14, *Lessonia variegata;* 15, *Pomatoceros cariniferus;* 16, *Hormosira banksii;* 17, *Ostrea heffordi;* 18, *Ulva reticulata;* 19, *Cystophora torulosa;* 20, *Macrocystis pyrifera;* 21, *Pyura pachydermatina; broken columns,* species present in patches; *unbroken columns,* species forming continuous belts. Note that the vertical range of *Melaraphe cincta* in very exposed places extends beyond the range of the figure. The very exposed region is at Little Panapui. It will be noted how striking differences are in the algae of the infralittoral fringe, under varying degrees of exposure.

of zone boundaries vague and variable. Naturally, there are considerable differences between the biota of the sheltered shores behind the Otago Peninsula and that of the coast at Little Papanui Bay, which is exposed to the open ocean.

This account of the Otago Peninsula area is based on the excellent work of Dr. E. J. Batham, Director of the Portobello Marine Station in Otago Harbour.

Little Papanui Bay

Little Papanui Bay, on the exposed seaward coast of the Otago Peninsula, consists of a half-mile stretch of sandy beach with cliffs at its north and south ends. In the north at the base of the cliffs, there are boulders and wave-cut platforms. Nearby, there is a reef of rocks running out into the sea. There is a similar arrangement of rocks at the southern end of the bay, but they face the sun and are well protected by the cliff. The zonation of the bay is diagrammed in Figure 11.42.

The Supralittoral Fringe. The black discoloration produced by lichens is dense but patchy on southward facing rocks. The more common species of *Melaraphe* is

M. cincta, which has a wide vertical range. *M. oliveri* is only locally abundant, and is more or less confined to splashed areas—quite in contrast to its distribution further north. A notable animal of this zone is *Notoacmea pileopsis sturnus*, which occurs crowded into crevices. *Bostrychia arbuscula* extends fairly high into this zone in shaded places damp with spray.

The Midlittoral Zone. Typically, *Chamaesipho columna* and *Elminius plicatus* form a substratum here, along with the small mussel *Modiolus neozelanicus* and the larger *Mytilus edulis* var. *aoteanus*. These mussels vary in dominance, according to exposure and the aspect of the rock, and both of them favour crevices. *Chamaesipho* flourishes best on flat-surfaced rocks that face the sun, and can tolerate violent surf. *Elminius*, however, is most successful in shaded areas, and is densest on vertical faces in association with mussels. Other animals characteristic of much of this zone are species of *Cellana*, *Lepsiella*, and *Sypharochiton*. In pools, and in places of fresh-water seepage at upper levels, *Ulva* and *Enteromorpha* are to be found.

Bostrychia arbuscula abounds in the middle part of this zone, and may become dominant on some shaded south faces. Here, too, is the familiar turf of short algae, the most common constituents of which are scattered tufts of *Apophloea lyalli*, *Porphyra* (which girdles boulders), and the seasonal species *Scytothamnus fascicularis* and *Scytosiphon lomentaria*.

There are additional algal species in the lower midlittoral, including *Splachnidium rugosum*, *Colpomenia sinuosa*, and *Leathesia difformis*, all of which are familiar on South African shores. The lower midlittoral is populated by the usual widespread New Zealand invertebrates, including *Mytilus edulis* var. *aoteanus* (many of these mussels carry *Chamaesipho columna* and *Notoacmea parviconoides* on their shells), *Anthopleura aureoradiata*, *Patelloida corticata*, and *Pyura suteri*.

The Infralittoral Fringe. In most wave-swept areas, *Durvillea antarctica* (mixed with *D. willana* at lower levels) is dominant; codominant is the large mussel *Mytilus canaliculus*. Among these, there is the usual varied population of animals (sponges, ascidians, bryozoa, hydroids, sea urchins, anemones, and so forth), and a number of smaller algae, including encrusting pink corallines and *Codium*.

For further details of the biota of this area, see Batham (1956, 1958).

Otago Harbour

The shores inside Otago Harbour are far more sheltered than those on the outer coast of the Otago Peninsula. The water movement in the harbour is due partly to tidal currents (which are stronger in some places than in others) and partly to waves in windy weather.

The following paragraphs give only a very brief summary of the biota of this more sheltered region, with various comparisons between it and more exposed areas nearby (Figure 11.42).

The Supralittoral Fringe. At the highest levels of the fringe, there are patches of blackening caused by two species of *Verrucaria*. In addition to *Melaraphe cincta* and *M. oliveri*, which vary in abundance from site to site, the notable animals are the ridged winkle *Risselopsis varia* and, in places, the green limpet *Notoacmea pileopsis sturnus*. At night, numerous amphipods—species of *Orchestia*—scamper about with the isopods so characteristic of the supralittoral fringe (here, the species is *Ligia novae-zealandiae*).

The Midlittoral Zone. Clumps of *Bostrychia arbuscula* are common on sunny faces, with dark patches of *Lichina* among them; in shade, however, *B. arbuscula* forms a continuous belt. Molluscs are plentiful, the four commonest being the limpet *Cellana ornata*, the chiton *Sypharochiton pelliserpentis*, the periwinkle *Melagraphia aethiops,* and the tiny black turret *Zeacumantus subcarinatus*. Other familiar species are the pulmonate limpet *Siphonaria zelandica*, the whelk *Lepsiella scobina albomarginata*, and the pulmonate slug *Onchidella patelloides*. In areas where the currents are strong, the barnacles *Elminius plicatus* and *Chamaesipho columna*, the mussel *Mytilus edulis* var. *aoteanus*, and the oyster *Ostrea heffordi* occur densely, but *Mytilus canaliculus* and the small *Modiolus neozelanicus* are virtually absent. The serpulid *Pomatoceros cariniferus* is abundant in some places, and patches of the ascidian *Pyura suteri* occur here also, as do various sponges and *Corallina officinalis*.

In the lowest part of the midlittoral, there is a sharply marked, dense algal belt, whose chief components are *Hormosira banksii* and *Ulva reticulata*. In places, *Cystophora torulosa* may be a major component. Under and among the algae are molluscs, anemones, brittle stars, serpulids, burrowing worms, compound ascidians, and sponges, the last two forming a much greater part of this varied population than they do of the corresponding population at Little Papanui Bay.

The Infralittoral Fringe. This is well defined, as *Macrocystis pyrifera* and *Pyura pachydermatina* cannot thrive here for very long above the level of low water of spring tides (Figure 11.43). There is an extensive and varied assortment of sponges, ascidians, bryozoa, amphipods, and polychaets growing around and under the holdfasts of the big algae. Many epiphytes grow on *Macrocystis* fronds, including such hydroids as *Obelia geniculata* and the feathery tufts of *Plumularia setacea* var. *optima*, swarming caprellids, and tiny white spirals of *Spirorbis*. Such populations are strongly reminiscent of those on shores of similar temperature on the coasts of South America, central California, south and west Cape Province in South Africa, and Tasmania.

The papers of Batham (1956, 1958, 1965) should be referred to for information on the seasonal fluctuations of animals and algae, the biota of pools and under stones, and details of the variations in biota under exposed and sheltered conditions.

The characteristic changes in the biota of the east and west coasts of South Island in comparable latitudes can be studied in detail from relevant New Zealand literature. These changes are of a very similar nature to those found between the east and west coast of North Island, the most prominent being the differences in algae in the lower midlittoral zones and infralittoral fringes of the two coasts.

FIGURE **11.43**

A rock near the Marine Station on the Otago Peninsula. The top of the rock lies just below the mean water level. At the top of the rock, there is a zone of barnacles with oysters below; below this, there is a belt of mixed lower-midlittoral algae and, finally, *Macrocystis* and *Pyura* of the infralittoral fringe, which are exposed only at extreme low water of spring tides. [Photograph by E. J. Batham.]

DIFFERENCES BETWEEN THE TRANSITIONAL WARM-TEMPERATE AND MIXED COLD-TEMPERATE REGIONS OF NEW ZEALAND

There are certain marked differences between the ecology of the transitional warm-temperate shores of New Zealand and that of its mixed cold-temperate shores. The most conspicuous of these are shown in Figure 11.44 and Table 11.4.

Conspicuous features of the far south are the thriving kelp beds of *Durvillea antarctica* and *D. willana*, an abundance of *Lessonia variegata*, and a luxuriant growth of *Bostrychia arbuscula*. *Apophloea lyalli* takes the place that is occupied on northern shores by *A. sinclairii*; similarly, the stouter *Xiphophora condrophylla* var. *maxima* replaces the *X. c.* var. *minus* of warmer water (Figure 11.45). A number of molluscs are particularly abundant in southern areas, including *Benhamina obliquata*, *Mytilus edulis* var. *aoteanus*, and *Aulocomya maoriana*; *Cellana redimiculum* is also a southern species. Two of the

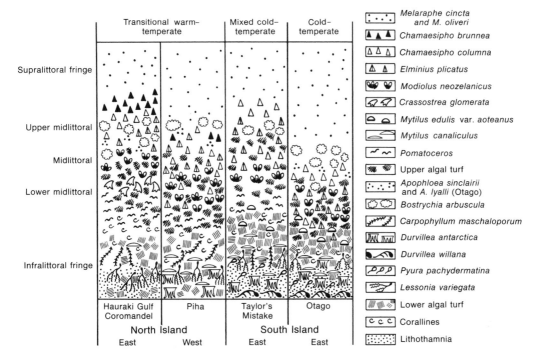

FIGURE **11.44**

The type of zonation found on characteristic coasts of New Zealand. Only a selection of typical species has been used, in order to keep the figure uncomplicated and easily intelligible.

TABLE **11.4**

Conspicuous differences between the biota of the mixed cold-temperate shores of New Zealand and that of its transitional warm-temperate shores

Mixed cold-temperate shores	Transitional warm-temperate shores
Apophloea sinclairii absent.	*A. sinclairii* forms a belt in the midlittoral.
Melanerita melanotragus of limited occurence, and only as far as Cook Strait.	*M. melanotragus* a prominent feature.
Benhamina obliquata important south of Cook Strait.	Absent.
The mussels *Aulacomya maoriana* and *Mytilus edulis* var. *aoteanus* are typical.	Both absent.
No conspicuous algal turf in the infralittoral fringe.	A well-defined turf present.
No belt of red algae present.	A belt of *Pachymenia himantophora, Gigartina alveata*, and *G. marginifera* present.
Brown algae fairly high on the shore, with *Cystophora* a most prominent feature.	Large brown alge present at low levels, with three species of *Carpophyllum* important (only *C. maschalocarpum* extends to South Island).

FIGURE **11.45**

Lessonia variegata and *Xiphophora chondrophylla* var. *maxima* on semiexposed rocks in Little Papanui Bay, New Zealand, at extreme low water of a spring tide. On extremely exposed faces, these species are replaced by *Durvillea antarctica* and *D. willana*. Compare with Figure 10.39 (p. 271), which shows the algal mat of the infralittoral fringe at Chepu, Chiloé Island, Chile. [Photograph by E. J. Batham.]

three northern species of *Carpophyllum, Chamaesipho brunnea,* and the belt-forming sabellariid tube worms do not reach as far south as mixed cold-temperate waters, and *Pomatoceros cariniferus* is near its southern limit at the overlap.

The coast at the extreme south of South Island adjacent to Stewart Island is included in the section dealing with the island of the transitional subantarctic region.

KERMADEC ISLANDS—TROPICAL

The Kermadec Islands, a dependency of New Zealand, lie roughly 500 miles northeast of Auckland (lat. 30°30′5″S, long. 178°30′W; see Figure 11.3). Myer Island, described below, is one of the group. It is made of volcanic rock, which erodes to a very uneven surface with sharp ridges and points. Small pools are common in the rock from the lower midlittoral downwards.

Dr. M. Miller has kindly provided us with the information on which this account is based.

The Supralittoral Fringe

The supralittoral fringe of Myer Island appears to be mainly empty, except for outliers of *Melanerita melanotragus* from the upper part of the zone below.

The Midlittoral Zone

Melanerita melanotragus marks the upper limit of the midlittoral zone. There is a little belt of a species of *Chaetangium*, and the usual concentration of *Balanus*, which continues down less densely throughout the whole zone. *Cellana creticulata* mixes with the upper *Balanus*, below which there is a belt of *Laurencia* spp. intermingled with patches of *Cladophoropsis membranacea*. In the lower midlittoral, a band of *Sargassum fissifolium* terminates in a strip of *Colpomenia sinuosa*, and *Balanus* again becomes more concentrated. The usual encrusting corallines then take over, studded with the giant limpet *Penepatella kermadecensis*. Other important animals are *Siphonaria carleyensis* and *Neothais smithii*. Except for *Melanerita melanotragus*, all the gastropods mentioned are endemics.

The Infralittoral Fringe

The top of the infralittoral fringe is indicated by a growth of larger algae, mainly *Pterocladia capillacea* and *Delisera fimbriata*. At the lowest limit of the fringe, a thick growth of coral can be seen.

It would appear that the algae here are of a general tropical type with little affinity with those of New Zealand. The origins and affinities of the animals are various, representing New Zealand, Australian, and Indo-Pacific species.

CHATHAM ISLANDS—MIXED COLD-TEMPERATE

The Chatham Islands lie 422 miles east-southeast of Wellington at lat. 44°20′S, long. 176°W. Though they lie on the line of the Subtropical Convergence, the shore waters of these islands appear to have temperatures varying roughly from 5°C to 16°C—the range of mixed cold-temperate conditions, certainly not of subtropical ones.

As might be expected, the shore biota proves to be an interesting compromise between those of New Zealand's North and South islands. For instance, the North Island alga *Carpophyllum plumosum* is common, and grows alongside the cold-water South Island algae *Apophloea lyallii* and *Pachymenia lusoria*. *Chamaesipho brunnea*, the one dominant barnacle, extends throughout the midlittoral zone. Listed in Table 11.5 are some notable New Zealand animals that are absent or of rare occurrence on the Chatham Islands. There seem to be no grapsid crabs and no anomurans. The algae have much in common with the New Zealand species.

The following is a very general summary of the zonation of a typical stretch of shore in the Chatham Islands.

The Supralittoral Fringe. The distribution of lichens and *Melaraphe* in the supralittoral fringe here is quite typical of many Australasian shores.

The Midlittoral Zone. This is dominated throughout only by *Chamaesipho brunnea* and *Cellana strigilis chathamensis,* which is quite distinct from the New Zealand species. There are two superimposing algal bands—*Bostrychia arbuscula* above and *Apophloea lyallii* below—a common arrangement in New Zealand.

In the lower midlittoral, additional animals are *Aulacomya maoriana, Elminius plicatus,* and *Siphonaria zelandica.* There is a belt of the alga *Splachnidium rugosum;* further seawards, there is a narrower belt of *Pachymenia lusoria* with lithothamnia and corallines beneath it. In more sheltered places, these larger species may be replaced by an algal mat consisting mainly of *Adenocystis utricularis, Bryopsis* sp., various corallines, *Gigartina* (on the outer margins), *Hormosira banksii, Leathesia difformis, Myriogloria lindauerii, Polysiphonia* sp., and *Ulva* sp.

The Infralittoral Fringe. Beginning in the lowest part of the midlittoral, and extending into the infralittoral fringe in exposed areas, *Durvillea antarctica* is dominant. On semiexposed platforms, it may be replaced by *D. caepestipes; D. willana,* however, is not recorded. There is the usual layer of lithothamnia and corallines. In less exposed places, *Xipophora chondrophylla* var. *maxima* may make a narrow band above the *Durvillea;* in very sheltered places, it may be replaced by *Lessonia variegata,* with the addition of *Carpophyllum plumosum* and *Macrocystis pyrifera* in the lowest part of the fringe.

Subantarctic features of the biota are *Cellana strigilis* and *Aulacomya maoriana* as the dominant limpet and the dominant mussel, respectively, and the striking alga *Landsbergia myricaefolia.*

THE SOUTH COAST OF SOUTH ISLAND, STEWART ISLAND, AND ADJACENT ISLANDS— TRANSITIONAL SUBANTARCTIC

The transitional subantarctic waters of New Zealand begin on the west of South Island near West Cape, and near the Otago Peninsula on the east. They surround the south coast and Stewart Island, and extend southward and eastward to include Snares Islands, Auckland Islands, Campbell Island, Bounty Islands, and Antipodes Islands (see Figure 11.3).

The South Coast of South Island and Stewart Island

The biota of the extreme south coast of South Island and on Stewart Island, although very similar to that of the Otago Peninsula, has a number of subantarctic species not found at the latter locality, which lies on the borderline of a cold-temperate area.

TABLE **11.5**
Some notable animals of New Zealand that
are absent or rare* in the Chatham Islands

Benhamina obliquata	*Mytilus edulis* var.
Chamaesipho columna	*aoteanus**
*Crassostrea glomerata**	*Melanerita melanotragus*
Elminius modestus	*Pollicipes spinosus*
Modiolus neozelanicus	*Pomatoceros cariniferus*
Mytilus canaliculus	*Siphonaria australis**

*Rare species are marked with an asterisk.

TABLE **11.6**
Some important algae and animals of Stewart Island
and the south coast of South Island

Algae	Algae (contd.)
NEW ZEALAND ENDEMICS	CIRCUMPOLAR SUBANTARCTIC SPECIES (contd.)
Apophloea lyallii	*Chaetomorpha darwinii*
Brogniartella australis	*Desmarestia firma*
Hymenena costata	*Halopteris funicularis*
	Macrocystis pyrifera
NEW ZEALAND SUBANTARCTIC ENDEMICS	*Phycodrys quercifolia*
Codium dimorphum	*Schizoseris davisii*
Delesseria crassinerva	*Scytothamnus fasciculatus*
Myriogramme crispata	
Pachymenia lusoria	Animals
	TYPICAL SUBANTARCTIC GENERA
CIRCUMPOLAR SUBANTARCTIC SPECIES	*Costokidderia*
Adenocystis utricularis	*Gaimardia*
Ballia scoparia	*Kidderia*
Chaetangium fastigiatum	*Margarella*

Significant features of this area are: (1) an intensive banding of lichens in the supra-littoral fringe; (2) in many places, a dense and extensive algal growth that extends downwards from the top of the midlittoral throughout the intertidal zone; and (3) a molluscan fauna dominated by the mussel *Aulacomya maoriana* and the characteristic limpet *Cellana strigilis*. Table 11.6 lists some of the important animals and algae of this district.

Adjacent Islands

The islands adjacent to the south coast of South Island are Campbell Island, which is roughly 320 miles south of Stewart Island, and four small island groups. Two of these groups, Snares Islands and Auckland Islands, lie roughly 75 miles and 150 miles south of South Island, respectively; the other two groups, Bounty Islands and Antipodes Islands, lie roughly 415 miles east-southeast and 450 miles southeast of the Otago Peninsula, respectively. These islands lie between lat. 52°30′S and lat. 47°43′S, and between long. 179°E and long. 166°13′E. They have many affinities with New Zealand, but the greater the distance from the mainland, the weaker these affinities become; on the most remote islands, the subantarctic constituents of the biota predominate.

The islands are treated here as a whole, and details of the variations of specific islands are not usually enumerated. Generally speaking, their intertidal zonation is unremarkable.

The Supralittoral Fringe.   This is banded with lichens. *Hildenbrandia lecannellieri* makes red incrustations in some places, and littornids are typically lacking.

The Midlittoral Zone.   This is mostly dominated by molluscs, including *Notoacmea pileopsis sternus, Kerguelenella flemingi,* and *Cellana strigilis.* In some places, there may be seasonal patches of such algae as *Bostrychia* and *Porphyra.* From the lower midlittoral, a dense growth of algae extends down into the sea; its constituents are almost identical with those of the lower midlittoral algal belt of the Chatham Islands (see p. 344). Common animals in this area are *Mytilus edulis* var. *aoteanus, Aulocomya maoriana,* the chitons *Plaxiphora aurata cambelli* and *Onithochiton subantarcticus,* various trochids of the genus *Margarella,* and the limpet *Patinigera terroris.*

The Infralittoral Fringe.   The low-level algal growth of the islands are typical of these waters: the large algae—*Durvillea antarctica, Lessonia variegata,* and *Macrocystis pyrifera*—have an undergrowth of such wide-ranging subantarctic forms as *Ballia callitricha,* and species of *Iridaea* and *Desmarestia.*

These islands have a few notable characteristics: (1) an absence of littornids, except for a few *Melaraphe cincta* on the Auckland Islands; (2) no barnacle belt, except for a reduced one on the Auckland Islands; and (3) a lack of many New Zealand belt-forming animals and seaweeds (some of these, conspicuous by their absence, are listed in Table 11.7).

TABLE **11.7**
Some of the belt-forming animals and algae
of New Zealand conspicuously absent on the islands
adjacent to the south coast of South Island

Animals	Algae
Cellana ornata	*Carpophyllum maschalocarpum*
Cellana radians	*Cystophora* spp.
Lepsiella albomarginata	*Ecklonia radiata*
Lepsiella scobina	*Glossophora kunthii*
Melaraphe aethiops	*Hormosira banksii*
Mytilus canaliculus	*Splanchnidium rugosum*
Patelloida corticata	*Zonaria augustata*
Pomatoceros cariniferus	
Sypharochiton pelliserpentis	

MACQUARIE ISLAND—SUBANTARCTIC

At lat. 54°29′S and long. 158°58′E, Macquarie Island lies roughly 850 miles southeast of Tasmania, Australia, to which it belongs. It is discussed in this section, rather than in the section on Australia, because of its greater proximity to New Zealand.

The island supports a fairly well-marked endemic element; the influence of the subantarctic component is very pronounced, and species common to New Zealand are extremely few.

The entire coastline is fully exposed. Temperatures range from 2.8°C to 7.2°C—a range typical of subantarctic regions. Great depths of water separate it from all adjacent land masses.

It is somewhat surprising that Kenny and Haysom (1962) state that the Stephenson zonation pattern is difficult to correlate with the zonation of Macquarie Island—especially so because their descriptions of the island conjure up an image of the shores of the Port Nolloth district of South Africa. Perhaps their impression of noncorrelation may be due to the fact that, in their list of references, only one paper on South Africa has been cited.

The following account of the biota of this region is general and brief. A careful and detailed account of the biota can be found in Kenny and Haysom's paper.

The Supralittoral Fringe

The supralittoral fringe has a strongly marked lichen zone in which *Verrucaria* is conspicuous. *Hildenbrandia* is common, but a littorinid population is lacking.

The Midlittoral Zone.

From the upper midlittoral, a dense mat of *Porphyra* extends into the lowest part of the infralittoral fringe. Here, the species is *P. umbilicalis* (which corresponds to the *P. capensis* of South Africa), with an admixture of *Ulva lactuca, Enteromorpha intestinalis,* and *Cladophora* in suitable damp places (algae of these last three genera are also found in association with *Porphyra* in South Africa). Farther down, other algae appear, including *Rhizoclonium* and *Chaetangium.* Below this belt, there is a zone that is bare but for closely cropped patches of algae (typically, the *Chaetangium* mentioned above and *Acrosiphonia lateralis*) and a scattering of the limpet *Siphonaria lateralis.* In the lower midlittoral, the biota increases: *Rhodymenia* spp. becomes dominant and conspicuous, and may extend to lower levels, making a cover for a varied population of small animals. *S. lateralis* is dominant here, but other gastropods, including *Nacella delesserti* and *Macquariella hamiltoni,* are also present. As in other areas of like climate and continuous wave action, the lower midlittoral and the infralittoral fringe are not clearly delimited in most places.

The Infralittoral Fringe

The infralittoral fringe has an encrusting layer of corallines with many species of red algae. The commonest species are *Ballia callitricha, Ceramium rubrum, Delesseria* sp.,

Desmarestia rossi, Iridaea sp., *Plocamium coccineum,* and *Schizoseris* sp. The holdfasts of some algae provide living quarters for a varied population of numerous small animals. The infralittoral zone below is a tossing mass of kelp (of which *Macrocystis* and *Durvillea antarctica* are conspicuous constituents) that effectively checks the full force of the surf from breaking on the shore.

That littorinids, balanoids, and certain other species (such as tube-making polychaets, ascidians, and so forth) are not found in any quantity, or may indeed be lacking, does not seem to invalidate our overall scheme of zonation. Here are represented the same widespread divisions: a supralittoral fringe, a midlittoral zone, and an infralittoral fringe. In these regions, the infralittoral fringe has the particularly extensive and fine growth of algae so characteristic of colder shores.

SELECTED REFERENCES

Barnes, E. H., 1963. The biogeography and intertidal ecology of the Australasian coasts. *Oceanogr. Mar. Biol. Annu. Rev.* 1: 341-404.

Batham, E. J., 1956. Ecology of southern New Zealand sheltered rocky shore. *Trans. Roy. Soc. N. Z.* 84(2): 447-465.

————, 1958. Ecology of southern New Zealand exposed rocky shore at Little Papanui, Otago Peninsula. *Trans. Roy. Soc. N. Z.* 85(4): 647-658.

————, 1965. Rocky shore ecology of a southern New Zealand fiord. *Trans. Roy. Soc. N. Z. Zoology* 6(21): 215-227.

Bennet, Isobel, 1960. Intertidal zonation of the exposed rocky shore of Tasmania and its relationship with the rest of Australia. *Aust. J. Mar. Freshwater Res.* 11(2): 182-221.

————, and Elizabeth C. Pope, 1953. Intertidal zonation of the exposed rocky shores of Victoria, together with a rearrangement of the biogeographical provinces of temperate Australian shores. *Aust. J. Mar. Freshwater Res.* 4(1): 105-159.

Beveridge, W. A., and V. J. Chapman, 1950. The zonation of marine algae at Piha, New Zealand, in relation to the tidal factors. *Pac. Sci.* 4(3): 188-201.

Carnhan, J. A., 1952. Intertidal zonation at Rangitoto, New Zealand, *Pac. Sci.* 6(1): 35-46.

Chapman, V. J., 1950. The marine algal communities of Stanmore Bay, New Zealand. *Pac. Sci.* 4(1): 63–68.

Cranwell, L. M., and L. B. Moore, 1938. Intertidal communities of the Poor Knights Islands, New Zealand. *Trans. Roy. Soc. N. Z.* 67: 375–407.

Dakin, W. J., 1953. *Australian Seashores.* Sidney: Halstead Press.

——, Isobel Bennett, and Elizabeth Pope, 1948. A study of certain aspects of the ecology of the intertidal zone of the New South Wales coast. *Aust. J. Sci.* (B)1(2): 176–230.

Deacon, G. E. R., 1967. Personal communication.

Dell, R. K., 1962. New Zealand marine provinces—do they exist? *Tuatara* 9(1): 18–23.

Dellow, V., 1950. Intertidal ecology at Narrow Neck Reef, New Zealand, *Pac. Sci.* 4(4): 355–374.

——, 1955. Marine algal ecology of the Hauraki Gulf, New Zealand. *Trans. Roy. Soc. N. Z.* 83(1): 1–91.

Endean, R., W. Stephenson, and R. Kenny, 1956. The ecology and distribution of intertidal organisms on certain islands off the Queensland coast. *Aust. J. Mar. Freshwater Res.* 7(3): 317–342.

——, R. Kenny, and W. Stephenson, 1956. The ecology and distribution of intertidal organisms on the rocky shores of the Queensland mainland. *Aust. J. Mar. Freshwater Res.* 7(1): 88–146.

Fairbridge, R. W., 1950. Recent and Pleistocene coral reefs of Australia. *J. Geol.* 58(4): 330–401.

Fischer, P. H., 1940. Notes sur les peuplements littoraux d'Australia. *Mem. Soc. Biogeogr.* 7: 288–313.

Fleming, C. A., 1962. New Zealand biogeography: a paleontologist's approach. *J. Biol. Soc. Victoria Univ. Wellington.* 110: 53–108.

Guiler, E. R., 1950. The intertidal ecology of Tasmania. *Pap. Proc. Roy. Soc. Tasmania* 1949: 135–201.

——, 1951. Notes on the intertidal ecology of the Freycinet Peninsula. *Pap. Proc. Roy. Soc. Tasmania* 1950: 53–70.

——, 1952a. The ecological features of certain sheltered intertidal areas in Tasmania. *Pap. Proc. Roy. Soc. Tasmania* 86: 1–11.

——, 1952b. The intertidal ecology of the Eaglehawk Neck Area. *Pap. Proc. Roy. Soc. Tasmania* 86: 31–61.

——, 1953a. Further observations on the intertidal ecology of the Freycinet Peninsula. *Pap. Proc. Roy. Soc. Tasmania* 87: 93–95.

——, 1953b. Intertidal classification in Tasmania. *J. Ecol.* 41: 381–384.

——, 1954. The intertidal zonation at two places in southern Tasmania. *Pap. Proc. Roy. Soc. Tasmania* 88: 105–118.

——, 1955. Australian intertidal belt-forming species in Tasmania. *J. Ecol.* 43: 138–148.

——, 1958. Long-term changes in intertidal zonation in Tasmania with special reference to the mollusca. *J. Malacol. Soc. Aust.* 3: 59–67.

——, 1960a. Notes on the intertidal ecology of Trail Harbour, Tasmania. *Pap. Proc. Roy. Soc. Tasmania* 94: 57–62.

——, 1960b. The intertidal zone-forming species on rocky shores of the east Australian coast. *J. Ecol.* 48(1): 1–28.

——, D. L. Serventy, and J. H. Willis, 1958. The Fisher Island Field Station—with an account of its principal fauna and flora. *Pap. Proc. Roy. Soc. Tasmania* 92: 165–183.

Hodgkin, E. P., 1959. Catastrophic destruction of the littoral fauna and flora near Fremantle, January 1959. *West. Aust. Natur.* 7(1): 6–11.

————, L. Marsh, and G. G. Smith, 1959. The littoral environment of Rottnest Island. *J. Roy. Soc. West Aust.* 42(3).

Kenny, R., and N. Haysom, 1962. Ecology of rocky shore organisms at Macquarie Island. *Pac. Sci.* 16: 245–263.

Knox, G. A., 1953. The intertidal ecology of Taylor's Mistake, Banks Peninsula. *Trans. Roy. Soc. N. Z.* 81(2): 189–220.

————, 1954. The intertidal flora and fauna of the Chatham Islands. *Nature* 174: 871.

————, 1957. General account of the Chatham Islands 1954 expedition. *N. Z. Dep. Sci. Ind. Res. Bull.* 122: 7–37.

————, 1960. Littoral ecology and biogeography of the southern oceans. *Proc. Roy. Soc. London B Biol. Sci.* 162: 577–624.

————, 1963. The biogeography and intertidal ecology of the Australasian coasts. *Oceanogr. Mar. Biol. Annu. Rev.* 1: 341–404.

Macnae, W., 1968. A general account of the fauna and flora of mangrove swamps and forests in the Indo–West-Pacific region. *Advan. Mar. Biol.* 6: 73–270.

Moore, L. B., 1961. Distribution patterns in New Zealand seaweeds. *Tuatara* 9(1): 18–23.

Morton, J., and M. Miller, 1968. *The New Zealand Sea Shore.* London and Auckland: Collins.

Oliver, W. R. B., 1923. Marine Littoral Plant and Animal Communities in New Zealand. *Trans. N. Z. Inst.* 54: 495–545.

Orr, A. P., and F. W. Moorhouse, 1933. Variations in some physical and chemical conditions on and near Low Isles Reef. In *Scientific Reports of the Great Barrier Reef Expedition, 1928–1929,* vol. 2, pp. 87–98. London: British Museum (Natural History).

Rochford, D. J., 1957. The identification and nomenclature of the surface water masses in the Tasman Sea (data to the end of 1954). *Aust. J. Mar. Freshwater Res.* 8: 369–413.

Steers, J. A., 1937. The coral islands and associated features of the Great Barrier Reefs. *Geogr. J.* 89(1, 2).

Stephenson, T. A., 1947. The constitution of the intertidal fauna and flora of South Africa, III. *Ann. Natal Mus.* 11: 207–324.

————, Anne Stephenson, G. Tandy, and M. A. Spender, 1931. The structure and ecology of Low Isles and other reefs. In *Scientific Reports of the Great Barrier Reef Expedition, 1928–1929,* vol. 3, pp. 17–112. London: British Museum (Natural History).

Stephenson, W., R. Endean, and Isobel Bennett, 1958. An ecological survey of the marine fauna of Low Isles, Queensland. *Aust. J. Mar. Freshwater Res.* 9(2): 261–318.

Womersley, H. B. S., and S. J. Edmonds, 1958. A general account of the intertidal ecology of South Australian coasts. *Aust. J. Mar. Freshwater Res.* 9(2): 217–260.

12

THE BRITISH ISLES

by J. R. Lewis

INTRODUCTION

The British Isles are approximately 700 miles long and, lying in the northeastern part of the North Atlantic Ocean, are subject to cool-temperate climatic conditions. In summer, the temperature gradient runs roughly north and south, with a bias towards milder and moister conditions on the west and northwest coasts than on the east coast. In winter, the Atlantic influence is stronger and the pattern changes, with the gradient roughly from east to west and the lowest sea temperatures in the southeast. The biological consequences of these conditions can be summarised briefly as (1) an uplift of the zonation of lower-shore species on northwestern coasts, and (2) greater northward extension of warm-water species in the west.

The seasonal variation in sea temperatures is roughly 7°C in the northern districts and up to 12°C in parts of the enclosed Irish Sea and on the southeastern coasts most subject to continental influences. Sea ice is exceedingly rare, and can be neglected as an ecological factor. The seasonal variation, together with the wide spacing of the isotherms and the consequent lack of sharp temperature barriers, produces a littoral population of eurythermic species in which Arctic, Boreal, and Lusitanean forms exist side by side. Spring and summer are the periods of maximum growth, and the rapid increase then in the amounts of algae, both annual and perennial, can effect considerable changes in the appearance of some shores.

The range of spring tides varies from 2 to 40 feet, but values between 12 and 20 feet are more common over most of the coastline. Tides are semidiurnal, and around much of the Atlantic coasts one low tide during the spring tide period occurs in the late morning or early afternoon.

The physical topography of the coastline and of the shores themselves is exceedingly variable, with the most extensive areas of rock occurring on the Atlantic coasts. These are often rugged and precipitous, but in Northern Scotland, Orkney, and parts of western Ireland, some areas with sedimentary formations have extensive platforms that slope gently seawards towards the incoming swell. The mountainous coasts of Ireland and Scotland are much indented, with many small bays and larger, fingerlike extensions of the sea ("lochs" or "loughs") that penetrate many miles inland. Together with the channels lying between the many flanking islets, these lochs form areas of very sheltered shoreline, the total extent of which probably exceeds that of the open coasts. So sheltered are these shores that stones and small shingle, which would be unstable and barren on open coasts, here serve equally with rocky outcrops for the attachment of sessile species. Yet, because these lochs are on coasts subject to the force of Atlantic waves, a distance of only a few miles—even a few hundred yards, in extreme instances—may separate localities subject, respectively, to extreme shelter and to extreme exposure. Along such sharp gradients, sudden changes in littoral populations and zonation patterns can be observed. Similar physical and biological transitions exist elsewhere, but around the North Atlantic they are rendered especially conspicuous by the presence of large, midshore algae (Fucaceae), the distribution and luxuriance of which are closely correlated with the degree of shelter. Although many other physical and biological factors influence local distribution, they appear to be subordinate to the effect of changes in the amount of wave action. Accordingly, throughout this chapter, the shore populations will be described in terms of the changes along the gradient from exposed to sheltered shores.

ZONATION

The general pattern of zonation follows closely that outlined in Chapter 2. Thus, the supralittoral fringe* is dominated by "black" lichens, dark microphytes, and littorinid snails; the midlittoral zone by various combinations of barnacles, mussels, limpets, snails, and brown and red algae; and the infralittoral fringe by laminarian algae, except where the substratum is insufficiently stable for such large plants. As indicated in Chapter 2, the placing of the boundary between the supralittoral fringe and midlittoral zone presents some practical difficulties which are best examined at the outset. Taking the "upper limit of barnacles in quantity" as the upper limit of the midlittoral zone, we find that where *Chthamalus stellatus* predominates (in southwestern areas generally, and

---

*For the sake of uniformity the author of this chapter has adopted the terminology used throughout the remainder of this book.

in exposed situations in the west and northwest), the "barnacle line" is higher—both absolutely, and relative to other species—than it is in areas where *Balanus balanoides* is present alone (on the north and east coasts, and in sheltered areas of the west and northwest). Consequently, some conspicuous zone-forming plants of narrow vertical range (the lichen *Lichina pygmaea* and the high-level fucoids *Pelvetia canaliculata* and *Fucus spiralis*) lie largely within the midlittoral zone on "*Chthamalus* shores" and partly or completely within the supralittoral fringe on "*Balanus* shores" (Lewis, 1955; see Figure 12.1). There are also intermediate areas where moderate amounts of *Chthamalus* lie above a well-defined *Balanus* line, so that even the selection of the barnacle line itself becomes arbitrary. Such complications are the inevitable consequence of using key indicator species to define zones, but, because no alternative method is either defensible or more practicable, they must be borne in mind when reference is made to species living at this level of the shore.

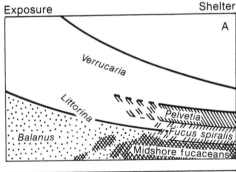

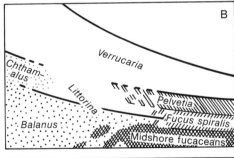

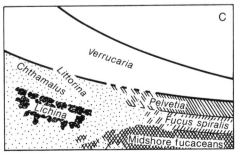

FIGURE **12.1**

The supralittoral fringe on various British shores: *A, Balanus* shores; *B,* places on northwest coasts where *Chthamalus* is confined to exposure; *C, Chthamalus* shores. [From Lewis (1964).]

The Supralittoral Fringe

The supralittoral fringe contains fewer species and a more uniform population than any other level of the shore. Its upper limit is placed, arbitrarily, at the junction between the black lichens and a band of orange or grey (or orange and grey) lichens that occurs on the majority of the rocky coasts of Britain. On extremely exposed cliffs, especially in northwestern districts, the upper limit is typically a diffuse transitional area at the top of a zone, which may be 30-70 feet deep, that lies entirely above the theoretical reach of the tides. Conversely, in very sheltered and sunny places, the junction of the two lichen belts is often very sharp, and the whole zone, which is compressed to a mere 2-3 feet, lies completely within reach of the spring tides. In the latter type of habitat, there is a close relationship between the depth of the zone and the tidal range; elsewhere, such considerations are overshadowed by the greater influence of variations in wave action (see Figure 12.14).

Flora. The vegetation comprises lichens and algae, and although there is much overlapping of ephemeral algae into the lichen belt, these two components tend to occupy the upper and lower levels, respectively. Among the lichens, two species predominate—*Verrucaria maura*, throughout much of the zone, and *Lichina confinis*, towards its upper limit. Although *L. confinis* is somewhat restricted by its preference for sunny situations, the two species occur under all conditions of wave action; together, they are primarily responsible for the blackening that characterises so much of the supralittoral area. Only on the more friable substrata (sandstone, chalk, soft limestone, or shale), or on unstable stony beaches, are they scarce or absent.

Closer to the barnacle line, *Verrucaria* thins out perceptibly and, depending upon whether algae are present or not, there may be a conspicuous bare region between the black lichens and the barnacle belt. The extent to which these levels do support algae varies with exposure, geographical location, and the time of year, so that the uniformity that characterised the lichen belt is reduced.

In wave-swept places, especially on ledges and shallow slopes, the algal growth takes the form of a very fine adherent layer dominated by cyanophyceans (*Calothrix* spp., in particular) and, more locally, filamentous green and red algae (*Ulothrix*, *Urospora*, and *Bangia*). Superimposed on this are the larger red alga *Porphyra umbilicalis* and species of the chlorophycean genus *Enteromorpha*, which may locally and temporarily obscure all others. This entire population is most abundant in the spring, and indeed by midsummer in the south and south west generally, and on sunny vertical faces elsewhere, it has either largely disappeared, or later generations of *Porphyra* or *Enteromorpha* have "migrated" downshore to leave the lower supralittoral almost completely bare again. However, the milder climatic conditions and greater moisture of northwestern Atlantic coasts permits a much longer period of growth, and under the most favourable conditions the cyanophyceans and *Porphyra* persist throughout the year. On surf-swept gentle slopes, this algal belt is itself subdivided to give a supralittoral fringe with a downshore sequence of four major organisms: lichens, cyanophyceans, *Porphyra*, and *Enteromorpha* (Figure 12.2). Here, too, may be the most exposure-tolerant of the large brown algae, *Fucus spiralis* f. *nanus,* and the rarer *F. distichus anceps.* Both

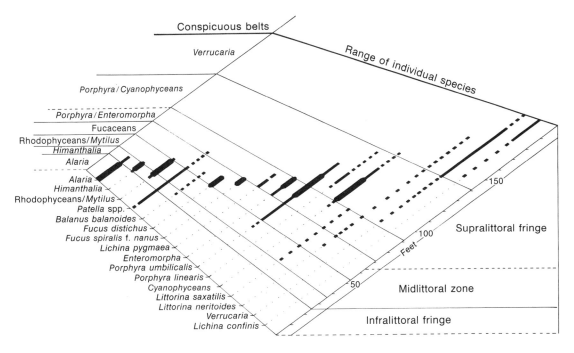

FIGURE 12.2

An extensive, unbroken slope in Orkney in April, showing a complex supralittoral fringe with several algal belts between the *Verrucaria* and barnacle zones. Note the great width of these belts compared with the remainder of the shore below. Only a few species of the lower and middle shore are depicted. [From Lewis (1964).]

occur among the *Porphyra/Enteromorpha* population, and their presence in quantity can give these high levels an appearance quite different from those of shores that are steeper or more southerly, and which, by contrast, are relatively barren for most of the year.

In less exposed places this algal belt changes it character completely. Most of the previously mentioned species decrease in abundance, even in spring, and are replaced by the more common high-level fucoids. The uppermost, *Pelvetia canaliculata*, is the more tolerant of wave action, and so may be found among the thinning *Porphyra* and cyanophyceans; more often, however, it occurs alone, a narrow belt among the lowest *Verrucaria*, well above the barnacle line on a *"Balanus"* shore, and astride or mostly above it in a *"Chthamalus"* area. Where conditions are less boisterous, or gentler slopes exist, it is joined immediately below by the more typical form of *Fucus spiralis*, and the two species occupy neat, narrow belts of roughly the same depth that persist into the most sheltered areas. Here, they form the upper fringe of a cover of fucoid algae that continues down to the laminarian zone, and the upper limit of *Pelvetia* now displaces the barnacle line as the most conspicuous biological horizon below the top of the *Verrucaria* belt. As with all the inhabitants of the supralittoral fringe, the absolute level occupied by these algae varies with the exposure. At its most wave-swept limits, *Pelvetia* may lie several feet above the theoretical high-tide level, whereas, in extreme shelter, it is well within the reach of spring tides. Even here, however, there are several consecutive days in each tidal cycle when it is not submerged, and loss of water reduces the fronds to a very brittle condition.

Associated with the denser growths of these fucoid algae is a distinctive subordinate flora comprising the red algae *Catenella repens* and *Bostrychia scorpioides,* and a wide variety of cyanophyceans not usually found on exposed coasts (including *Calothrix* spp., *Plectocnema, Microcoleus,* and *Rivularia*). *Catenella* is the most widespread. It first appears on fairly open shores in damp clefts, on shaded faces, or around the bases of *Fucus spiralis,* and is joined by the other species only where conditions are very sheltered. Together, they form a dark, shrubby or felted layer about 1 cm thick at most, and where most abundant—in the muddy and perhaps brackish backwaters of the indented western coasts—they may locally displace *Pelvetia* and *F. spiralis.*

The vegetation of the supralittoral fringe consists, therefore, of a wide belt of black lichens succeeded below by an algal belt that is perennial in shelter but of short duration on exposed shores (except on those of very exposed northerly coasts). One further plant should be mentioned here: Among the upper barnacles on exposed and sunny faces, there are conspicuous black patches of another species of *Lichina, L. pygmaea.* In *"Balanus"* areas, the uppermost patches tend to extend above the barnacle line, and when the spring growth of filamentous algae has died away, *L. pygmaea* may thus intrude into what would otherwise be the bare region at the bottom of the supralittoral fringe.

Fauna. In the British Isles, the characteristic *Littorina* population of the supralittoral fringe consists of two species, *L. neritoides* and *L. saxatilis.* Neither is restricted to these high levels (indeed, they may locally be more abundant in the upper part of the midlittoral zone), but having greater ability than other animals to withstand long periods without submersion or wetting by spray, they extend well upshore and are typically the only conspicuous animals of the lower and middle *Verrucaria* belt. *L. neritoides* usually has the greater vertical range, and because of its ability to live on very exposed shores with a wide spray zone, it can reach the greater absolute height, being found as much as 60 feet above high-tide level on the Atlantic coasts.

Littorina neritoides occupies the exposed shores and *L. saxatilis* occupies the sheltered shores, but there is a considerable area of overlap between them on moderately exposed coasts. In extreme shelter, *L. saxatilis* is most abundant on stable shingle, and seldom extends above the level of *Pelvetia.*

On open coasts both species appear to need the protection afforded by crevices and pitted surfaces, by the compact thallus of *Lichina pygmaea,* or by the small spaces and empty shells among barnacles. Consequently, their very local distribution varies greatly; cracks across otherwise smooth, bare rock may support concentrations of several hundreds or even thousands of individuals, while elsewhere, the snails may be scattered throughout an area of, for example, pitted limestone.

On most shores, the term *"Littorina* zone" adequately conveys the zoological character of these levels, for other animals are absent or difficult to find. Mites occur here, as at all levels, and the deep cracks harbour the thysanuran *Petrobius maritimus* and the isopod *Ligia oceanica.* Closer to the barnacle line, an overlap of essentially midshore species may occur, with *Patella vulgata* and *Littorina littorea* especially in shaded or slow-drying places. On stable shingle, *Ligia* and various amphipods accompany *Littorina saxatilis*, but they are not usually seen unless the stones are moved.

The Midlittoral Zone

At one extreme, the midlittoral zone is dominated by barnacles or mussels (or both), and at the other, by exceptionally heavy growths of long-fronded fucoid algae. Between the two are various overlapping or intermediate populations, and indeed, the infinite variety of local topography and degrees of exposure results in many stretches of shoreline supporting a mosaic of communities that can be rather bewildering to one without prior experience of the main types of populations and their characteristics where they occur alone. Accordingly, this zone is described primarily in terms of the distinctive communities dominated by barnacles, mussels, and fucoids.

Barnacle-Dominated Areas. Barnacles are among the most widespread of littoral animals. Although they tend to be most abundant on exposed shores, especially verticals or steep slopes, their ability to live under most physical conditions suggests that their occurrence in quantity is controlled less by their own requirements than by those of the competing mussels and fucoid algae. As these competitors generally favour the very exposed and sheltered sites, respectively, barnacles tend to dominate the areas of intermediate exposure. However, as mussels appear rather erratic in distribution, and do not occupy all exposed sites (especially if they are steeply sloping), barnacles can dominate under the most severe exposure conditions that exist. Equally, the fucoid algae may be lacking from verticals, even in extreme shelter, or from gentle slopes on open coasts where grazing molluses abound, so that one cannot be too dogmatic about where barnacle-dominated communities will or will not occur.

Where they are abundant, the barnacles can extend from their very sharp upper limit—the barnacle line—down to within a foot of the topmost laminarians. Within this cover, the actual species present vary with the local habitat and the geographical location. *Balanus balanoides* is almost ubiquitous, becoming very scarce only in the extreme southwest; *Chthamalus stellatus*, on the other hand, predominates in the southwest, but is absent from the North Sea coasts and the eastern half of the English Channel (see Figure 12.12; for further discussion of the distribution of barnacles, see Crisp and Southward, 1958; Moore and Kitching, 1939; Southward, 1967; and Southward and Crisp, 1954b, 1956). Where the two occur together on western coasts, *Chthamalus* is uppermost; towards the north, it occupies a decreasing proportion of the shore and becomes increasingly restricted to the most exposed and southwesterly-facing situations, where it forms a belt 1–3 feet high at the top of the barnacle zone (see Figure 12.4). As its geographical limits are reached, or in less exposed sites, it dies out as a thin scatter above a well-defined "*Balanus* line" and among the developing *Pelvetia*.

In moderately exposed sites around southwest England and Wales (but not in Ireland), a third species of much larger size, *Balanus perforatus*, occupies a belt 2–3 feet high immediately above the laminarians, or forms isolated patches in shaded or damp places at slightly higher levels. Since the late 1940s, the Australasian barnacle *Elminius modestus* has been establishing itself in harbours and estuaries and along the less exposed open coasts, mainly at the expense of *Balanus balanoides*. As the process of its dispersal and colonisation is greatly assisted by its ability to live on the undersides of boats, its geographical distribution here, and in northwestern Europe generally, is expanding irregularly. At the time this chapter was written, the species was apparently absent from most of Scotland, the Atlantic coasts of Ireland, and the more exposed districts of southwest England and south Wales (Crisp, 1958).

The overlapping local requirements of these species and their different geographical ranges mean that a complete cover of barnacles may comprise all four species in some parts of southwest England and south Wales, generally two species (*B. balanoides* and *C. stellatus*) on the exposed western coasts of Ireland and Scotland, and only *B. balanoides* in northeast Scotland and those sheltered west-coast areas not yet colonised by *Elminius*.

The macroscopic subsidiary fauna of a barnacle shore includes limpets (*Patella*) and dogwhelks (*Thais*), together with topshells (*Gibbula*) and anemones (*Actinia*) more locally, and the lower part of the population of *Littorina neritoides* and *L. saxatilis* that begins in the supralittoral fringe. Indeed, if we exclude such forms as nematodes, which may abound in the interstices of a dense barnacle cover, these littorinids are usually the most abundant species, but because they occupy spaces between barnacles and within their empty shells, they do not have the visual impact that similar or lesser numbers have higher up in the pits and crevices of the *Verrucaria* belt. Both species are most plentiful among the upper barnacles, but both can be found as low as there are barnacles to provide shelter. The individuals tend to be very small and may, therefore, be young specimens moving upshore; in the case of *L. saxatilis*, however, they may represent a variety peculiar to this habitat.

The limpets are much more conspicuous, and, as with the barnacles, there are local and geographical differences in the species present (Crisp and Southward, 1958; Evans, 1947b). *Patella vulgata* is ubiquitous and spans the entire zone, except on extremely exposed shores, where its lower limit rises relative to those of other species. *P. aspera* is fairly widespread geographically, but it is restricted to lower levels or pools in wave-swept places, rising farther upshore the more exposed or damp the conditions. The third species, *P. depressa*, found only in southwest England and Wales (see Figure 12.11), is less restricted locally than *P. aspera*, but has a more limited vertical range than *P. vulgata*.

Although the numbers of limpets may commonly range up to 300 per square metre, the individuals among barnacles tend to be fairly randomly distributed. By contrast, the other conspicuous mollusc of these habitats, *Thais lapillus*, congregates in large numbers during winter in pools and clefts or around the bases of large stable boulders, and only in summer do the individuals become well scattered. Even then, however, they largely avoid vertical faces subject to strong wave action, unless protecting ledges and clefts are present. Being less tolerant of desiccation than *Balanus balanoides* or *Chthamalus*, they are usually confined to the middle and lower levels of the zone. Here, their predation of the barnacles and mussels may reach such high proportions that the midshore may be largely denuded of these animals (and repopulated by them) annually.

The remaining large animals are much more local in their distribution. *Gibbula umbilicalis* is more numerous on sheltered shores, so its presence in wave-swept areas

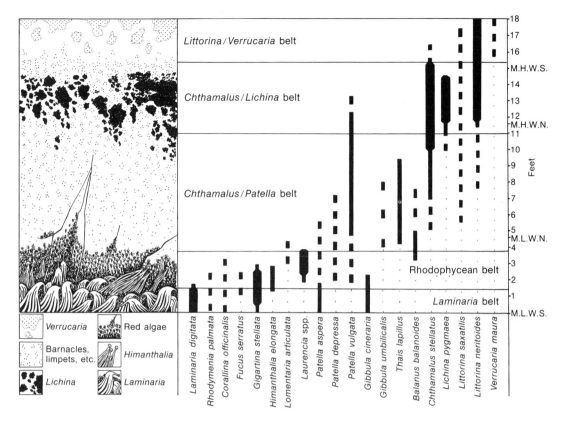

FIGURE **12.3**

A barnacle-dominated face near Hope Cove, South Devon, typical of many exposed and south-facing areas of the Channel coast. [From Lewis (1964).]

is dependent upon there being protecting clefts and depressions. It, too, is most abundant in the southwest, and is absent from North Sea coasts. *Actinia equina*, although primarily a pool-living species, is remarkably tolerant of desiccation. Thus, it not only hangs from the sides of shaded gullies, but also lives on open slopes where small depressions form temporary pools as the tide recedes.

By definition, barnacle-dominated shores lack algae in quantity, but on virtually none of them are algae completely absent. Most commonly, there is an algal belt (predominantly of rhodophyceans) immediately above the laminarian zone (Figure 12.3). At one extreme, on steep and very exposed faces, this zone may consist of little but encrusting lithothamnia; at the other, it may be a close, compact turf of such density that all barnacles, limpets, and other organisms are excluded. The number of algal species in this population is high, but the following are especially common: *Corallina officinalis,* *Gigartina stellata, Laurencia pinnatifida, Lomentaria articulata, Membranoptera alata,* *Plumaria elegans,* and *Rhodymenia palmata*. These and the many others vary greatly in their habitat requirements: *Corallina,* for example, abounds in exposed sites, *Lomentaria* on shaded vertical faces, and *Rhodymenia* in shade in southern localities but

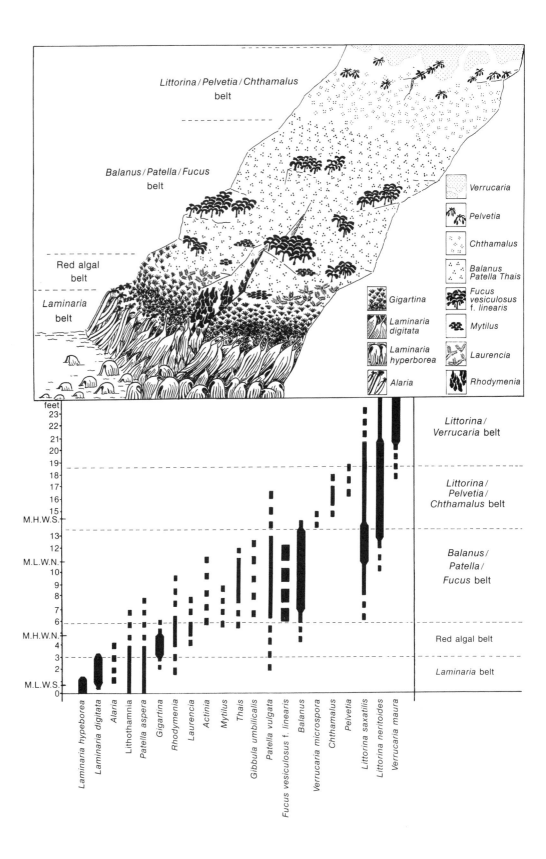

FIGURE 12.4

A type of barnacle-dominated slope that is very common on moderately exposed shores of northwestern Scotland and northwestern Ireland. [From Lewis (1964).]

on well-splashed open slopes and platforms further north. Where the rock is near sand, there are other, more specialised, communities of *Rhodochorton* spp., *Gelidium* spp., the chlorophycean *Cladophora rupestris,* and the phaeophycean *Cladostephus* spp. (For further discussion of the algae of the rhodophycean belt, see Börgesen, 1908; Cotton, 1912; Gibb, 1938, 1939; Lewis, 1953; and Rees, 1935.)

Within this belt, considerable stratification occurs. *Laurencia* spp. and the brown *Leathesia difformis* are uppermost, *Gigartina* and *Lomentaria* occupy the middle levels, and *Rhodymenia, Corallina,* and many others may overlap well into the laminarian zone. The depth of the belt responds sharply to any factor influencing desiccation: thus, on shaded gullies in northern Scotland, or on slow-draining platforms on open shores, the topmost *Laurencia* may extend one-third to one-half of the way up the midlittoral zone, whereas on steep, sunny faces that are unsplashed at low tide, the entire belt may rise only a foot or so above the laminarians.

Superimposed upon these turf-forming species are a few larger algae—the browns *Himanthalia elongata* and *Bifurcaria bifurcata,* and species of the chlorophycean genus *Codium.* Apart from *Himanthalia,* these algae are largely restricted to southwestern districts of England and Ireland and are relatively rare; *Himanthalia,* however, is widespread and characteristic of this low-level community. It occurs in a variety of local habitats, provided they are not excessively exposed, but it is undoubtedly most abundant on platforms and gentle slopes, where its reproductive fronds, 2–3 feet long, overlie the red algae and completely transform the appearance of the lower shore.

This belt of red algae and *Himanthalia* may be the only macroscopic vegetation on exposed barnacle shores, but two other species may intervene more locally, namely *Lichina pygmea* (Figures 12.3, 12.5A, and 12.6), and *Fucus vesiculosus* f. *linearis* (Figure 12.4). The lichen may overlap the lower part of the supralittoral fringe (p. 356), but on steep, sunny faces, its characteristic black patches are abundant among the upper barnacles, and in southwestern England, they may even extend down towards the middle of the zone. The very compact, branching thallus harbours large numbers of *Littorina* spp. and the very small bivalve *Lasaea rubra.* The *Fucus* is a wave-tolerant form of small size (6–9 inches) lacking the distinctive vesicles of typical *F. vesiculosus.* It may vary in quantity from solitary plants to the dense ankle-deep patches that characterise many northern shores, and usually it occupies the middle levels of the barnacle zone, only rarely overlapping into the red algal belt. Dense growths reduce the barnacle cover, and the shaded underlying rock then supports *Actinia, Gibbula umbilicalis,* and upshore extensions of *Patella aspera, Corallina, Laurencia,* and *Rhodymenia.*

Summarizing briefly, the barnacle-dominated areas have a virtually complete cover of barnacles, limpets, and littorinids (plus *Thais, Actinia,* and *Gibbula*), which generally gives way to a rhodophycean/*Himanthalia* belt at low levels. Additionally, the upper and middle levels may support belts of *Lichina,* or *Fucus vesiculosus* f. *linearis,* or both. When all three plant communities are luxuriantly developed, "barnacle dominance" becomes rather a misnomer.

Mussel-Dominated Areas. The common mussel, *Mytilus edulis*, was not mentioned in the previous section, but small numbers occur commonly in crevices or on ledges among the barnacles. In many localities, however, they spread out onto the open surface, and may form extensive carpets that displace the barnacles from all but the upper part of the midlittoral zone. Although this development occurs mainly in very exposed places, and is more common on gentle slopes than on vertical faces, the variation between apparently similar sites is such that physical factors are clearly not the sole influences. Recent unpublished work suggests that the almost erratic local distribution of *Mytilus* simply reflects stages in irregular cycles of settlement, competition with barnacles, predation, denudation, and resettlement, which vary in phase from one site to another. Mussels may occur as a thin scatter among the barnacles, as irregular patches that tend to coalesce towards low levels, as a virtually unbroken sheet down from a sharp upper limit, or as isolated belts or patches at any level within the species' potential range (Figure 12.6), the latter pattern probably representing the long-lived survivors of more extensive populations (Seed, 1969).

Nevertheless, whatever the variation, maximum occupation of rocky shores by *Mytilus* occurs on very exposed and typically slow-draining positions, mainly in northern and western districts. These physical conditions, accentuated by the water retained among the mussels, are at once reflected in a general uplift of distributional limits. *Patella aspera*, for example, rises towards the upper limit of *Mytilus*, while *P. vulgata* and the barnacle-inhabiting *Littorina* spp., although not entirely displaced from the lower levels, become largely confined to the narrow barnacle belt at the top of the zone. The response of the rhodophycean/*Himanthalia* belt is most interesting. The upper limit of *Himanthalia* rises slightly—in tidal terms—but the belt persists in the lower shore until it dies out in places of increased exposure or steepness. Among the red algae, the less wave-tolerant species give way to *Corallina* and *Gigartina* especially, and to some *Rhodymenia* and several less conspicuous species, and most of this population rises upshore to overlie almost all the mussels. Thus, whereas *Gigartina* and *Rhodymenia* normally occupy a lower level than *Himanthalia*, they rise well above it on mussel-dominated slopes. Indeed, most of the red algae on these exposed *Mytilus* shores is midlittoral in distribution (Figure 12.7).

Parallel with this change is the development of a new midshore population of red algae that is almost entirely restricted to the surface of *Mytilus* shells. This population is dominated by two small, bushy species, *Ceramium shuttleworthianum* and *Callithamnion arbuscula*, and downshore extensions of *Porphyra umbilicalis* from the supralittoral. Thus, where *P. umbilicalis* persists throughout the year at high levels (as in northern districts), it has two areas of abundance separated by a gap astride the upper barnacles, but elsewhere, the summer die-back of the supralittoral population leaves this upper *Mytilus* belt as the main area for summer *Porphyra*.

The combined effect of the uplift of low-level species and the development of this higher-level population is to overlay the *Mytilus* with a bewildering array of mixed algal communities, with mosaics of single species growths, and with very local, irregular patterns of zonation within the underlying barnacle/mussel division of the shore (Lewis, 1954, 1964). Indeed, the only regular algal feature is the low-level position of the *Himanthalia* belt—where that species is present. Further complication is added by the

A

FIGURE **12.5**

A, a moderately exposed shore on the Sound of Jura, western Scotland, with a midshore belt of barnacles and limpets fringed above by a belt of *Lichina pygmaea:* in the lower shore, an irregular belt of rhodophyceans and *Fucus serratus* precedes the infralittoral population of *Laminaria digitata* and *L. hyperborea. B,* a severely exposed face near Kerry, southwest Ireland, showing the lower part of the *Verrucaria* zone, a narrow belt of barnacles (*Chthamalus*) and the main midlittoral population of *Mytilus.* Small amounts of *Lichina* occur among the barnacles, and some rhodophyceans (especially *Porphyra*) overlie the mussels. The uppermost laminarian algae are just visible above the water line.

B

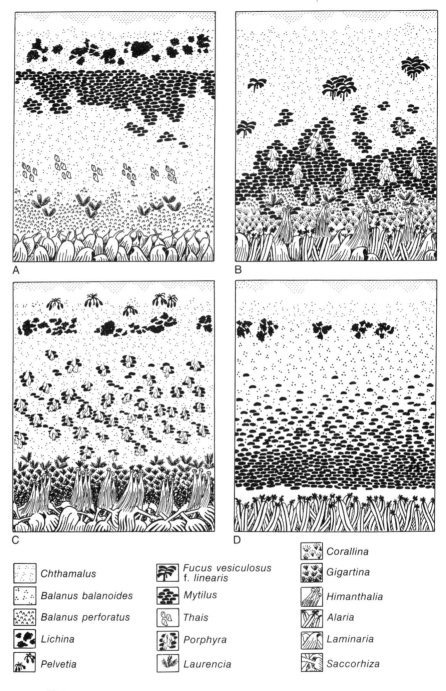

:::: Chthamalus	🌿 Fucus vesiculosus f. linearis	🌿 Corallina
::: Balanus balanoides	🐚 Mytilus	🌿 Gigartina
🔺 Balanus perforatus	🐚 Thais	🌿 Himanthalia
🔲 Lichina	🌿 Porphyra	🌿 Alaria
🌴 Pelvetia	🌿 Laurencia	🌿 Laminaria
		🌿 Saccorhiza

FIGURE **12.6**

Patterns of *Mytilus* distribution: *A*, St. Ives, June 1960; *B*, near Pendeen Head, Cornwall, June 1960; *C*, near Derrynane, County Kerry, July 1959; *D*, near Stoer Bay, Sutherland, August 1952. The space below the mussels in *D* is occupied by lithothamnia. *Patella* spp. and *Littorina* spp. have been omitted. [From Lewis (1964).]

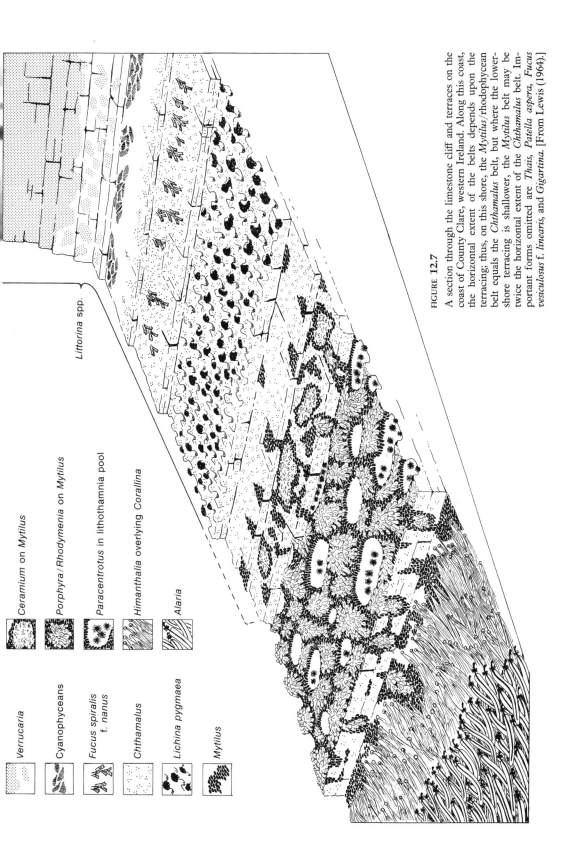

Verrucaria

Cyanophyceans

Fucus spiralis
f. nanus

Chthamalus

Lichina pygmaea

Mytilus

Ceramium on Mytilus

Porphyra/Rhodymenia on Mytilus

Paracentrotus in lithothamnia pool

Himanthalia overlying Corallina

Alaria

Littorina spp.

FIGURE **12.7**

A section through the limestone cliff and terraces on the coast of County Clare, western Ireland. Along this coast, the horizontal extent of the belts depends upon the terracing; thus, on this shore, the *Mytilus*/rhodophycean belt equals the *Chthamalus* belt, but where the lower-shore terracing is shallower, the *Mytilus* belt may be twice the horizontal extent of the *Chthamalus* belt. Important forms omitted are *Thais, Patella aspera, Fucus vesiculosus* f. *linearis*, and *Gigartina*. [From Lewis (1964).]

presence of *Fucus vesiculosus* f. *linearis,* which can be very abundant on the flatter reefs, especially those of northern Scotland; but again, there is no regularity in its distribution from one site to another.

A further interesting feature of *Mytilus* shores, compared with barnacle areas, is the rich fauna inhabiting the water-retaining interstices between the mussels: the anemones *Sagartia* and *Metridium,* small colonies of hydroids and sponges, and small crabs and other crustaceans. In the west of Ireland, the pool-living urchin *Paracentrotus lividus* is not uncommon in the dampest mussel patches.

Fucoid-Dominated Areas. The progressive replacement of barnacle- or mussel-dominated communities as exposure decreases begins with either the first appearance of *Fucus vesiculosus* f. *linearis,* or with the increasing abundance of this species in districts where it is already present. Generally, this occurs first on slopes that lack other fucoids at other levels (apart from those districts with *F. spiralis* f. *nanus* and *F. distichus anceps;* see p. 354), but as the process develops, *Pelvetia* gradually appears in the supralittoral fringe, and *Fucus serratus* starts to mingle with the low-level *Himanthalia.* As the longer and bladdered form of *F. vesiculosus* replaces *F. vesiculosus* f. *linearis,* the gap between it and *Pelvetia* will support the first *F. spiralis,* and *F. serratus* will displace *Himanthalia* almost entirely. Next, *Ascophyllum nodosum* starts to appear in the flatter or more protected places among the *F. vesiculosus,* and the process culminates in very sheltered bays and lochs with luxuriant narrow belts of *Pelvetia* and *F. spiralis* surmounting a wide midshore belt of long-fronded *Ascophyllum* (up to 12 ft) with a narrow belt of *F. serratus* just above the laminarians (Figure 12.8).

The stages in this succession do not, however, always appear to follow the general plan. Because of the very erratic distribution of *Fucus vesiculosus* f. *linearis* (which is associated, presumably, with such matters as dispersal and limpet grazing), the midshore may remain empty until the normal *F. vesiculosus* appears, or one may find areas of the two types of plant separated spatially and without the usual transitional forms. Moreover, because all of these algae prefer flat to steep surfaces, very local changes in topography at different levels can vary the order in which the species appear along a transect from a headland to a bay. Thus, a flat lower shore can have *F. serratus* and *Ascophyllum,* while upper levels that are steeper or have loose stones retain an exposed facies and remain bare (Figure 12.9). Alternatively, it may be the upper levels that are flatter, provide local shelter, and become occupied first, so the sequence outlined above can only be seen where uniform topography exists along the exposure gradient.

FIGURE **12.8**

Three stages in the progressive occupation of the shore by fucoid algae: *A,* moderate exposure (Sutherland, northwest Scotland); *B,* moderate shelter (Bantry Bay, southwest Ireland); *C,* extreme shelter (Anglesey, northern Wales). In *A,* from the water line upwards, the following belts can be recognized: rhodophyceans and *Fucus serratus;* barnacles and scattered clumps of short *Fucus vesiculosus;* the *Verrucaria/Littorina* zone, in the middle part of which there is a belt of *Pelvetia canaliculata;* and orange and grey lichens. In *B,* the high belt of orange and grey lichens is followed below by a very narrow belt of *Verrucaria,* dense but narrow belts of *Pelvetia* and *Fucus spiralis,* and a wide midshore belt of *Fucus vesiculosus* and barnacles. In *C,* the belts of *Pelvetia* and *Fucus spiralis* are succeeded below by long-fronded drapes of *Ascophyllum nodosum.*

A

B

C

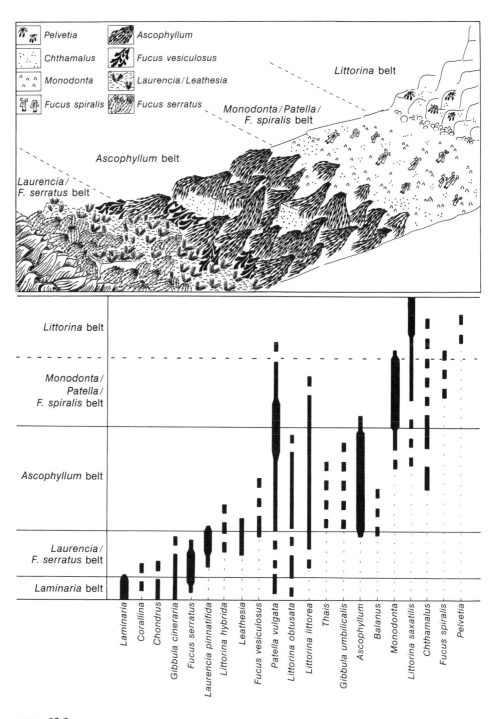

A flat shore on the open coast near Bantham Bay, South Devon, where abrasion in the upper levels disrupts the algal sequence and produces a rather barren area dominated by *Patella*, *Monodonta*, and *Fucus spiralis*. [From Lewis (1964).]

Wherever the overall degree of shelter permits a continuous cover of fucoids, the relative proportions of the midlittoral zone occupied by *Ascophyllum, Fucus vesiculosus,* and *F. serratus* vary greatly. On open coasts, *Ascophyllum* prefers the flatter areas, but in greater shelter, it hangs in long festoons from vertical faces and large boulders. *F. vesiculosus* fills the steeper positions and outcrops near sand on open coasts, but in sheltered places it is displaced, through competition with *Ascophyllum*, to the smaller loose-lying stones that do not provide adequate anchorage for the latter. *F. serratus* responds to shade or slow-draining positions by rising upshore appreciably and raising the lower limit of the other two species; but in extreme shelter, the *F. serratus* belt on steep faces may be almost eliminated by the masses of *Ascophyllum* hanging down into the laminarian zone. On sheltered shingle, there is a wide area of overlap between *F. vesiculosus* and *F. serratus,* and it is not uncommon to find scattered plants of *F. spiralis* well below the normal level. Thus, the combined effects of wave action and the topography of the shore determine the actual pattern of zonation in any particular transect (Figure 12.10).

Two other midshore variations occur more locally: In estuaries and small outlets of streams, the fresh-water flow may be marked by the presence of *Fucus ceranoides,* and on extremely sheltered shingle beaches subject to alternating salinities, there may be free-living forms of the fucoids. Of these, *Ascophyllum nodosum* f. *mackaii* is by far the most common (Gibb, 1957), and the presence of extensive midshore beds of this highly distinctive plant in the lochs of Scotland and Ireland marks the end point in the succession from exposed to sheltered conditions.

Turning to the smaller algae, we find that the shade of the first growths of midshore fucoids encourages *Laurencia, Leathesia,* and other members of the red algal belt to extend patchily upshore, but under very dense growths of *Ascophyllum* and *Fucus serratus,* most of these algae are eliminated and the underlying rock supports only lithothamnia and small growths of shade-tolerant species. The removal of *Ascophyllum* is at once followed by the redevelopment of the "turf" (especially of *Laurencia, Leathesia, Cladophera rupestris,* and *Chondrus crispus*), and even of *Himanthalia,* if the site is not too distant from a permanent population of this alga.

Many changes in the fauna accompany the increase in fucoids. First, there is the loss of such open-coast species as *Littorina neritoides, Patella aspera, P. depressa, Balanus perforatus,* and (typically) *Mytilus,* with it associated flora. Of those that remain, the barnacles are fewer and persist mainly on minor elevations and faces. In northwestern areas, *Chthamalus stellatus* may be absent even where there is no full cover of fucoids, leaving *B. balanoides* as the only barnacle present. *Patella vulgata* and *Littorina saxatilis* remain in somewhat smaller numbers, but both now attain a larger maximum size, and the former in particular achieves a prominence it lacked where the barnacle cover was heavier. *Thais lapillus* persists where barnacles still exist, but never in the dense aggregations that are so typical of exposed coasts.

The changes are not, however, only a matter of reduction. Low-level and flat outcrops of rock adjacent to sand support the massed tubes of the polychaete *Sabellaria alveolata* in southwest England and Wales, and these may locally displace the belt of *Fucus serratus* and red algae. The topshell *Gibbula umbilicalis* becomes plentiful throughout the middle of the zone, and is joined by *G. cineraria* and *Monodonta lineata* in the lower and upper

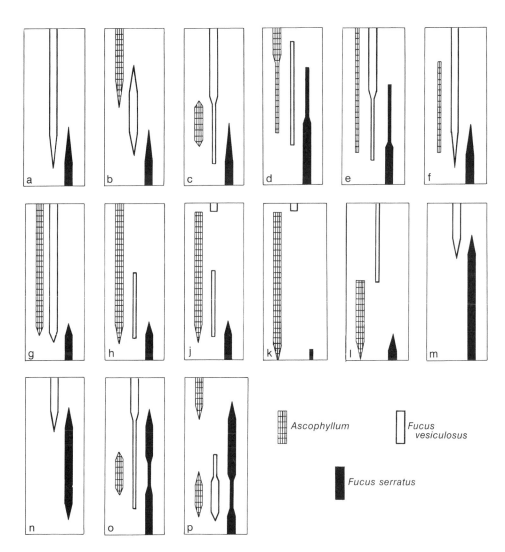

FIGURE **12.10**

Some of the variations in the proportions of the midlittoral zone occupied by *Asocphyllum, Fucus vesiculosus,* and *F. serratus: a,* open coast, maximum degree of exposure under which full cover of fucoids develops; *b,* moderately sheltered open-coast boulder beach (southwestern Anglesey, Wales) or very stable stony beach with rocky upper levels (Menai Straits, Wales; Loch Sunart, Scotland); *c,* irregular open coast with midshore depressions (northwestern Donegal, Ireland); *d,* flat limestone terraces, slow-draining and fairly sheltered (northeastern Anglesey, Wales); *e,* extensive flat shore of rocky outcrops and boulders (Berwick-shire, Scotland); *f,* stable shingle with scattered boulders (Menai Straits, Wales); *g,* boulders and shingle throughout (Menai Straits, Wales; Loch Creran, Scotland); *h,* steep rocky slopes or beaches of large boulders (very widespread in lochs and sheltered bays); *j,* steep rocky slopes in extremely sheltered backwaters (Outer Hebrides; Menai Straits, Wales; Loch Laxford, Scotland); *k,* steep rocky slopes (many Scottish lochs, but especially in the Outer Hebrides); *l,* stony upper levels with boulders and rocky outcrops below (Loch Linnhe, Scotland; rather uncommon elsewhere); *m,* steep open coast with heavy shade (northern Anglesey, Wales); *n,* steep open coast with heavy shade, becoming steeper and more exposed at low levels (northern Anglesey, Wales); *o,* shaded and very irregular slope (northern Anglesey, Wales); *p,* north-facing jetty with boulder and shingle below northern Anglesey, Wales). Further variation not shown here is caused by the occasional descent of *Fucus spiralis* to these levels and the occasional replacement of *F. serratus* by *Himanthalia.* [From Lewis (1964).]

levels, respectively (see Figure 12.9). *Littorina saxatilis* is joined by *L. obtusata,* mainly on the fucoids, and by large numbers of *L. littorea,* a large winkle that abounds in flat and otherwise sparsely occupied sites and may range from the topmost fucoids down to the laminarian zone. All these grazing molluscs, together with *Patella vulgata,* are particularly common on flat, open-coast sites lacking many barnacles and, at the same time, surprisingly bare of fucoids: it may well be that their grazing and wandering activities are inimical to barnacle larvae and fucoid germlings alike. There nevertheless comes a point along the transition from exposure to shelter where the fucoids do establish themselves in quantity and few of these gastropods exist. Where long-fronded *Ascophyllum* drapes boulder and outcrops, *L. saxatilis* is confined to the level of *Pelvetia* and *Fucus spiralis,* and *L. obtusata* is common on the weed itself, but the underlying rock will have only an occasional *Patella* or *L. littorea.*

In the shingle and stony areas lacking *Ascophyllum,* however, the fauna is richer and more conspicuous among the sparser *Fucus vesiculosus* and *F. serratus.* Much depends on the stability of the shingle and the amount of silt present, but such areas can have a light cover of barnacles on the upper surfaces, large numbers of *L. littorea,* some *Gibbula umbilicalis* and *G. cineraria,* and a very plentiful downshore extension of *L. saxatilis.* Indeed, *L. saxatilis* is more abundant in many such areas than at higher levels.

Mytilus edulis requires special mention here because, although it dies out on most sheltered open coasts, it may reappear in quantity in the lochs again. The individuals are fewer, but much larger, and they can occur on stable shingle, beneath short *Ascophyllum,* or on the lower parts of steep faces. The latter may unaccountably lack midlittoral algae, even in extreme shelter. At once, *Balanus balanoides* reappears (with or without *Mytilus* at lower levels), and, as is true of so many other species, it is again notable that the individuals, although fewer, are very much larger than those on open coasts. Such *Balanus* or *Balanus/Mytilus* sites are not common in extreme shelter, but where they do occur they stand out in marked contrast with the adjacent, extensive areas of *Ascophyllum* or *Fucus.*

The Infralittoral Fringe

Flora. The flora of the infralittoral fringe is characteristically dominated by the laminarians, but the range of exposure and of substrate conditions is such that other plant forms may be more important locally. Most of the permanently submerged laminarian "forest" consists of *Laminaria hyperborea,* and only the upper fringe of this population is uncovered at the lower spring tides. Above this species on open coasts, two species predominate—*Alaria esculenta,* in very exposed situations, and *Laminaria digitata* elsewhere. They form a continuous narrow belt, typically not more than 1-2 feet deep; where oceanic swell is continuous, however, the area occupied by *Alaria* widens appreciably, partly because *L. hyperborea* retreats to lower levels, but mainly because the upper limit of *Alaria* rises. Indeed, in such situations in northern districts where desiccation is less, the uppermost *Alaria* may reach a level comparable to that of the upper barnacles only a short distance away.

Insofar as it can be ascertained, the subordinate flora of the most severely exposed sites appears to consist of little but a continuous coating of encrusting lithothamnia;

elsewhere, there is typically some *Corallina officinalis* and *Gigartina stellata,* the latter extending down from the midshore mussel communities where it is more abundant. As *Laminaria digitata* replaces *Alaria,* the undergrowth becomes more variable and luxuriant, and commonly includes such species as *Ceramium* spp., *Chondrus crispus, Cladophora rupestris, Cystoclonium purpureum, Delesseria sanguinea, Dictyota dichotoma, Membranoptera alata, Plocamium coccineum, Plumaria elegans, Polysiphonia* spp., and *Rhodymenia palmata.* Some of these, together with other red algae, also occur as epiphytes on the stipes of *Laminaria hyperborea,* where they are often arranged in a definite order that presumably reflects the light gradient among the larger plants. (For further discussion of the subordinate flora of the infralittoral fringe, see Börgesen, 1908; Cotton, 1912; Gibb, 1938, 1939; Lewis, 1953; and Rees, 1935.)

On southwestern coasts, certain large algae provide local novelty, especially the laminarian *Saccorhiza polyschides,* but also species of *Codium* and *Cystoseira, Bifurcaria bifurcata,* and unusually low-level growths of *Himanthalia elongata.* Very locally, any of these may be dominant, but usually there are just small patches or solitary plants among the normally occuring *Laminaria.*

In sheltered lochs, *Laminaria digitata* and *L. hyperborea* lose their predominant position. This stems partly from the greater abundance of boulders, stones, and shingle, which, although apparently unsuitable for those two species, are sufficiently stable in sheltered waters to support a wide variety of other algae, both large and small. *L. hyperborea* dies out first; even on rocky faces, its place is taken by *L. saccharina, Halidrys siliquosa,* and (locally) by *Codium* and *Cystoseira* (the latter in Ireland but not in Scotland), all of which find adequate anchorage on bedrock, boulders, and stones alike. *L. digitata* persists further and, even on small boulders, may still form an upper fringe to the *L. saccharina*/*Halidrys* population. However, as the substrate deteriorates further, it too is replaced by *L. saccharina* and *H. siliquosa* among which may also be found some *Fucus serratus,* extending to lower levels than it does on open coasts.

Parallel changes occur among the smaller species. Bedrock and large boulders still have a turf comparable with that of the open coasts, but here it is dominated by *Chondrus crispus* and *Furcellaria fastigiata,* with *Corallina officinalis* and *Cladophora rupestris,* and these may be sufficiently abundant locally to exclude all but isolated plants of the larger species. In areas of shingle, this turf gives way to larger but more delicate, lightly attached forms, all of which float erect in the still waters to produce a buoyant layer of very soft vegetation. Common genera here are *Asperococcus, Ceramium, Chylocladia, Cystoclonium, Dictyota, Enteromorpha, Laurencia, Lomentaria, Mesogloia, Scytosiphon, Spermatochnus, Stilophora,* and *Ulva* (see Lewis, 1957a, 1957b; Lewis and Powell, 1960; and Rees, 1935).

Where shingle is replaced by coarse gravel and mud, all the larger species die out, except for strands of *Chorda filum,* and the smaller algae become very patchy. Locally, however, the increasing barrenness is replaced by blanketing, woolly layers of fine filamentous rhodophyceans—*Bonnemaisonia hamifera* and *Rhodochorton floridulum*—or by extensive meadows of the eelgrass *Zostera.* Thus, as wave action ceases to be an important factor, the flora is determined by the nature of the substratum; depending upon the variability of the latter, the population may remain very uniform for miles of loch shores, or patches of all the above types may occur together in one small area.

There is, however, one type of habitat in which water movement—in the form of strong tidal streams, rather than waves—reappears as the controlling influence (Bassindale et al., 1948; Lewis, 1957a, 1957b; Lewis and Powell, 1960; and Rees, 1935). Such areas occur at constrictions in long, narrow lochs, or in the channels between small islands, and as more of the indented western coasts are investigated, more examples come to light. In some, the tidal stream flows like a gentle river, and *Laminaria saccharina* and *Halidrys siliquosa* persist as the dominant plants, and may attain a very much larger size than elsewhere. They are sometimes joined by that "typical" open coast alga *Himanthalia elongata*, which here may occur sublittorally with its unusually long fronds (up to 20 feet) floating up among the other two species. Where the tidal flow is very strong and turbulent "rapids" result, the character of the population assumes an open-coast facies with the return of luxuriant plants of *Laminaria* and, less commonly, *Saccorhiza polyschides*. In the subordinate flora, the first casualties of faster water flow are the lightly attached buoyant forms (see above), but finally, the *Chondrus/Furcellaria* turf gives way to a more open scattering of *Gigartina stellata*, *Rhodymenia palmata*, *Corallina officinalis*, and *Cladophora rupestris* on an underlying cover of encrusting lithothamnia.

Fauna. Just as the laminarians visible on the shore are but the upper fringe of a more extensive infralittoral population, so the majority of the animals here are more abundant below tidal levels. Although some littoral species (*Patella aspera*, for example, and *Gibbula cineraria*) lie astride this level, most of the animals dealt with in earlier sections die out above or among the upper laminarians, and the fauna changes from one of large numbers of relatively few species to one of small numbers of very many species, few of which achieve the high densities and recurrent prominence of the midlittoral barnacles and mussels.

The richness of the fauna depends, therefore, upon the extent to which these sublittoral animals can penetrate the tidal area in any particular habitat, and although there is almost always some increase in the total number of species towards low-water mark, the greatest variety and abundance occurs in sites that are not only protected from excessive desiccation, but are also sheltered, or fairly free from sediment, or both.

Thus, the open rock surface of very exposed *Alaria* shores supports little but *Patella aspera* and irregular downward extensions of *Mytilus edulis*. The sublittoral element is to be found in clefts or on overhanging surfaces, and consists primarily of a few hardy sponges (such as *Halichondria* spp.), anemones (such as *Sagartia* spp. and *Tealia felina*), hydroids (*Tubularia* spp.), polyzoans (*Mucronella*, for example), or compound ascidians (such as *Botryllus* and *Botrylloides*).

With less wave action and the change to *Laminaria* spp., with their more varied subflora, the fauna diversifies in various ways. On the open rock surface, *Patella* and *Mytilus* give way to a scattering of barnacles (*Verruca stroemia* and *Balanus crenatus*), tubiculous polychaetes (*Pomatoceros triqueter* and *Spirorbis* spp.) and a greater variety of small hydroids, polyzoans, and compound ascidians. Crawling over the surface may be chitons, cowries, nudibranchs, the topshell *Calliostoma zizyphinum*, and the lowermost midlittoral molluscs including *Patella vulgata, Littorina littorea,* and (in some places) *Thais lapillus.*

The algae, especially the basal parts of perennial plants, now provide attachment for hydroids and polyzoans, and the stipes and blades of *Laminaria* often support the limpet *Patina pellucida* and the topshell *Gibbula cineraria.*

Nevertheless, the total number of animals on open slopes remains small, and for richer collecting on open coasts, one must turn either to wave-cut and shaded overhangs or to boulder shores. In the former, the emphasis is entirely upon sedentary suspension-feeding forms, and in some limestone districts (in parts of South Pembrokeshire, for example), an impressive number of sponges, hydroids, anemones, polyzoans, and compound ascidians can be found. Where the topography and orientation are most favourable, such populations may extend well upshore into the lower midlittoral, displacing the more usual algae. The other rich habitat—stable boulders—owes its richness to a wide range of habitats in a small area: upper and lower surfaces, shaded overhangs, algae, pools, and organic debris around the boulder bases. Such areas not only have most of the aforementioned species, but are further enriched by nereid worms, scale worms, crabs, nudibranchs, and echinoderms—but of course, most of these are not strictly equatable with the fauna of emergent rock surfaces with which we are concerned.

One other feature of this zone on open coasts is the upshore migration into it by certain infralittoral animals. For some animals—for nudibranchs generally, and the tectibranch *Aplysia punctata*—this movement is very intermittent, but in the case of *Echinus esculentus* and *Asterias rubens,* there is greater regularity, with the former appearing in late spring and summer and the latter in autumn and winter. *Asterias* moves onto exposed slopes in very great numbers, and when it reaches the lower midlittoral, destroys many of the lower mussels.

In the sheltered lochs, the fauna may be very much richer than it is in most open-coast sites, although the sediment that characterises many sheltered places may result in the rock fauna itself being greatly depleted locally. Where mud is plentiful, and a dense *Chondrus/Furcellaria* turf covers all the available surface, most of the animals are to be found on the large algae, *Laminaria saccharina, Halidrys,* and *Codium.* Presumably because dislodgement is rare in these quiet waters, this population is typically much richer than that on *L. digitata* on open coasts, and the delicate feathery covering of hydroids and polyzoans is commonly joined by more robust forms, such as solitary ascidians (*Ciona intestinalis* and *Ascidiella* spp.) and the anemone *Anemonia sulcata,* any of which may occur at the rate of several dozen per plant (Ebling et al., 1948; Lewis, 1957a, 1957b; Lewis and Powell, 1960).

In the mud-free areas, there is a general increase in species, but the details of the fauna vary greatly with the type of substratum and the degree of water flow. On the sides of large boulders in still or slow-moving waters, for example, the smaller suspension feeders are joined by solitary ascidians (*Ascidia* spp., especially), and some boulder areas have large numbers of echinoids (*Echinus esculentus* generally, and *Paracentrotus lividus* in southwestern Ireland), whose grazing habits appear to be responsible for the local absence of all algae except lithothamnia (Kitching and Ebling, 1961). Where small stable stones and clean gravel predominate and algae are lacking, the entire emphasis switches to a thin scatter of tubeworms (*Spirorbis* spp., *Pomatoceros triqueter,* and *Serpula vermicularis*), the limpets *Acmaea tessulata* and *A. virginea,* occasional banks of *Mytilus edulis,* and large numbers of *Psammechinus miliaris* locally, the whole making a low-level counterpart to the barnacle/gastropod population of stable midlittoral shingle (Lewis, 1957a).

As the rate of water flow increases in the vicinity of tidal rapids, there is initially an increase in the suspension-feeding element, but with severe turbulence and the return of "open coast algae," there is a dramatic fall in the fauna living on the stipes and blades of algae, and even on the top and sides of boulders there are typically few animals but encrusting sponges and the hardier hydroids and polyzoans. The richest areas are probably those of moderate tidal current, where *Halidrys* and *Laminaria saccharina* still persist and support their characteristic fauna, and where, with a reduction in the smaller algae, the boulders support a dense, tangled mat of large hydroids and polyzoans, pendulous sponges, and sheets of compound and colonial ascicians (the actual species present vary greatly from one channel to another). Details of these faunas are inappro-pirate here, and can be obtained elsewhere (see, for example, Ebling et al., 1948, 1960; Lewis, 1957a, 1957b; Lewis and Powell, 1960; Lilly et al., 1953; and Sloane et al., 1957): suffice it to say that they are probably the richest in the British Isles.

CONCLUSION

The variation in the littoral populations of the British Isles has a number of causes. On the geographical scale, the climatic and hydrographic differences between north and south and between east and west, referred to in the introduction to this chapter, permit a considerable number of warm-water species to extend to various points in the south-west, west, or northwest, but exclude them from the coasts of the North Sea and the eastern half of the English Channel (Crisp and Southward, 1958). Although such adverse local factors as unsuitable substrata probably hasten the decline of these species in the eastern half of the English Channel, the general correlation between distribution and winter isotherms suggests that low temperatures are the controlling factor—indeed, it was these southern species on western coasts that suffered most severely from the unusually cold winter of 1962–63 (Crisp, 1964).

The greater variety of species on the western coasts does not, of course, radically change the character of a shore. A barnacle-dominated area in the southwest is super-ficially similar to one on North Sea coasts, but has three or four species of barnacles instead of one (or possibly two), three limpets (*Patella* spp.) as opposed to two, three topshells instead of one, and so on for most groups of organisms except the fucoid algae. Amongst this southern component (see Figure 12.11) are such conspicuous or abun-dant species as *Chthamalus stellatus, Balanus perforatus, Patella depressa, Gibbula um-bilicalis, Monodonta lineata, Anemonia sulcata, Sabellaria alveolata, Paracentrotus lividus, Bifurcaria bifurcata,* all species of *Cystoseira,* and a large number of red algae. By contrast, there are few conspicuous northern species entirely confined to North Sea coasts, this element of the population being almost ubiquitous in the British Isles. The principal exceptions are *Balanus balanoides,* which does become rare in the extreme

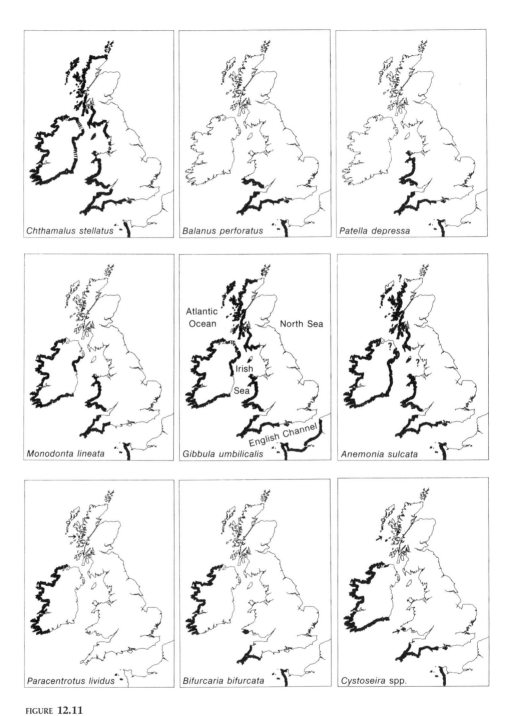

FIGURE **12.11**
Maps showing, in heavy outline, the geographical areas of the British Isles in which certain southern species will probably occur, provided local conditions are favourable. [From Lewis (1964).]

southwest (Southward, 1967; and Southward and Crisp, 1958), and *Fucus distichus anceps*, which is absent from most western and all North Sea coasts, and so constitutes a rather special case (Powell, 1957).

Climatic differences do, however, influence the appearance of the shore in another way, for the mild, moist summers of all northern and some western coasts appear to contribute greatly both to the persistence there throughout the summer of the high-level *Porphyra*/*Enteromorpha*/cyanophycean belt, and to the occupation of the exposed mid-littoral zone by species, especially red algae, that are generally restricted to low levels in the south (Lewis, 1954). The combined effect of these two phenomena is to give exposed northern shores a much greater cover of small algae, even in midsummer, than exists on comparable shores in the south. Sheltered areas lack this difference, for the fucoids are dominant in all regions. The considerable local differences in the distribution and zonation of the three midlittoral fucoid species are responses to very local factors (such as wave action, aspect, and substratum) that tend to obscure any geographical differences that may exist.

Geographical variation, however, reflects not only climatic differences, but also the geographical distribution of particular types of habitat. For example, the most varied and richest populations of the infralittoral fringe occur in the lochs and tidal rapids, habitats that are completely restricted to western coasts. Because similar types of populations recur on the similarly indented coasts of southwestern Norway and western Sweden, we may reasonably infer that the comparative poverty of this zone on the North Sea coasts of Britain generally is due, in part, simply to the lack of such habitats.

A further illustration of this "chance" geographical influence is provided by the *Porphyra*/cyanophycean and *Mytilus*/rhodophycean populations. As mentioned earlier (p. 354), these are favoured by cool, moist summers of northern districts, but they also require very exposed, gentle slopes for maximum development; such habitats are much more common in Orkney, northern Scotland, and western Ireland than elsewhere. Indeed, in the more northern of these areas, the climatic influence probably only accentuates the already favourable conditions of topography and exposure. In the west of Ireland (County Clare, especially), we see the interesting combination of exposed platforms, cool summers, and very mild winters, and the consequent uplift of rhodophyceans imparts a "northern" character to shores that, paradoxically, still support a number of the southern, warm-water species.

A more common form of variation, demonstrated on all coastlines to some degree, is that associated with differences in exposure to wave motion. As the entire descriptive account in this chapter is based on this variation, little more need be said. However, it is important to emphasise that, although we have the obvious differences between wave-swept headlands and land-locked bays, the "exposed" or "sheltered" character of a shore is not solely a matter of its position relative to open seas and oceanic swell (Figure 12.12). How and with what force the waves break against a shore is largely determined by the very local topography of that shore, and in general, the more broken and extensive it is from high to low levels, the less severe the conditions will be. Consequently, wide, broken shores or areas of large, stable boulders will, even on Atlantic coasts, provide many areas occupied by sheltered-shore species, or will show a gradient of increasing shelter from the seaward to the landward margins.

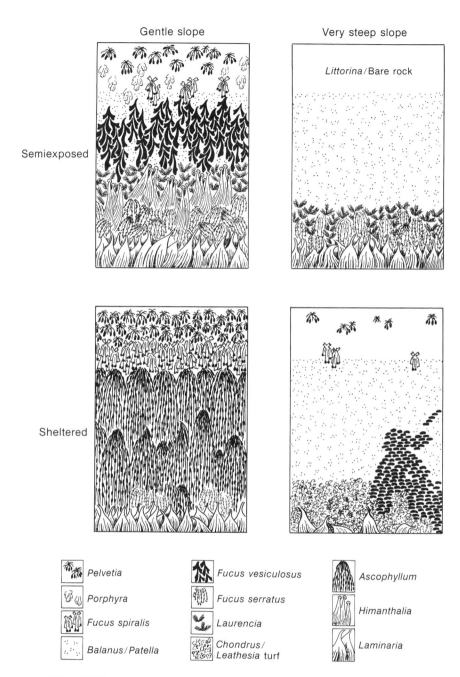

FIGURE **12.12**

Adjacent gentle and very steep slopes with markedly different populations. The semiexposed sites are at Baligill Point, northern Sutherland, and the sheltered sites in Loch Craignish, Argyll. The slope at Baligill Point shows an unusual variation with *Himanthalia* above *Fucus serratus*. [From Lewis (1964).]

The very striking differences between exposed and sheltered shores stem primarily from the presence in Britain of the large intertidal fucoids and their restriction to, or greatest luxuriance in, shelter. Where such species are occasionally absent from sheltered sites (some vertical surfaces, for example), barnacles and mussels reappear; the differences between exposure and shelter are then a matter of degree rather than of kind. Undoubtedly, the distribution of fucoids reflects the inability of some species to survive under wave-beaten conditions, but it is far from clear if the direct effects of wave action explain the absences in all cases. It is perhaps significant that, where the same species occur in New England and Nova Scotia (see Chapter 9), on shores lacking genuinely intertidal limpets, these algae live under more exposed conditions than they appear able to tolerate in Britain. Although there may well be different exposure tolerances in populations on either side of the Atlantic, it is equally probable that the absence and apparent inability to tolerate exposure in Britain is largely a consequence of the high density of grazing limpets on exposed shores here (Burrows and Lodge, 1949; Jones, 1948; and Southward, 1956).

In a similar way, the apparent inability of "open-coast species" to live in sheltered situations is partly explicable in terms of their inability to settle or survive in the presence of the long-fronded fucoids. Where the latter are scarce, or have been removed, a number of exposed shore species, including *Balanus balanoides*, *Chthamalus stellatus*, *Mytilus edulis*, *Lichina pygmaea*, and *Himanthalia elongata*, may reappear in what is typically the "wrong" habitat. However, there are also some exposed-shore species (including *Littorina neritoides*, *Patella aspera*, *Fucus distichus anceps*, *Alaria esculenta*, and a variety of red algae) whose absence from shelter, even when competition is nil, suggests that some stage in their life cycle is very sensitive to the physical regime of sheltered environments.

The varying patterns of zonation described in previous sections of this chapter are, to a large extent, a reflection of local or geographical changes in population. There remains the variation in the level occupied by the same zones and species from one site to another; in this context, the range of environmental conditions in Britain is sufficient to produce quite dramatic changes (see Burrows et al., 1954; Evans, 1947a, 1947b; Lewis, 1953, 1954, 1955, 1957a; and Moyse and Nelson-Smith, 1963). Although the absolute limits reached by particular species probably varies somewhat from year to year as a result of chance sets of physical or biotic conditions operating at one stage during the previous season or seasons, the general level occupied relative to theoretical tidal levels is fairly constant in any one situation, and is a response to the balance between a wide range of local and geographical, physical and biotic factors. Changes from one site to another in, for example, slope and drainage, aspect, shading, or water-retaining species, can produce sudden changes in level of several feet (Figure 12.13), especially among lower-shore species. These effects, however, are usually slight compared with those of increasing exposure, of regular subjection to continuous waves and oceanic swell where cliffs and steep slopes rise out of deep water. All such areas exhibit a widening of zones and a raising of upper limits, which, under severe conditions, may bring the *Alaria* belt above theoretical mean tide level, the barnacle line 10–15 feet above mean high water of spring tides, and the uppermost *Littorina/Verrucaria* another 40–50 ft. higher still (Figure 12.14). Much of this elevation affects the supralittoral

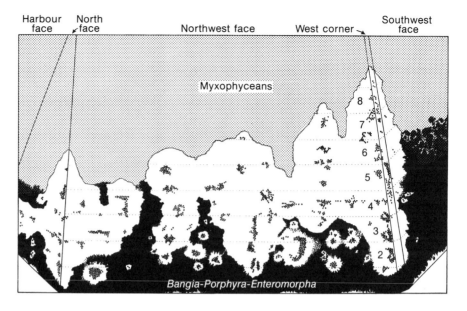

Harbour face | North face | Northwest face | West corner | Southwest face

Myxophyceans

8
7
6
5
4
3
2

Bangia-Porphyra-Enteromorpha

FIGURE 12.13

The influence of aspect and wave splash on *Patella vulgata* at the end of a breakwater. Note the grazing areas, the general increase in the height of the upper limit of limpets towards the most wave-splashed west corner, and the failure of limpets to round this corner on to the sunny southwest face. The numbered zones (*2-8*) are 2 feet deep [From Lewis (1964).]

fringe, the lower margin of which in exposure may be several feet above the level of the upper margin in nearby shelter. The species of the upper shore are, of course, adapted to long periods out of water, and are therefore better able to take advantage of spray to increase their vertical range, and also to withstand the intermittent calm periods that occur even on exposed coasts. Cool, moist summers will clearly be beneficial (Burrows et al., 1954), and although exact comparisons are exceedingly difficult to make, the greatest elevation certainly occurs in northern districts.

If, finally, we combine all the geographical, climatic, seasonal, and local features together to assess the range of variability on British coasts, the following would rank as the extreme types of conditions and populations:

1) Very enclosed bays and lochs with rock and shingle: a very narrow supralittoral fringe entirely within tidal reach; a midlittoral completely dominated by fucoids (*Ascophyllum* on rock, the free-living forms on shingle in Scotland and Ireland); and an infralittoral rich in algae and sedentary suspension-feeding animals, but poor in laminarians.

2) Exposed steep faces in southwestern England in summer, facing south and unsplashed at low tide: a wide but sparse supralittoral lacking all algae; a midshore covered almost exclusively with barnacles, limpets, and littorinids down to the narrow rhodophycean belt; and a "typical" forest of *Laminaria digitata* and *L. hyperborea*.

3) Slow-draining slopes and platforms shaded by cliffs and subject to oceanic swell, in northern Scotland or western Ireland in winter and spring: a very wide supralittoral with successive high-level algal belts; a midlittoral dominated by *Mytilus*/rhodophycean communities; and a very sparse infralittoral of *Alaria* and lithothamnia.

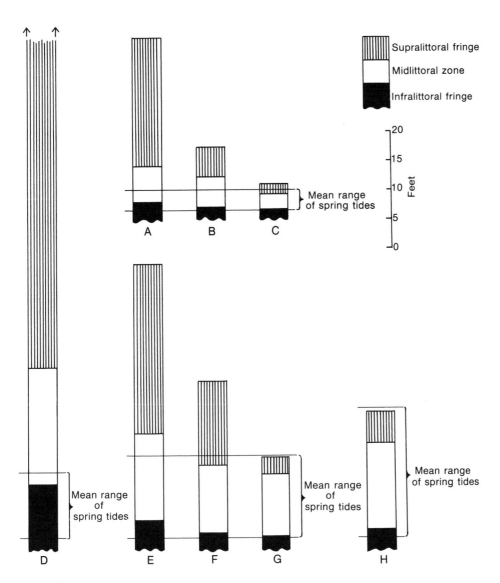

FIGURE **12.14**

The varying influence of tidal range and wave action upon the actual range and the relative proportions of the three major zones: *A–C*, Kintyre, Scotland (*A*, the most exposed site accessible; *B*, a site of moderate exposure; *C*, the maximum shelter on the open coast); *D*, an exceedingly exposed Atlantic site at Kerry Head, County Kerry (measurements are approximate, and the upper limit of the *Verrucaria* zone lies in a wide spray-extended area); *E–G*, the southwest coast of Holy Island, Anglesey (*E*, the most exposed site accessible; *F*, a site of moderate exposure; *G*, in the shelter of Rhoscolyn Bay); *H*, the extreme shelter of the Menai Straits, Anglesey. [From Lewis (1964).]

SELECTED REFERENCES

[This chapter is essentially a summary of the author's more extensive account of the British Isles (Lewis, 1964), to which readers are referred for more details and for a bibliography of more than 450 entries. The literature cited below is a selection of papers that particularly amplify the necessarily brief description given here.]

Anand, P. L., 1937. An ecological study of the algae of the British chalk-cliffs, I and II. *J. Ecol.* 25: 153–188, 344–367.

Barnes, H., 1958. Regarding the southern limits of *Balanus balanoides* (L.) *Oikos* 9: 139–155.

Bassindale, R., and J. H. Barrett, 1957. The Dale Fort marine fauna. *Proc. Bristol Nat. Soc.* 29: 227–328.

———, F. J. Ebling, J. A. Kitching, and R. D. Purchon, 1948. The ecology of the Lough Ine rapids with special reference to water currents, I. Introduction and hydrography. *J. Ecol.* 36: 305–322.

Börgesen, F., 1903. The marine algae of the Shetlands. *J. Bot. (London)* 41: 300–306.

———, 1908. The algae-vegetation of the Faeröese coasts. In *Botany of the Faeröes*, vol. 3. Copenhagen.

Burrows, E. M., 1958. Sublittoral algal population in Pt. Erin Bay, Isle of Man. *J. Mar. Biol. Ass. U. K.* 37: 687–703.

———, E. Conway, S. M. Lodge, and H. T. Powell, 1954. The raising of intertidal algal zones on Fair Isle. *J. Ecol.* 42: 283–288.

———, and S. M. Lodge, 1949. Notes on the inter-relationships of *Patella, Balanus* and *Fucus* on a semi-exposed coast. *Rep. Mar. Biol. Sta. Port Erin* (62): 30–34.

Colman J., 1933. The nature of the intertidal zonation of plants and animals. *J. Mar. Biol. Ass. U. K.* 18: 435–476.

Connell, J. H., 1961a. The effects of competition, predation by *Thais lapillus*, and other factors on natural populations of the barnacle, *Balanus balanoides*. *Ecol. Monogr.* 31: 61–104.

———, 1961b. The influence of interspecific competition and other factors on the distribution of the barnacle, *Chthamalus stellatus*. *Ecology* 42: 710–723.

Cotton, A. D., 1912. Marine Algae. Clare Island Survey. *Proc. Roy. Ir. Acad.* 31(15): 1–178.

Crisp, D. J., 1958. The spread of *Elminius modestus* Darwin in north-west Europe. *J. Mar. Biol. Ass. U. K.* 37: 483–520.

———, ed., 1964. The effects of the severe winter of 1962–63 on marine life in Britain. *J. Anim. Ecol.* 33: 165–207.

———, and E. W. Knight-Jones, 1954. Discontinuities in the distribution of shore animals in North Wales. *Rep. Bardsey Bird Field Obs.* 1954: 29–34.

———, and A. J. Southward, 1958. The distribution of intertidal organisms along the coast of the English Channel. *J. Mar. Biol. Ass. U. K.* 37: 157–208.

———, and ———, 1959. The further spread of *Elminius modestus* in the British Isles to 1959. *J. Mar. Biol. Ass. U. K.* 38: 429–437.

de Valera, M., 1962. Some aspects of the problems of the distribution of *Bifurcaria bifurcata* (Velley) Ross on the shores of Ireland. *Proc. Roy. Ir. Acad.* 62: 77–100.

Ebling, F. J., J. A. Kitching, R. D. Purchon, and R. Bassindale, 1948. The ecology of the Lough Ine rapids with special reference to water currents, II. The fauna of the *Saccorhiza* canopy. *J. Anim. Ecol.* 17: 223–244.

————, M. A. Sleigh, J. F. Sloane, and J. A. Kitching, 1960. The ecology of Lough Ine, VII. Distribution of some common plants and animals of the littoral and shallow sublittoral regions. *J. Ecol.* 48: 29–53.

Evans, R. G., 1947a. The intertidal ecology of Cardigan Bay. *J. Ecol.* 34: 273–309.

————, 1947b. The intertidal ecology of selected localities in the Plymouth neighbourhood. *J. Mar. Biol. Ass. U. K.* 27: 173–218.

————, 1949. The intertidal ecology of rocky shores in South Pembrokeshire. *J. Ecol.* 37: 120–139.

Fischer-Piette, E., 1936. Etudes sur la biogéographie intercotidale des deux rives de la Manche. *J. Linnean Soc. London Zool.* 40: 181–272.

Gibb, D. C., 1938. The marine algal communities of Castletown Bay, Isle of Man. *J. Ecol.* 26: 96–117.

————, 1939. Some marine algal communities of Gt. Cumbrae. *J. Ecol.* 27: 364–382.

————, 1950. A survey of the commoner fucoid algae on Scottish shores. *J. Ecol.* 38: 253–269.

————, 1957. The free-living forms of *Ascophyllum nodosum* (L.) Le Jol. *J. Ecol.* 45: 49–83.

Jones, N. S., 1948. Observations and experiments on the biology of *Patella vulgata* at Port St. Mary, Isle of Man. *Proc. Trans. Liverpool Biol. Soc.* 56: 60–77.

Kain, J. M., 1960. Direct observations on some Manx sublittoral algae. *J. Mar. Biol. Ass. U. K.* 39: 609–630.

Kitching, J. A., 1935. An introduction to the ecology of intertidal rock surfaces on the coast of Argyll. *Trans. Roy. Soc. Edinburgh* 58: 351–374.

————, and F. J. Ebling, 1961. The ecology of Lough Ine, XI. The control of algae by *Paracentrotus lividus* (Echiniodea). *J. Anim. Ecol.* 30: 373–384.

————, J. F. Sloane, and F. J. Ebling, 1959. The ecology of Lough Ine, VIII. Mussels and their predators. *J. Anim.* Ecol. 28: 331–341.

Knowles, M. C., 1913. The maritime and marine lichens of Howth. *Sci. Proc. Roy. Dublin Soc.* 14: 79–143.

Lewis, J. R., 1953. The ecology of rocky shores around Anglesey. *Proc. Zool. Soc. London* 123: 481–549.

————, 1954. The ecology of exposed rocky shores of Caithmess. *Trans. Roy. Soc. Edinburgh* 62: 695–723.

————, 1955. The mode of occurrence of the universal intertidal zones in Great Britain; with a comment by T. A. and Anne Stephenson. *J. Ecol.* 43: 270–290.

————, 1957a. Intertidal communities of the northern and western coasts of Scotland. *Trans. Roy. Soc. Edinburgh* 63: 185–220.

————, 1957b. An introduction to the intertidal ecology of the rocky shores of a Hebridean island. *Oikos* 8: 130–160.

————, 1964. *The Ecology of Rocky Shores.* London: English University Press.

————, and H. T. Powell, 1960. Aspects of the intertidal ecology of rocky shores in Argyll, Scotland, I. General description of the area. *Trans. Roy. Soc. Edinburgh* 64: 45–74.

Lilly, S. J., J. F. Sloane, R. Bassindale, F. J. Ebling, and J. A. Kitching, 1953. The ecology of the Lough Ine rapids with special reference to water currents, IV. The sedentary fauna of sublittoral boulders. *J. Anim. Ecol.* 22: 87–122.

Moore, H. B., 1936. The biology of *Balanus balanoides,* V. Distribution in the Plymouth area. *J. Mar. Biol. Ass. U. K.* 20: 701–716.

————, 1940. The biology of *Littorina littorea,* II. Zonation in relation to other gastropods on stony and muddy shores. *J. Mar. Biol. Ass. U. K.* 24: 227–237.

————, and J. A. Kitching, 1939. The biology of *Chthamalus stellatus* (Poli). *J. Mar. Biol. Ass. U. K.* 23: 521-541.

Morton, J. E., 1954. The crevice faunas of the upper intertidal zone at Wembury. *J. Mar. Biol. Ass. U. K.* 33: 187-224.

Moyse, J., and A. Nelson-Smith, 1963. Zonation of animals and plants on rocky shores around Dale, Pembrokeshire. *Field Stud.* 1(5): 1-31.

Naylor, G. L., 1930. Notes on the distribution of *Lichina confinis* and *L. pygmaea* in the Plymouth district. *J. Mar. Biol. Ass. U. K.* 16: 909-918.

Powell, H. T., 1957. Studies in the genus *Fucus* L., II. Distribution and ecology of forms of *Fucus distichus* L. amend. Powell in Britain and Ireland. *J. Mar. Biol. Ass. U. K.* 36: 663-693.

Rees, T. K., 1935. The marine algae of Lough Ine. *J. Ecol.* 23: 69-133.

Seed, R., 1969. The ecology of *Mytilus edulis* L. (Lamellibranchiata) on exposed rocky shores, II. Growth and mortality. *Oecologia* (*Berlin*) 3: 317-350.

Sloane, J. F., R. Bassindale, E. Davenport, F. J. Ebling, and J. A. Kitching, 1961. The ecology of Lough Ine, IX. The fauna and flora associated with undergrowth-forming algae in the Lough Ine area. *J. Ecol.* 49: 353-368.

————, F. J. Ebling, J. A. Kitching, and S. J. Lilly, 1957. The ecology of the Lough Ine rapids with reference to water currents, V. The sedentary fauna of the laminarian algae in the Lough Ine area. *J. Anim. Ecol.* 26: 197-211.

Smith, A. L., 1908. Lichens. Clare Island Survey. *Proc. Roy. Ir. Acad.* 31(14): 1-14.

Smith, J. E., and G. E. Newell, 1955. The dynamics of the zonation of the common periwinkle (*Littorina littorea* (L.)) on a stony beach. *J. Anim. Ecol.* 24: 35-56.

Southern, R., 1915. Marine Ecology. Clare Island Survey. *Proc. Roy. Ir. Acad.* 31(67): 1-110.

Southward, A. J., 1953. The ecology of some rocky shores in the south of the Isle of Man. *Proc. Trans. Liverpool Biol. Soc.* 59: 1-50.

————, 1956. The population balance between limpets and seaweeds on wave-beaten rocky shores. *Rep. Mar. Biol. Sta. Port Erin.* 68: 20-29.

————, 1967. Recent changes in the abundance of intertidal barnacles in south-west England: a possible effect of climatic deterioration. *J. Mar. Biol. Ass. U. K.* 47: 81-96.

————, and D. J. Crisp, 1954a. The distribution of certain intertidal animals around the Irish coast. *Proc. Roy. Ir. Acad. Sect. B.* 57: 1-29.

————, and ————, 1954b. Recent changes in the distribution of the intertidal barnacles *Chthamalus stellatus* Poli and *Balanus balanoides* L. in the British Isles. *J. Anim. Ecol.* 23: 163-177.

————, and ————, 1956. Fluctuations in the distribution and abundance of intertidal barnacles. *J. Mar. Biol. Ass. U. K.* 35: 211-229.

————, and J. H. Orton, 1954. The effects of wave-action on the distribution and numbers of the commoner plants and animals living on the Plymouth breakwater. *J. Mar. Biol. Ass. U. K.* 33: 1-19.

Walton, C. L., 1915. The distribution of some littoral Trochidae and Littorinidae in Cardigan Bay. *J. Mar. Biol. Ass. U. K.* 10: 114-122.

13

CONCLUSIONS

THE PRINCIPAL VARIATIONS IN ZONATION

The ocean, however inanimate, produces an impression of unrestrained passion. It never sleeps, but is in ceaseless turmoil. This statement may seem exaggerated to anyone who has lived beside a more or less sheltered shore somewhat removed from the open ocean, but if one lives on an oceanic coast, nothing is more impressive than the ceaseless activity of the sea. Really calm days are rare, and even when the surface appears, at a casual glance, to be smooth, there is usually a greater or lesser swell, which, causing no "white horses" on the surface, explodes into violent action when it strikes the coast. For the creatures living between tidemarks in such places, turmoil is the normal background. Some of them appear not merely to tolerate it, but to prefer it: actively moving, highly aerated water seems to be vital to them, and its impact does not destroy them. Others cannot withstand too much water movement of a violent type, and shun it to varying degrees. We emphasise the roaring activity of the surf here because it is one of the leading causes of major variations in the zonation between tidemarks. We have learnt to expect one type of zonation on exposed, wave-beaten coasts, and another in sheltered inlets far from the open sea. Not only this, but on any hundred yards of coastline, if the topography is such that some of the individual rocky faces are exposed to strong wave action and others sheltered from it, very marked variations will occur. Other influences that control zonation will be mentioned as this chapter proceeds.

The boundaries of the major zones do not lie at exactly the same relative levels everywhere. They are characterised by different species in different parts of the world, and it would be too much to expect that every species of barnacle, for example, dominating the upper part of the midlittoral should cease to be abundant at exactly the same level in relation to the tides and the other zones. We know also that the broad changes that can be followed along seashores from the tropics towards the poles involve gradual changes in level for the organisms of the shore; and again, that the absolute levels of given zone boundaries vary greatly in accordance with differences in exposure to wave action and other environmental features (Figure 13.1). This being so, we can reasonably expect the upper limit of any zone to occur in a fairly constant relative

FIGURE **13.1**

A comparison of the widespread features of zonation with an example that complicates them. A coast is shown on which smooth granite spurs are exposed to considerable wave action. On the middle spur, some of the widespread features are summarized and the following succession is shown: *A,* supralittoral fringe (= *Littorina* zone), blackened below by myxophyceans; *B,* midlittoral (balanoid zone), occupied by barnacles above and lithothamnia below; *C,* infralittoral fringe, dominated in this case by laminarians growing over lithothamnia. On the other spurs, (*foreground and background*) the actual zonation from the Atlantic coast of Nova Scotia is shown. Here, the simplicity of the basic plan is complicated by maplike black patches in the supralittoral fringe, consisting of *Codiolum, Calothrix,* and *Plectonema;* the existence of a strongly developed belt of *Fucus* (mostly *F. vesiculosus* and *F. edentatus,* in this example) occupying a large part of the midlittoral zone and overgrowing the uppermost barnacles; and a distinct belt of *Chondrus crispus* growing over the lower part of the midlittoral zone and largely obliterating the belt of lithothamnia, which, on the middle spur, extends above the laminarians.

position everywhere, but we can also expect its *exact* level to vary over a limited vertical distance. It will always be found within a strip bearing a fairly constant relation to some particular tidal level (the upper edge of the infralittoral fringe, for example, will never occur near high water), but not always at exactly the same level within that strip.

The following paragraphs attempt to give a preliminary idea of the kind of variation which may be expected.

The Supralittoral Fringe

The supralittoral fringe, in many places, is populated by myriads of small snails (mainly species of *Littorina* and *Tectarius*) and by rapidly-running isopods (*Ligia*), and may have no vegetation other than encrusting microphytes; it varies from this mean, however, in both directions. On the Florida Keys, for example, it becomes highly developed and organised. Here, the snails include about fifteen species and show a zonation among themselves; there is an invasion of flowering plants from above and of small mosslike marine algae from below; and the zone is divided into three well-marked subzones, in each of which the rock is differently coloured and supports a different assortment of species. However, at Marineland, in northern Florida (about 300 miles north of the Keys), snails are almost lacking in this zone. Careful searches over an eight-day period revealed only fourteen individuals of *Littorina* on this shore; and these were not found in the supralittoral fringe but in the upper edge of the midlittoral. On other shores, littorinids are completely absent from the supralittoral. This does not make it impossible to identify the limits of the zone: these can be recognised either from the distributions of other animals, or from those of plants.

The Black Zone. This band may occupy the supralittoral fringe almost exactly, extend somewhat above it, or (most commonly) occupy its lower part only, perhaps overlapping into the upper midlittoral. At one extreme, it forms a very black, sharply marked, continuous band that may be so straight that it appears to have been painted between ruled lines. At the other extreme, it disappears. Between these extremes, it shows various intermediate conditions; in some places, its boundaries are vague, and in others, it exists in the form of discontinuous patches. The cause of the blackening of the rock surface is commonly a film or crust of more or less microphytic myxophyceans or lichens adapted for life at a particular level. This does not imply that there are none of these organisms at other levels—in fact, there are myxophyceans in all the intertidal zones and below tidemarks, and lichens throughout the tidal belt and far above it—but the black zone is caused by a concentration of particular species belonging to these groups in a belt situated, as has been described, at high level. The lichens are examplified by several species of *Verrucaria*. The myxophyceans belong to various genera and species; for example, we collected about twenty species in this zone in North America. All discoloration of the rock is not necessarily caused by myxophyceans and lichens, but may also be due to small chlorophyceans or to encrusting rhodophyceans, such as *Hildenbrandia*. In special places, such as on the English chalk cliffs, other variations may occur. Further, although we have referred to the discoloration broadly as a "blackening," which it frequently is, the tint may vary from blackish to grey, green, brown, or other shades. Some accounts dealing with this vegetation are those of Anand (1937), Berner (1931), Chapman (1946), Cranwell and Moore (1938), du Rietz (1925), Feldmann (1938), Fritsch (1931, 1945), Geitler (1930–32), Ginzberger (1925), Grubb (1936), Johnson and Skutch (1928), Kylin (1937), and Smith (1921).

The Midlittoral Zone

Where the midlittoral zone is most clearly developed, barnacles form a dense continuous sheet on the rock; in most places, the sheet has a rather sharp upper limit, which forms one of the most distinct landmarks on the shore. Individual barnacles straggle above this limit, but, while these are visible from a slight distance, the upper boundary of the main mass of the barnacles often forms a conspicuous line, which may be visible miles away. In some places, the barnacles continue downwards uninterrupted (though it need not be the same species all the way down) almost to the infralittoral fringe, either fading off gradually, or ceasing with a distinct lower limit; in a very few places, they continue below low-water level of spring tides. On many shores, other organisms interfere with or suppress them in the lower parts of the zone, and such interference may extend to moderately high levels. The competing organisms may be turflike growths of short algae, a mass of living polychaet tubes, a sheet of zoanthids, or various other sedentary forms; but there are influences other than competition that tend to reduce the numbers of barnacles at the lower levels. On many shores, there is a fairly sharp line between an upper midlittoral subzone dominated by barnacles and (perhaps) limpets, and a lower subzone, which may support a patchwork of algal turf, polychaets, and other forms competing with the barnacles. In other places, no such sharp line can be seen; in yet others, there are more subzones than two (three is a common number). There are also regions where the barnacles are reduced in number, or even absent over the entire zone. In such places, the extent of the zone is indicated by other organisms of similar distribution. It should not be supposed that barnacles are necessarily absent from the zones below the midlittoral. The large *Balanus titinnabulum*, for example, may occur near or even below low water, and other species are invariably found below tidemarks. But the barnacles that characterise the midlittoral belong to a selection of species adapted for life between tidemarks, especially at the middle levels, and are not the same as the low-level or offshore species. Moreover, the population of the infralittoral fringe is not typically dominated by barnacles even where these are present. Finally, there are barnacles characteristic of caves and overhangs, and these sometimes complicate the picture.

The Infralittoral Fringe

Some indication of the variations to be found among the populations of the infralittoral fringe has already been given (Chapter 2). The fringe is usually recognisable, though there are abnormal regions in which it departs in many ways from the standard, commonly in the reduction of the population due to some special local feature. The wide flats laden with calcareous sediment that occur at this level along the Florida Keys serve as an example of such features. On the other hand, where the fringe is well developed, it may, like the higher zones, include subzones, as in New South Wales, where a *Pyura* belt sometimes occurs above a kelp subzone.

The Lithothamnia Zone. This belt, which extends upwards from below tide-marks through the whole infralittoral fringe and into the lower midlittoral, tends to be carpeted throughout by encrusting coralline algae, wherever these can get a footing. The number of species involved is considerable, and these change from one part of the world to another; several species may occupy a single shore. It is not easy to give a precise account of the distribution of these algae, because they are not easy to identify, and specialists able to discriminate them accurately have been rare. During our survey of the South African coast, we estimated that not fewer than sixteen species occur there between tidemarks, and possibly many more. These species show a subsidiary zonation among themselves. Nodular and branching (but unjointed) forms may be present, as well as encrusting ones. On shores where these lithothamnia are well developed (and this applies by no means only to coral reefs), they tend to form a conspicuous and more or less continuous pinkish covering over the rock of the lower levels. This covering may (on open rock) have an upper limit so sharp that it becomes the most conspicuous landmark on the shore. The lithothamnia may also encrust things other than the rock itself. In South Africa, for example, they cover the shells of *Patella cochlear* so persistently that it is very difficult to find a shell of this animal that is not so overgrown. On other shores, the growth of lithothamnia is weaker and less conspicuous, and tends to become intermittent. On yet others, this growth is almost suppressed, and the Florida Keys provide a good example of this. Here, these algae have only a restricted distribution and occur in strongly subnormal amounts in the fringe (as distinct from their occurrence below tidemarks). This restriction seems to be connected, on the one hand, with a superabundance of sediment and, on the other, with a reduction in wave action, which allows much of the intertidal rock to become very hot and dry when exposed to the air.

The Occurrence of Other Zones

We have seen that the widespread zones vary from place to place, and that each of the three principal ones may be divided into subzones, which are more local in distribution than the major zones. The complexities of the whole problem are still further increased by the fact that, on some coasts, there exist other, special zones—zones that occur in one region alone, and are "special" in the sense that they cannot readily be classified as subzones, but are more of the nature of additional zones superimposed on the widespread basic pattern. A good example of such a zone is the belt inhabited by *Patella cochlear* on the South African coast. This belt is a spectacular, well-marked feature of the shore, and is present along more than a thousand miles of coastline; but there is no evidence that it occurs elsewhere. It is situated between the lower midlittoral and the infralittoral fringe. Similar zones may be seen in New Zealand. The well known fucoid zones of northwestern Europe may also be mentioned here. In Britain, zonation

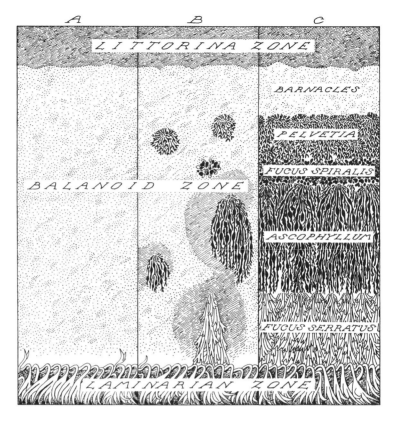

FIGURE **13.2**

Three of the possible types of zonation met with on the British coasts (for example, on the steep rock faces in Argyllshire). In *A*, the three widespread zones described in the text (*Littorina* zone, balanoid zone, and the infralittoral fringe, represented here by a laminarian zone) are alone developed; in *B*, the balanoid zone is colonised by patches of brown algae at different levels (the lower ones surrounded by areas of bare rock more or less free from barnacles); in *C*, a full series of fucoid zones is superimposed on the balanoid zone, overshadowing the barnacles. In this diagram only part of the width of the *Littorina* and laminarian zones is included. The relative widths of the other zones are accurate for the places chosen. For *C*, a locality without *Fucus vesiculosus* and *Porphyra* was selected. [Reproduced by the courtesy of the Linnean Society.]

on the shore was described for some time in terms of the species of conspicuous brown algae belonging to this group that so commonly form bands along the shore, one above the other (Figure 13.2). These species include *Pelvetia canaliculata, Fucus spiralis, F. vesiculosus, F. serratus,* and *Ascophyllum nodosum.* It has been realised more recently that the basic zonation in Britain conforms to the same pattern as that of the rest of the world, and that these fucoid zones, conspicuous though they may be, are a feature restricted to a certain part of the northern hemisphere, and are not widespread, even in Europe. They could be regarded as subzones of the midlittoral, but this classification does not fit them exactly, because the two uppermost zones, those of *Pelvetia* and of *F. spiralis,* lie in the upper midlittoral on some shores and above it on others. It is perhaps more accurate, therefore, to regard them as additional local zones superimposed on the basic pattern.

LOCAL ABSENCES OF IMPORTANT SPECIES

It is a well known feature of seashores that some common organism, ecologically important and perhaps characteristic of a zone or subzone, will be present in force on one part of a coast, less common on another, and absent from yet another. Such fluctuations are often readily explicable in terms of local conditions. As we have already seen, strong wave action will encourage one species and discourage another; consequently, in some sheltered spots, an animal that characterises a zone in rough places may no longer be present. In addition to this, however, there are other absences of which much less is known. These seem to belong to different categories from the ones just mentioned, and some of them can hardly be explained in terms of our present knowledge.

Before visiting North America from 1946 to 1948, we had supposed (along with many other marine biologists) that, on any rocky coast subject to average environmental conditions, there would be a fairly standard set of plant and animal forms represented. In other words, if any locality were free from restricting influences—such as those that prevail, for example, in an estuary—one could expect to find species of periwinkles, acorn barnacles, crabs, limpets, anemones, whelks, and other common forms of shore life. There are certain ecological "niches," such as those tenanted on many shores by limpets, that one might expect to find always occupied by these animals—although it is well known that, in different parts of the world, they are filled by different species. These species may belong to different genera or different groups, but they are still "limpets." Thus, many limpets are prosobranchs, while others (*Siphonaria* and *Gadinia*) are pulmonates; *Patella* is the leading genus in some parts of the world, *Acmaea* in others. But it is so general a feature of seashores that limpets of some kind are common on open rock surfaces, that it is difficult to believe that, on some shores, there are none. Yet this seems to be the case. On the southeast coast of the United States, for example, the common limpets of Florida are species of *Siphonaria;* but during our stay in North Carolina and South Carolina, we could not find a single open-rock limpet of any sort. There are under-stone limpets there (*Crepidula* and *Diodora*), but these belong to quite a different ecological sphere, the cryptofauna. Farther north again, *Acmaea testudinalis* becomes plentiful on open rock, but even within its range, there are shores that lack it; in any case, it appears to be a less dominant limpet than the British and South African species of *Patella.*

There are rocky shores on which periwinkles, whelks, crabs, anemones, and barnacles are lacking, or nearly so. On some, this absence can be explained on the lines indicated above, or may perhaps be due to a lack of currents capable of bringing larvae from suitably inhabited areas. On other shores, however, these explanations are inapplicable.

We recommend to future workers a fuller study of such absences as likely to lead to interesting discoveries. We do not consider that the instances that we have observed are, as yet, fully established, because it is no easy matter to certify the complete absence of some species from a locality, especially during a short period of investigation. At the same time, the apparent absences we have noted cannot be put down to defective observation. After working on seashores for more than 25 years, two experienced observers cannot spend several weeks examining a district, keeping a watch for limpets all the time, and (having found thousands the month before a few hundred miles away) not see a single one—unless limpets, at that place and season, are at least very scarce. Moreover, our observations have received confirmation from those of other authors, even though the amount of available information is still very limited.

THE BOUNDARIES BETWEEN THE ZONES

It will have become evident from the earlier parts of this chapter that a particular species, although it may occur chiefly in one zone, may also extend in small numbers into another. More than this, we find examples of plants or animals whose distribution overlaps completely some zone boundary. Such a form may be equally common, for example, in the lowest part of the supralittoral fringe and the uppermost part of the midlittoral. If we forget the major zones for a moment and think instead of the vertical distribution of individual species between tidemarks, we find that a picture emerges that is quite different from the simple pattern of a succession of striplike zones, one above the other, that has engaged our attention so far. Figure 9.6 (p. 169) illustrates the distributions of twenty common plant and animal species (or groups of species) in relation to the principal zones on the Florida Keys. This figure shows clearly how these distributions form an overlapping and fairly continuous series from above high-water mark to below low water. A similar diagram can be constructed for almost any coast, and the basic features revealed in this one are of general occurrence. One would expect that, with so much overlapping, the shore, when viewed as a whole, would show no particularly noticeable zones. We know already that this is not the case, and that despite the overlapping, a fairly simple series of zones is, in fact, visible to the eye. Why is this so? Mainly because some of the species concerned, when spread out over the rock in quantity, make conspicuous coloured bands, while others do not; partly because some of them are much more abundant than others; and partly because some are more readily visible than others. Even so, there is no avoiding the fact that even visible zones sometimes overlap. This makes matters very difficult when one attempts to define boundaries between zones for descriptive purposes. Unless we can define these in such a way as to minimise the difficulties due to overlapping, we cannot provide ourselves with a framework that proves serviceable when we need to describe the arrangement of plant and animal communities within the zones. The only solution seems to be to select, as descriptive boundaries between the main zones, certain lines that both serve this purpose and correspond with natural realities.

We have found, in practice, that it is serviceable to define the upper limit of the midlittoral as the line along which barnacles cease to be abundant. For the infralittoral fringe, we can define the upper limit, similarly, as the line along which laminarians, *Pyura*, or some other characteristic organisms cease to be abundant; this line serves a double purpose, as it represents the lower limit of the midlittoral as well as the upper limit of the infralittoral fringe. Again, the upper limit of barnacles (in quantity) marks the lower limit of the supralittoral fringe as well as the upper limit of the midlittoral. The upper boundary of the supralittoral fringe is indicated by different things in different places—in some places, it coincides with the lower edge of the growth of maritime lichens (as distinct from intertidal ones), or that of maritime flowering plants; in other places, it is indicated by the upper limit of *Littorina*. It will be noticed that all these limits refer to organisms and not to tidal levels—in fact, the only tidal level suitable for use in this connection is extreme low water of spring tides, which sets a convenient lower boundary to the infralittoral fringe. Even this, as knowledge advances, may be superseded by a limit defined in terms of organisms, because it is possible that the transition line from this "fringe" to the permanently submerged region below low water actually lies a little *below* extreme low water of springs tides.

It may strike a reader unaccustomed to shore work that this method of defining the boundaries of the zones is rather arbitrary and not very precise. This impression is probably misleading. The boundaries, as we have said, really do exist on the shore, and protracted study and discussion have led to the conclusion that, even though they are not the only ones that might be employed for such purposes, they are the best available. As to precision, the whole system of zonation is too fluid and variable to permit anything more exact. If the scheme outlined here is adopted, the major zones become defined as strips, not overlapping, but separated by linear boundaries. When one writes an account of a whole shore, they act as essential lines of reference in relation to which the overlapping distributions, referred to at the beginning of this section, can be described.

THE RELATION BETWEEN ZONES AND TIDES

In the foregoing section, we drew attention to the fact that the boundaries between the principal zones had been defined in terms of organisms, rather than in terms of tidal levels. It now becomes necessary to explain this choice in more detail.

It has always been a seductive idea that, as there is a relation between tides and zones, we ought to be able to demonstrate that some particular zone lies neatly between two significant tidal levels. It would be most convenient, for example, if the midlittoral could be defined as the belt between the level of extreme high water of neap tides and the level of extreme low water of neap tides. This idea has appealed so strongly to the minds of certain authors, and has seemed so "scientific" (sometimes a fatal lure), that they have persisted with attempts to apply it, even after it has become apparent that such attempts can hardly succeed. We should have liked to devise such a system ourselves, but have given it up for a number of reasons.

In the first place, shores can be found on which the relation between zones and tidal levels is quite different in places only a few yards apart. A measured example is shown in Figure 10.7 (p. 212). The figure refers to a small island known as Brandon Island, situated in Departure Bay, British Columbia. The two columns in the figure illustrate the zonation at points on the opposite sides of this island, less than 100 yards apart. Both sides are equally subjected to relatively slight wave action, and are not, therefore, sharply differentiated from one another on this account. They are sharply distinguished, however, in that the southern side is sloping and sunny, while the northern side is cliffed and shady. It is evident from the diagram that, at the points measured, the upper limit of the main population of barnacles (line *D-K*) lies more than two feet higher up on the shaded side than on the sunny side. This means that the upper boundary of the midlittoral is situated at quite a different tidal level at the first place than at the second. The same applies to the upper limit of the main growth of *Fucus* (line *E-L*) and to that of the densest part of the black zone (line *C-J*). Other differences are that the whole barnacle belt on the shady side is deeper than it is on the sunny side; there is an additional zone on the sunny side (between points *G* and *H*) that is not distinguishable on the shaded side; and although the upper limit of the infralittoral fringe is at the same level on the two sides, the population of this zone is by no means the same on the one side as on the other. The existence of shores such as this seems to us to show conclusively that, even were there no other reasons for the belief, the zone boundaries vary so strongly in relation to the tide levels that no attempt to define one in terms of the other can be expected to succeed.

There are other reasons for the same belief. What is a neap tide, and what is a spring tide? The answer to this apparently simple question covers a multitude of complications. Typically, the amplitude of the tide changes quite gradually from one day to another. The highest spring tide of a particular series is followed by one that is a little lower, by a second a little lower still, and so on, until a point is reached at which high-water level is minimal for the period in question, and may be described as high water of a neap tide. Therefore, if we are to use tidal levels in connection with zones, we are faced with two possibilities. We can choose the highest spring occurring within a given period, determine the level reached by the water on a shore at this time, and describe this level as "extreme high water of springs." Alternatively, we can work out, over a suitable period of time, the average level reached by spring tides and the average reached by neap tides, and identify these levels on the shore. But the answer will vary according to exactly how many "spring" or "neap" tides we use when computing our averages.

Are we to use the highest single tide in each fortnight, over a year or more, as the basis for our average for spring tides? Or should it be the two highest, or the three highest, or how many? How do we decide (seeing that the series is continuous) where a "spring" tide stops and a "neap" tide begins? Of course, we can (and do) make an arbitrary choice, for purposes of our man-made definitions—but how do we know which choice is the relevant one as far as organisms are concerned? Each choice will lead to a different figure, and each will lead to the identification of a different level on the shore as "mean high water of spring tides," or whatever the level in question may be. This means that our conceptions of "mean" levels, at least in this particular connection, are purely arbitrary. Again, which of these two kinds of levels—means or extremes—is the important one for organisms? We cannot be sure, but it is likely that both are effective. If so, we have to determine not only what are the effects of the extremes, but also which of the possible mean levels are the most vital ones, and this is no easy matter.

Whatever may be the outcome of the complexities outlined in the preceding two paragraphs, it seems clear that, when we do identify some particular level on a seashore as presenting "mean high water of neap tides" (or any other level), this takes us little nearer our goal. For we then discover that, in certain places, the imaginary level we have identified coincides with some real natural boundary, but that in other places it does not. For practical purposes, this in itself makes the application of any system of coincidences between tide levels and zone boundaries almost impossible, particularly when we add difficulties of the kind encountered at Brandon Island. We can see, of course, that there is often a relation, if not a coincidence, between some boundary and a given tidal level. For instance, the upper limit of a particular zone may occur 6 inches above some specific tidal level; but as relationships of this kind are not constant from one shore to another, they seem to have little value for purposes of definition.

The points emphasized in the preceding paragraphs are by no means surprising, if we realise that the popular idea that zonation is caused by tides is itself untrue. Vertical stratification or zonation of plant and animal species is very widespread in the world, even in places where there is no tidal action. There is a zonation on high mountains from the summit to sea level, another in tropical rain forests between the canopy and the ground, and so on. Zonation occurs, both above and below the water level, round static waters such as ponds and lakes, where there is no tidal change. It is true that the surface level in such bodies of water can rise during heavy rain or fall because of evaporation, and that this variation can affect the zonation; but it is an irregular rise or fall, unlike that of the tides, and hardly affects the present argument.

There seems to be no doubt that the starting point for the development of zones near a body of water is the existence of an interface between water and air. Below the surface of the water there is a gradient in light penetration and a gradient in the depth of the water (effective for certain plants that can or cannot grow up from particular depths to the surface), and these can be accompanied by other gradients, such as a change in sedimentation with increasing depth. Zonation results from these gradients. Above the water surface, there is further zonation, controlled by whatever factors are effective in a particular place. Round a pond, the soil may be very wet at first, marshy beyond this, drier beyond the marshy belt, and so on; the distribution of plants will follow these changes in soil moisture. If the substratum is rocky, it may still be affected by the proximity of a body of water; there may be spray, or moisture caused by evaporation, rendering the rocks fringing the pond or lake less arid than those further away; or animals from the water may frequent these rocks, within a certain distance of the water itself. In a marine area with no tide, zones would immediately develop corresponding to our supralittoral and infralittoral zones, each with its subzones. If we add considerable and steady wave action (but still no tides), we produce a third zone, the midlittoral, related to the average amplitude of the waves. We therefore have the beginnings of intertidal zonation without any tide at all.

That this argument is not fanciful is well illustrated by the state of the tides round the Mediterranean. While that sea is not, as generally imagined, literally tideless, there are many parts of it where the tidal amplitude rarely exceeds 1 foot, and where the sea level varies considerably more in accordance with irregular changes in barometric pressure and in the force and direction of the wind than it does as a result of tidal action (a good graphic illustration of this is given by Feldmann, 1938, p. 31). On the northwestern shore of the Mediterranean coast of France nearest to Spain, only the supralittoral fringe and the upper part of the midlittoral are represented; the middle and lower belts are missing, and the upper limit of various species of the algal genus *Cystoseira* indicates the low-water mark.

If we add the effect of tides to that of wave action, we may reasonably expect that the zonation will be strengthened and made more marked and varied. This expectation does, in fact, seem to be realised; and there is nothing against it, provided we appreciate that the tides have not created the system of zonation but have merely developed and modified it. Even here, we make no general statement to the effect that zonation is more marked where the tidal range is great than where it is small, because it is now well known that very marked zonations can occur in connection with small tidal ranges, and also that, in the Bay of Fundy, where the tidal range is enormous, the zonation is no more elaborate than anywhere else—in fact, it is less marked (in the places we have seen ourselves) than on the Atlantic coast of Nova Scotia, where the range is much less. If, indeed, any generalisation on this subject proves to be possible, it may be that the distinctness and complexity of the zonation is increased by lesser tidal ranges, rather than by greater ones.

We believe that the arguments just presented show conclusively that we cannot hope to define the boundaries of the intertidal zones in terms of tide levels. We must therefore define them in terms of organisms, as we have done throughout this book. When describing any particular zone, of course, we can state what its relation to tidal levels is, but we cannot expect its boundaries to coincide neatly with these levels.

It may be added that zones between tidemarks have no constant depths whatever. Tidal behaviour and tidal ranges vary so much in different parts of the world that no such constancy is to be expected; the depths of zones are controlled, not only by tidal phenomena, but by variations in the amount of wave action, distribution of sun and shade, and other factors. This means that, even on a single stretch of coast with the same tidal range, the zones may vary in depth from one rocky face to another. It is possible that, if enough measurements were made, and if a sufficient number of corrections were applied to them (for differences in tidal range, exposure to wave action and sunshine, and so forth), a constant proportion between the depths of the main zones could be demonstrated; this, however, has not been ascertained.

The widths of zones vary in relation to the slope of the rock on which they are found, apart from the factors that control their vertical depths. Thus, a zone that is a foot deep on a vertical wall may also occur on a nearly horizontal rocky platform where, although its vertical extent may still be only a foot, its width may be many yards.

THE CAUSES OF ZONATION

There are some influences at work on the shore that seem, quite obviously, to affect the zonation, and we may be reasonably confident that we can recognise these where they occur. There are other influences whose effects are more in doubt, and probably yet others whose effects have not yet been recognised at all.

In 1943, one of us (T. A. Stephenson) opened a discussion on zonation that had been organised by the Linnean Society. The published account of that introductory speech dealt with the causes of the vertical and horizontal distribution of organisms between tidemarks in South Africa. Although this published statement referred chiefly to one country, much of it was of wider application, and the following paragraphs represent a modified version of it, brought up to date and stated in more general terms.

The formal proof that the zonation observed on the shore is due to the interaction of a stated set of factors can only be obtained experimentally. But we must remember that what is visible on the shore is itself the result of a large-scale natural experiment: the experiment of exposing to an intertidal environment the thousands of eggs, spores, and larvae deposited or liberated into the water between tidemarks or flung against the rocks by the sea. This means that we may hope to solve the problem of zonation not only by interpreting laboratory experiments, but also by an interpretation of the facts observed on the shore—that is, by interpreting the results of the natural experiment. We can probably understand the results of the laboratory experiments more readily than observations made on the shore, because such experiments are small in scale and can deal with one variable at a time. Field observations, however, have the advantage that they record the results of experiments on a sweeping scale unattainable in the laboratory. Moreover, these results ensue from combinations of conditions that do, in fact, occur on the shore, whereas laboratory experiments inevitably take place in a somewhat abnormal environment, and cannot simulate natural conditions beyond a certain point. A third type of information may be gained by deliberate interference with the life of the shore by human agencies; but here again, the scale can never be as extensive as that of the interaction of natural forces on a whole coastline.

In the following paragraphs, single factors will be treated separately as far as possible, but it will become apparent to the reader over and over again that the distribution of any species between tidemarks tends to be due to several factors working together, rather than to a single, isolated cause. It is doubtful, therefore, whether much is to be gained by too great an insistence on the exact part played by each factor in a complex. For example, if it seems clear that the upper limit of a species is determined by its ability to resist desiccation, acting together with its responses to temperature and light, then at least we know something definite about that species, even if we are unable to disentangle the exact effects of each of the interacting variables.

The Nature of the Substratum

The Composition of the Rock. Rocks between tidemarks include granite, hard quartzitic sandstones, soft sandstones (calcareous or not), dolerite, conglomerates, limestones, slatelike rocks, and various other materials. We have already seen that the nature of the rock may exert a certain selective influence on the organisms that attempt to colonise it, in the sense that some rocks are susceptible to the activities of boring organisms and others are not, and certain minor variations in the zonation may also be attributed to it. Nevertheless, the main outlines of the zonation usually seem to be unaffected by it. Thus, where granite and hard sandstone occur intermingled on the same beach, the zonation is essentially the same on both, and may be the same again on slatelike rocks at no great distance away. At one locality in South Africa, where outcrops of granite and marble occur close to one another, there seem to be no marked differences in zonation between the two. The arrangement of organisms on areas of calcareous sandstone (that is, a sandstone in which the grains are bonded by a calcareous cement) often shows divergencies from the normal that, at first sight, seem to be connected with the composition of the rock, but which on close examination prove, almost certainly, to be due to other causes.

On the other hand, the nature of the rock may, in some places, produce an effect that is too noticeable to be classified as a "minor" one. Near La Jolla, California, for instance, there is a great deal of soft sandstone with a surface so porous that it holds moisture for some time after the receding tide has left it exposed to the air. (Although it may look dry, if one sits down on it, one's clothes absorb moisture from it.) Among the soft sandstone, however, there are areas of hard rock with a nonporous surface, which dry out at low water. The relative development of barnacles and lithothamnia on these two kinds of rock is striking. On the soft, moisture-holding rock, the crust of lithothamnia extends freely, on open surfaces, into the upper midlittoral. This is quite exceptional, and does not occur on the adjacent hard rock. The barnacles, on the other hand, form a normal population only on the hard rock, and are much reduced on the soft; with the result that the local population of limpets (*Acmaea digitalis* and others) on the soft rock is far more conspicuous at high levels in this district than it otherwise would be.

The Texture and Configuration of the Rock.   The texture of the rock surface is one consequence of its composition that seems to exert only minor effects on zonation, unless we consider that the friability of the surface of some soft rocks is an aspect of their texture. If the surface readily crumbles or breaks away, this reduces its value as a support for barnacles and other attached forms. In some exceptional cases, where the intertidal reefs are made of water-worn peat (as in North Carolina), barnacles can rarely remain attached for long enough to reach maturity. Where ice affects soft rocks, it may scrape away not only the organisms, but also the surface of the rock itself.

The configuration of the major rocky masses, on the other hand, has certain definite effects on zonation. Thus, granite tends to weather into smooth slopes and rounded boulders, hard sandstone into ledges and rectangular shapes, slate and shale into jagged ridges, and so on. These various shapes are effective in the sense that they may present the zones in a diagrammatic form (as on smooth slopes) or break them up into a kaleidoscopic series of fragments (as on a tumbled chaos of rocks), but they are also effective in their interactions with other factors. Thus, the shapes and the arrangement of the rocks interact with the waves and help to determine the type and degree of wave action to which any particular group of organisms is exposed. This, in its turn, causes modifications in the zonation. Other local effects due to the shapes of the rocks are the occurrence of caves, overhangs, and other shaded surfaces, which introduce variations in light-intensity. Shade, however, tends to encourage the cryptofauna and shade-loving flora and to restrict those species tolerant of sunshine. Its action thus splits any population into two sections, in each of which zonation is evident. The existence of shade provides a distinct habitat, rather than affecting the problem of zonation in general.

The Proximity of Sand. There is little doubt that any considerable body of sand affects the population of adjacent rocky areas. On some coasts, the sand shifts a good deal, so that a low rocky reef may be buried under the sand at one time, and emerged at another time. Submergence of this kind does not necessarily destroy everything on the reef, unless it lasts for a long time. On many shores, rocky spurs or masses run down to a sandy slope: here, the movement of the sand induced by tides and wave action may scour the lower parts of the rock and reduce or eliminate the population; but very often it seems to do little harm, except near the foot of the rocks where they pass below the sand. In places where the proximity of sand leads to a continual suspension of sand grains in the water at some states of the tide, so that the adjacent rocks are sprinkled with sand (but not buried in it—after all, waves can wash sand away as well as deposit it), the rock may carry a very dense population, but it will not necessarily be the same population it would be were there no sand in the neighbourhood. Excellent examples are seen in South Africa, where a mosaic of limpets (*Patella cochlear*) occurs at low levels on many rocks unaffected by sand. Where such rocks are flanked by sand, the limpets are commonly replaced by other communities, such as beds of mussels and sea anemones (*Bunodactis reynaudi*). The species of mussel differ in their response to the proximity of sand. *Mytilus meridionalis,* for example, seems to be favoured by it (beds of this mussel may even occur on a sandy foundation), but *M. crenatus* is inhibited by it.

There is a further complication, however: because proximity of sand and some degree of reduction in wave action often go together, the effects of the two are not easily disentangled. By this we do not mean that the effects attributed to sand in the preceding paragraph are really due to reduced wave action, but that there are shores on which the results produced are due partly to the one cause and partly to the other.

The Width of the Rocky Tract. This may affect the distribution of organisms such that species that tend to occupy its whole width where it is narrow may die out to landward where it is wide. Thus, at certain localities, barnacles are absent or reduced on inshore rocks at levels that they colonise freely to seaward.

Desiccation

There appears to be no doubt that the degree to which organisms are able to withstand desiccation affects their distribution between tidemarks. Desiccation may affect organisms in two ways: it may be intense, but of short duration, or less severe, but of repeated occurrence. Desiccation is intensified by increased temperature and air movement (and possibly supplemented by the effects of light), and it is mitigated by certain other agencies.

Extreme Desiccation of Short Duration. This is seen occasionally in South Africa when a hot, dry wind persists for two or three consecutive days in summer, coinciding with spring tides. It can result in the death of thousands of fronds (or whole plants) of certain algae of the lower levels (including *Hypnea spicifera, Gigartina radula,* and *G. stiriata*), while species characteristic of higher levels, such as *Chaetangium* *ornatum,* may be quite unharmed. Here, the fatalities appear to be due to the combined action of desiccation, heat, and light, but as the light is presumably not abnormally intense on such occasions, it is probably effective only because heat and drying render the tissues of the plants more susceptible than they normally are to its action. It seems a fair conclusion that *Chaetangium ornatum* can establish itself higher up than the other species mentioned because of its greater resistance to these conditions. It is easy to imagine that occurrences less drastic than those just described might destroy many animal larvae and young plants.

A comparable mishap, due to cold, is sometimes seen when, on such coasts as those of southern California, an unusually cold wind prevails for long enough to kill many intertidal algae; but here, the effect seems more likely to be due to low temperature than to desiccation.

Repeated Desiccation. The controlling effects of desiccation are probably most commonly exerted by means of repeated exposure of shore organisms to its action. Such exposure varies in length and frequency at different levels on the shore; it may last for several days at a time, because the highest levels are not submerged at high water of neap tides. That intertidal organisms do possess, in varying degrees, the ability to withstand prolonged desiccation is certain, both from field observations and experiments. Widespread over the Indo-Pacific tropics, for example, millions of oysters (*Ostrea cucullata*) inhabit levels on the intertidal rocks where they are baked in the sun for a large part of their lives and experience body temperatures of 45°C and higher. It can hardly be doubted that these oysters live where they do because they are resistant to high temperatures and protected against desiccation, but the reason why they do not extend to lower levels is to be sought elsewhere. It is equally clear that species without such powers cannot extend to high levels. Experiments have shown that certain snails and barnacles of the higher levels can be dried for long periods and yet survive, and that species from different levels may show graded reactions to desiccation. A particularly neat example of this has been provided by Broekhuysen (1940), who showed that a series of South African intertidal gastropods exhibited graded reactions that corresponded remarkably well, on the whole, with the order of their occurrence on the shore.

The Interaction of Desiccation with Other Factors. High temperatures and wind will increase the rate of desiccation and augment its total amount within a given period. It is suggested by experiments, such as those of Broekhuysen (1940), that the drying action of these agencies is of more importance than the actual temperature involved (at least for some organisms and in some latitudes). For example, Broekhuysen found that adult snails showed a graded resistance to desiccation even when the desiccation was unaccompanied by unduly high temperatures; and also that, although they showed a graded resistance to high temperatures unaccompanied by desiccation, their lethal temperatures (when submerged) were much higher than anything they were likely to experience (frequently, at any rate) in their natural habitats; and that temperature, therefore, hardly seemed likely to be acting as primary limiting factor in their vertical distribution. His experiments also suggest, however, that high temperatures may reduce the resistance of the snails to desiccation, and may thus become important in the warmest latitudes. Furthermore, Broekhuysen pointed out that the temperatures lethal to the young of the molluscs he studied had not been determined, and these may be lower than those lethal to the adults. The relation between the action of temperature and desiccation and that of light also needs further study. Organisms high up on the shore are presumably exposed to the same light intensities as those lower down (that is, when these are exposed to the air), and it is possible, as has already been noted, that the action of drying and heating, after a certain point, may render them susceptible to light intensities that normally have no harmful effects on them.

SELECTED REFERENCES

Anand, P. L., 1937. An ecological study of the algae of the British chalk-cliffs, I and II. *J. Ecol.* 25: 153–188, 344–367.

Berner, L., 1931. Contribution à l'étude sociologique des algues marines dans la Golfe de Marseille. *Ann. Mus. Hist. Nat. Marseille* 24(1): 3–84.

Broekhuysen, G. J., 1940. A preliminary investigation of the importance of desiccation, temperature and salinity as factors controlling the vertical distribution of certain intertidal marine gastropods in False Bay, South Africa. *Trans. Roy. Soc. S. Afr.* 28: 255–292.

Chapman, V. J., 1946. Algal zonation in the West Indies. *Ecology* 27(1): 91–93.

Cranwell, L. M., and L. B. Moore, 1938. Intertidal communities of the Poor Knights Islands, New Zealand. *Trans. Roy. Soc. N. Z.* 67: 375–407.

du Rietz, G. E., 1925. Die Hauptzüge der Vegetation der Insel Jungfrun. *Svensk Bot. Tidskr.* 19(3): 323–346.

Feldman, J., 1938. Recherches sur la végétation marine de la Méditeranée. *Rev. Algol.* 10: 1–339.

Fritsch, F. E., 1931. Some aspects of the ecology of fresh-water algae. *J. Ecol.* 19: 233–272.

———, 1945. *The Structure and Reproduction of the Algae*, vol. 2. Cambridge: Cambridge University Press.

Geitler, L., 1930–32. Cyanophyceae. *In* G. L. Rabenhorst, ed., *Kryptogamenflora von Deutschland, Österreich und der Schweiz*, vol. 14. Leipzig: Akademische Verlags-gesellschaft.

Ginzberger, A., 1925. Der Einfluss der Meerwassers auf die Gliederung der süd-dalmatischen Küstenvegetation. *Österr. Bot. Z.* 74(1): 1–14.

Grubb, V. M., 1936. Marine algal ecology and the exposure factor at Peveril Point, Dorset. *J. Ecol.* 24: 392–423.

Johnson, D. S., and A. F. Skutch, 1928. Littoral vegetation on a headland of Mt. Desert Island, Maine. Ecology 9(3): 188–215, 307–338.

Kylin, H., 1937. Über die Farbstoffe und die Farbe der Cyanophyceen. *Förhandl. Kungl. Fisiogr. Sällsk. Lund* 7(12): 1–28.

Smith, A. L., 1921. *Lichens.* Cambridge: Cambridge University Press.

ACKNOWLEDGMENTS

This book has inevitably involved assistance from many people, and it is with great pleasure and deep gratitude that I acknowledge my indebtedness to them. It is impossible to express my thanks to all these people by name, but some must have special mention. First and foremost, without the unfailing help, counsel, and encouragement of Professor C. F. A. Pantin (until his untimely death) and Sir Maurice Yonge, this work could never have been accomplished. I am most grateful to Sir George Deacon—always willing and prompt to give help with the knotty problems of tides—and to Dr J. R. Lewis, who wrote the chapter on the British Isles.

For comments on the manuscript and other aid, I give my most sincere thanks to Miss J. I. Campbell, Mr J. S. Colman, Dr H. E. Davenport, Miss M. Eates, Dr M. W. Holdgate, Miss E. H. Ramsden, Dr A. J. Southward (England); the members of the staff of the South Kensington Natural History Museum, London; Dr K. M. F. Scott (South Africa); Miss I. Bennett and Miss E. Pope (Australia); and Dr E. J. Batham, Dr V. Cassie, and Dr M. Miller (New Zealand). I am indebted to my brother, Mr H. B. Wood, and to Miss Celia Wood of Welwyn Garden City, for a number of beautiful figures; to Mr. J. Rickus and the staff of the photographic department of Glasgow University for fine photographic prints; to Mr C. Field, of Wynberg, Cape Town, for beautiful colour transparencies of the Cape taken specially for us; and to Miss C. Swan of Glasgow and Mrs H. Wareing of Aberystwyth for cheerfully undertaking the wearying task of typing the manuscript. I most happily thank all these helpful and generous people and any others I have failed to mention.

NAME INDEX

INDEX OF GENERA AND SPECIES

GENERAL INDEX